Structural Mathematical Modeling Applications in Technological Machines and Transportation Vehicles

Andrey Eliseev
Irkutsk State Transport University, Russia & Irkutsk National Research Technical University, Russia

IGI Global
PUBLISHER of TIMELY KNOWLEDGE

A volume in the Advances in Mechatronics and Mechanical Engineering (AMME) Book Series

Published in the United States of America by
IGI Global
Engineering Science Reference (an imprint of IGI Global)
701 E. Chocolate Avenue
Hershey PA, USA 17033
Tel: 717-533-8845
Fax: 717-533-8661
E-mail: cust@igi-global.com
Web site: http://www.igi-global.com

Library of Congress Cataloging-in-Publication Data

Names: Eliseev, A. V. (Andrey Vladimirovich), author.
Title: Structural mathematical modeling applications in technological machines and transportation vehicles / authored by Andrey Eliseev.
Description: Hershey, PA : Engineering Science Reference, [2023] | Includes bibliographical references and index. | Summary: "The monograph is devoted to the development of the basis of modern machine science. A wide class of problems of dynamics of mechanical engineering objects operating under conditions of intense dynamic loading is considered. The basis of the approach is the formation of mathematical models of technical objects that can be displayed using mechanical oscillatory systems with several degrees of freedom and concentrated parameters"-- Provided by publisher.
Identifiers: LCCN 2022055121 (print) | LCCN 2022055122 (ebook) | ISBN 9781668472378 (hardcover) | ISBN 9781668472385 (paperback) | ISBN 9781668472392 (ebook)
Subjects: LCSH: Mathematical models. | Machine design. | Vibration--Mathematical models.
Classification: LCC TA342 .E525 2023 (print) | LCC TA342 (ebook) | DDC 621.8/15015118--dc23/eng/20230103
LC record available at https://lccn.loc.gov/2022055121
LC ebook record available at https://lccn.loc.gov/2022055122

This book is published in the IGI Global book series Advances in Mechatronics and Mechanical Engineering (AMME) (ISSN: 2328-8205; eISSN: 2328-823X)

British Cataloguing in Publication Data
A Cataloguing in Publication record for this book is available from the British Library.

All work contributed to this book is new, previously-unpublished material. The views expressed in this book are those of the authors, but not necessarily of the publisher.

For electronic access to this publication, please contact: eresources@igi-global.com.

Advances in Mechatronics and Mechanical Engineering (AMME) Book Series

J. Paulo Davim
University of Aveiro, Portugal

ISSN:2328-8205
EISSN:2328-823X

Mission

With its aid in the creation of smartphones, cars, medical imaging devices, and manufacturing tools, the mechatronics engineering field is in high demand. Mechatronics aims to combine the principles of mechanical, computer, and electrical engineering together to bridge the gap of communication between the different disciplines.

The **Advances in Mechatronics and Mechanical Engineering (AMME) Book Series** provides innovative research and practical developments in the field of mechatronics and mechanical engineering. This series covers a wide variety of application areas in electrical engineering, mechanical engineering, computer and software engineering; essential for academics, practitioners, researchers, and industry leaders.

Coverage

- Computational Mechanics
- Nanomaterials and nanomanufacturing
- Medical Robotics
- Design and Manufacture
- Sustainable and green manufacturing
- Intelligent Sensing
- Intelligent Navigation
- Tribology and surface engineering
- Vibration and acoustics
- Manufacturing Methodologies

IGI Global is currently accepting manuscripts for publication within this series. To submit a proposal for a volume in this series, please contact our Acquisition Editors at acquisitions@igi-global.com or visit: https://www.igi-global.com/publish/.

The Advances in Mechatronics and Mechanical Engineering (AMME) Book Series (ISSN 2328-8205) is published by IGI Global, 701 E. Chocolate Avenue, Hershey, PA 17033-1240, USA, www.igi-global.com. This series is composed of titles available for purchase individually; each title is edited to be contextually exclusive from any other title within the series. For pricing and ordering information please visit https://www.igi-global.com/book-series/advances-mechatronics-mechanical-engineering/73808. Postmaster: Send all address changes to above address.

Titles in this Series

For a list of additional titles in this series, please visit: www.igi-global.com/book-series

Handbook of Research on Emerging Designs and Applications for Microwave and Millimeter Wave Circuits
Jamal Zbitou (LABTIC, ENSA of Tangier, University of Abdelmalek Essaadi, Morocco) Mostafa Hefnawi (Royal Military College of Canada, Canada) Fouad Aytouna (LABTIC, ENSA of Tetouan, University of Abdelmalek Essaadi, Morocco) and Ahmed El Oualkadi (LabTIC, ENSA of Tangier, University of Abdelmlek Essaadi, Morocco)
Engineering Science Reference • ©2023 • 542pp • H/C (ISBN: 9781668459553) • US $325.00

Trends, Paradigms, and Advances in Mechatronics Engineering
Mohamed Arezki Mellal (M'Hamed Bougara University, Algeria)
Engineering Science Reference • ©2023 • 299pp • H/C (ISBN: 9781668458877) • US $225.00

Handbook of Research on Aspects and Applications of Incompressible and Compressible Aerodynamics
Sathish K. Kumar (Nehru Institute of Engineering and Technology, India) and Naren Shankar Radhakrishnan (Vel Tech Rangarajan Dr. Sagunthala R&D Institute of Science and Technology, India)
Engineering Science Reference • ©2022 • 429pp • H/C (ISBN: 9781668442302) • US $335.00

Advances in Computational Approaches in Biomechanics
Pritam Pain (Haldia Institute of Technology, India) Sreerup Banerjee (Haldia Institute of Technology, India) and Goutam Kumar Bose (Haldia Institute of Technology, India)
Engineering Science Reference • ©2022 • 306pp • H/C (ISBN: 9781799890782) • US $260.00

Developing Charging Infrastructure and Technologies for Electric Vehicles
Mohammad Saad Alam (Aligarh Muslim University, Aligarh, India) Reji Kumar Pillai (India Smart Grid Forum, India) and N. Murugesan (India Smart Grid Forum, India)
Engineering Science Reference • ©2022 • 343pp • H/C (ISBN: 9781799868583) • US $240.00

Multidisciplinary Perspectives on Green Electromobility and Charging Stations
Yu Miao (East China University of Sciencer and Technology, China)
Engineering Science Reference • ©2022 • 305pp • H/C (ISBN: 9781799839972) • US $225.00

Electric Vehicles and the Future of Energy Efficient Transportation
Umashankar Subramaniam (Prince Sultan University, Saudi Arabia) Sheldon S. Williamson (Ontario Tech University, Canada) Mohan Krishna S. (Alliance University, India) and Febin Daya J. L. (Vellore Institute of Technology, India)
Engineering Science Reference • ©2021 • 293pp • H/C (ISBN: 9781799876267) • US $240.00

IGI Global
PUBLISHER of TIMELY KNOWLEDGE

701 East Chocolate Avenue, Hershey, PA 17033, USA
Tel: 717-533-8845 x100 • Fax: 717-533-8661
E-Mail: cust@igi-global.com • www.igi-global.com

Table of Contents

Preface

Ensuring the reliability of technological and transport machines, improving the dynamic quality and level of their competitiveness in modern market conditions is a continuous process of improving design and technical developments and finding new solutions. In the development of such approaches, it is essential to advance the development of the methodological basis and scientific directions in the field of machine science, the theory of mechanisms and machines, dynamics and strength, theoretical mechanics, oscillation theory, applied system analysis and mathematical modeling.

In recent years, mathematical modeling methods related to the development of circuit theory and graph theory, the theory of automatic control and applied system analysis have been rapidly developing. Structural mathematical modeling, based on the possibility of using special technologies for constructing mathematical models of objects and processes, taking into account the variety of dynamic connections, is represented in the modern literature by a large number of theoretical and applied works in the field of electronics, electromechanics, oscillation theory, mechatronics and automatic control theory.

Methods of structural mathematical modeling based on the development of the apparatus of the theory of automatic control are widely used in research, development and evaluation of the properties of technical objects, the design schemes of which can be represented by mechanical oscillatory systems of various physical nature. The methods of structural mathematical modeling are diverse, and generally compatible.

Structural mathematical modeling, which is based on the idea of considering mechanical oscillatory systems as some analogues of automatic control systems, has been developed in the works of Russian scientists M.Z. Kolovsky, V.L. Veitz, M.D. Genkin, I.I. Galperin, I.G. Shatalov, based on the use of dynamic analogies. A structural mathematical model of a mechanical oscillatory system with several degrees of freedom and concentrated parameters is considered as a block diagram of a dynamically equivalent automatic control system.

Such representations are quite convenient when solving problems of machine dynamics, when their dynamic state is estimated by determining the level of oscillatory processes or vibrations arising from the interactions of elements of mechanical systems.

In this regard, the problems of machine dynamics in the form of problems of vibration protection and vibration isolation of technical objects are of interest. It can be noted that the problems of vibration protection, in a certain sense, are significant for the study of machine dynamics, in general, when structural mathematical modeling highlights an object whose dynamic state is evaluated, a system of disturbing factors, dynamic connections of interacting elements, levels of oscillatory movements at various points are evaluated.

In the field of theory and practice of vibration protection, the works of K.V. Frolov, M.Z. Kolovsky, V.L. Veitz, I.I. Blekhman, V.A. Ivovich, M.D. Genkin, A.V. Sinev, B.G. Korenev, P.M. Alabuzhev, I.I.

Galiev, V.A. Nekhaev, etc. have gained fame. Among foreign scientists, the developments of D.P. Den-Gartog, Ch.E. Crede, C.M. Khamis, F.S. Tse, I.E. Morse, R.T. Hinkle, D.S. Snowdon, K.N. Tonga, etc. can be noted.

The proposed work reflects the results of research conducted for two decades under the guidance of Professor, Doctor of Technical Sciences, Honored Scientist of the Russian Federation S.V. Eliseev at the Irkutsk Polytechnic Institute (now Irkutsk National Research Technical University), the Irkutsk Computing Center SB RAS (now the Institute of System Dynamics and Control Theory SB RAS named after V.M. Matrosov), and also at the Irkutsk State Transport University.

In the mentioned publications, attention was paid to the development of the methodological foundations of the structural mathematical modeling method in applications to various problems of machine dynamics. Important for the purposes of practical application was the creation of a certain basis that ensures the implementation of technologies for structural transformation of mathematical models, which led to the disclosure of a number of new dynamic effects, the introduction of concepts of dynamic stiffness, quasi-springs, interpartial connections, the role and significance of the distribution of oscillation amplitudes, etc.

A great contribution to the development of the methodology of structural mathematical modeling was made by Prof., Doctor of Technical Sciences A.P. Khomenko, Yu.N. Reznik, A.I. Artyunin, N.K. Kuznetsov, V.I. Sobolev, P.A. Lontsikh, S.V. Belokobylsky, as well as assoc., Candidate of Technical Sciences Yu.V. Ermoshenko, A.V. Dimov, R.Y. Upyr, A.S. Logunov, D.N. Nasnikov, A.O. Moskovskikh, I.V. Fomina, A.N. Trofimov, M.V. Drach, E.A. Parshuta, E.V. Kaimov, I.S. Sitov, V.B. Kashuba, R.S. Bolshakov.

The materials of the monograph reflect the results of joint scientific work of the author of the monograph S.V. Eliseev with his colleagues who participated in the development of individual areas and applications.

The monograph is preceded by an overview of modern works on machine science, which reflects new research trends, problem-setting, innovations in methods and approaches. A great contribution was made by the developments of Ph.D. R.S. Bolshakov, E.A. Parshuta, E.V. Kaimova. It is necessary to note interesting results in detailing the features of the application of structural mathematical modeling methods of Ph.D. R.Yu. Upyr, M.V. Drach, A.S. Logunov, D.N. Nasnikov, I.V. Fomina, A.N. Trofimov.

The main scientific results in the mechatronics of vibration-proof systems and the development of methods of structural mathematical modeling and its applications were obtained by a team of employees of the Scientific and Educational Center for Modern Technologies, System Analysis and Modeling of the Irkutsk State Transport University.

The monograph is devoted to the development of the basis of modern machine science. A wide class of problems of dynamics of mechanical engineering objects operating under conditions of intense dynamic loading is considered. The basis of the approach is the formation of mathematical models of technical objects that can be displayed using mechanical oscillatory systems with several degrees of freedom and concentrated parameters.

The first chapter is devoted to the basic concepts of structural mathematical modeling methods in the dynamics of mechanical oscillatory systems applied to the problems of vibration protection of machinery and equipment. An overview of the problems of dynamics, vibration interactions of machine elements and mechanisms is given; the design and technical features of technological machines and the forms of their dynamic loading are considered; typical design schemes of transport vehicles, taking into account the suspension features, are presented. The concepts of lever connections, which are implemented in the mechanisms of technical objects, are introduced. Approaches to the construction of computational circuits in problems of machine dynamics, taking into account the limitations in the connections of movable elements, are presented.

In the second chapter, ideas about the dynamic states of elements of technical objects in conditions of vibration loads are developed. The features of using methods and devices for changing the state of vibration, taking into account additional restrictions, mechanisms and devices, are described. Some features of vibration protection systems, the principles of the formation of the element base are discussed. The features of the use of lever mechanisms in assessing the dynamic properties of technical systems are considered. Methods of dynamic synthesis of vibration protection, features of calculation and selection of system parameters are given.

In the third chapter, the provisions on the basics of assessing the dynamic states of technical objects are developed. The fundamentals of the structural approach in mathematical modeling in the problems of dynamics and vibration protection of technical objects are presented. The ideas about the possibility of simplifying the structural scheme of a technical object by introducing a generalized spring are introduced. Some forms of suppression of dynamic oscillations in systems with several degrees of freedom are listed. The features of the dynamics of three-mass vibration protection systems are considered. Approaches to simplification of structural schemes are proposed. Typical elements used for the formation of structural schemes are considered.

The fourth chapter shows the results of the development of generalized approaches to the problems of evaluation and control of the dynamic state of mechanical oscillatory systems under vibration loading. The design and engineering features of technical objects necessary for the construction of mathematical models within the framework of structural approaches are described. The issues of determining dynamic rigidity in systems with a rotating object of protection are considered. Possible directions of development of the structural theory of vibration protection systems are presented. The concepts of connecting blocks of typical elements in mechanical oscillatory systems are described in detail. Methods for determining the reactions of dynamic coupling with vibration effects are described. Representations of motion transformation devices in the structure of mechanical oscillatory systems are given. Ideas about the possibilities and forms of lever interactions in the system are developing. The concepts of articulated solids are introduced. Examples of connecting elements in a beam system with two degrees of freedom are considered. The possibilities of forming virtual connections in a mechanical oscillatory system are discussed.

In the fifth chapter, theoretical concepts of lever connections in mechanical oscillatory systems are developed. The main provisions on the possibilities and forms of lever interactions in the system of connections of typical elementary nodes are presented. Systems containing a lever of the first kind are considered. A new approach to assessing the possibilities of sequential connections of typical elements in constructive interpretations of vibration protection systems is proposed. The features of parallel connections of typical elements are shown. Some issues of assessing the states of dynamic vibration dampers of the lever type are considered. The issues of the connection of solids with the dynamic properties of vibration protection systems, taking into account lever connections, are discussed. Ideas about a virtual lever mechanism are being developed. The possibilities of equivalent representations of mechanical systems with angular vibrations of solids are discussed.

The sixth chapter is devoted to the development of theoretical ideas about the forms and features of the implementation of lever connections in mechanical oscillatory systems. Mathematical models that have been developed taking into account the features of lever connections are presented. The dynamic effects of introducing additional restrictions are considered. The concepts of levers in dynamic interactions of system elements are considered in detail. The possibility of using a typical element to form lever connections in dynamic interactions. The dynamic properties of vibration protection systems with lever connections are considered, taking into account the motion conversion device.

The seventh chapter lists possible directions for the development of structural mathematical modeling methods. The possibilities of using the characteristic frequency equation to evaluate the structure, dynamic rigidity and features of the interaction of system elements are considered. The possibilities of using additional inertial-type constraints in the problems of the dynamics of technical systems are discussed. The possibility of using the connectivity of the movement of elements and forms of external influences in mathematical models to evaluate interactions in chain structures is considered. The ideas about the forms of joint movements of the three mass elements of the oscillatory system are detailed, taking into account the influence of dynamic rigidity. The effects of changes in the reduced stiffness in the problems of dynamic damping of vibrations with the introduction of additional external influences are discussed. Questions arise about the role of the system in assessing the dynamic characteristics of the system.

In general, the monograph is of interest to specialists working in the field of modern machine science, dynamics and strength of machines, theoretical and applied mechanics, and can also be used as a textbook for training students of technical specialties in bachelor's and master's degree programs focused on mechatronics, robotics, and modern mechanical engineering.

Andrey Eliseev
Irkutsk State Transport University, Russia & Irkutsk National Research Technical University, Russia

Acknowledgment

The work on the proposed monograph was started under the guidance of my research supervisor Eliseev Sergey Viktorovich. I will give a brief scientific biography.

Eliseev Sergey born in 1940. In 1963 he graduated from the Omsk Institute of Railway Engineers. The thesis was defended in 1967 thesis "Study of the dynamics of the moving contact and the characteristics of the switching process under the action of mechanical factors". 1973 – Defense of doctoral dissertation at the Institute of Mechanics of the Ukrainian Academy of Sciences, Kyiv. Theme of doctoral dissertation "Theory of active vibration protection systems and its applications" in "Theoretical mechanics" in 1974. Scientific supervisor Professor M. Z. Kolovsky. He passed a scientific internship in 1974 at the Illinois Institute of Technology. Head Robert Klein, as well as at the Massachusetts Institute of Technology, supervisor Stefan Crandal. Trained: Higher Technical School (Paris), Imperial College (England), Germany, Poland, India, etc. Main places of work: Irkutsk Scientific Center – Irkutsk Computer Center (now the Institute of System Dynamics and Control Theory academician V. M. Matrosov). Worked in Irkutsk Polytechnic Institute (now Irkutsk National Research Technical University), SibExpoCentre and other organizations. From 2002 to 2021 - Professor Irkutsk State Transport University. Honored worker of science of the Russian Federation. Full member of the Russian Academy of engineering Sciences, full member of the Russian Academy of Engineering, full member of the Academy of quality problems, full member of the Tsiolkovsky Academy of cosmonautics, full member of the International Academy of engineering, corresponding member of the Russian Academy of Natural Sciences. Scientific specialization: mechanics and mechanical engineering. Over the past 10 years, his research interests have focused on the development of structural mathematical modeling methods. The principal feature of the method is the use of the analytical apparatus of the theory of automatic control to solve the problems of dynamics of various technical objects of transport and technical purpose operating under conditions of intense dynamic load, provided that mechanical oscillatory systems will be considered as the design schemes of technical objects.

To my profound sorrow, Sergey Viktorovich passed away in 2021.

I considered it my duty to do everything in my power to publish the materials of the prepared monograph.

Andrey Eliseev
Irkutsk State Transport University, Russia & Irkutsk National Research Technical University, Russia
2023

Introduction

The problem of dynamics of machines is complex, which implies the interconnectedness of the processes of studying the state of the technical object, determined by the structural and technical parameters of the machine, and the evaluation of its dynamic properties based on the use of computational schemes, as well as the need to solve many specific problems of dynamic interaction of machine elements within the mathematical modeling methods. Mechanisms in the structure of technical objects, including transport and technological machines, are very diverse. Most of the mechanisms are mechanical chains consisting of links in the form of solids connected by kinematic pairs. Detailed consideration of problems related to the definition of kinematic and dynamic parameters of the functioning of mechanisms, belong to the domain of the theory of mechanisms and machines. The problems of the dynamics of machines and mechanisms initiate the development of mathematical modeling methods and form different new directions in machine science, theoretical and applied mechanics.

Specific problems of machine dynamics, with all their diversity, are focused in many of their manifestations on improving the reliability of the machines, ensuring the safety of operation and competitiveness of technical objects that should show high performance and dynamic quality.

The operation of machines and mechanisms in conditions of intense dynamic heating is often accompanied by manifestations of vibrational processes, which arise both for internal reasons related to the imperfection of constructive and technical design, and during the interactions of working bodies of the machines with the external environment, including the implementation of various process technologies.

A detailed approach to assessing the dynamic capabilities of technical objects is based on the drawing out schematic diagrams of mechanisms. The elastic properties of the elements (links) of the mechanisms, their connections (or kinematic pairs) often cause the occurrence of vibrations in mechanisms; however, these processes in mechanisms can occur for other reasons, such as movement of the mechanism base, machine malfunction, the loading variability of the working bodies of machines, etc.

In recent decades, the problems of vibration of machines and mechanisms are considered as a fairly autonomous direction in the dynamics of technical objects. They have a developed methodological basis that determines the solubility of a wide variety of tasks, including vibration protection and vibration insulation of machines, equipment, and apparatus of various purpose.

In a certain sense, the tasks of vibration protection of machines and equipment can be identifies as significant in the dynamics of machines, since the ability to evaluate, control and change the dynamic states of technical objects under the action of perturbing factors of various nature is common in many cases of research, in one form or another.

The tasks of machine dynamics, focused on the problems of vibration protection and vibration insulation, are quite sufficiently represented by the scientific and technical literature of recent years.

Modern machine science that defines the theoretical basis of mechanical engineering, and problem solution related to the assessment of dynamic states are characterized by the widespread use of various methods of mathematical modeling. The specificity of solving the problems of assessing the dynamic states of technical objects is the use of mathematical models based on computational schemes in the form of mechanical oscillatory systems with one, two or more degrees of freedom, with lumped and distributed properties.

The use of mechanical oscillatory systems as computational schemes of technical objects operating under the conditions of intense dynamic loading that contributes to the manifestations of different forms of vibratory states, has its own inherent characteristics. Vibration processes, in many cases, are seen as oscillatiory motions of various objects; first of all - as small periodic motions relative to the position of the static equilibrium of the initial system or the relatively established dynamic state.

Consideration of small motions predetermines the possibilities of simplifying mathematical models, reducing them to linear forms, followed by interpretations of external periodic impacts as the sum of a number of harmonic processes. In this regard, the methods of the Fourier and Laplace integral transformations were highly developed and widely applied.

The representation of technical objects of transport or technological purpose, which are basically some mechanisms or groups of interconnected mechanisms, is a rather complicated and time-consuming task.

Mechanical oscillatory systems, considered as computational schemes of technical objects, consist of a certain (quite limited) set of standard elements. The most developed approaches in this regard are reflected in the theory of vibration insulation or the theory of vibration protection systems. Considered elementary standard links include, in the first place, mass-inertia elements (material points and solids), springs, dampers, and combinations of them. Mathematical models using these sets of elements, in their simplest versions, in dynamic states are displayed by systems of ordinary linear second-order differential equations with constant coefficients.

At the same time, taking into account the real properties of the mechanisms displayed by mechanical oscillatory systems implies attention to the fact that manifestations of various kinds of dynamic linking occur between the elements of technical objects in the initial systems. Of great importance are the lever linkages, the effects of the conversion of mass-inertia and kinematic parameters, implemented by special mechanisms (lever, screw non-locking, gear, etc.). Such mechanisms are included in the structures of mechanical oscillatory systems and are represented accordingly in mathematical models.

It is typical for modern machines to widely use links in the form of elastic elements, as well as links in the form of pneumatic, hydraulic and electromagnetic devices that are included into and operate in the structures of conventional mechanisms. The representation of the properties of these mechanisms in the structures of computational schemes in the form of mechanical oscillatory systems requires appropriate techniques and interpretations. This problem is further complicated by the fact that in some cases the standard elements of mechanical oscillatory systems are servoactuators made on the basis of hydro-, pneumatic and electrical automation. These mechanical oscillatory systems are automatic control systems (provided there are devices for collecting and processing state information). These complex vibration protection systems have been called active vibration protection systems; such systems are widely used to improve dynamic quality and improve the reliability of many technological and transport machines.

Thus, the evaluation of the dynamic properties of technical objects (in particular, transport and technological machines), using computational schemes in the form of mechanical oscillatory systems, requires that new forms of construction of mathematical models be developed. These models should take into account the potential occurrence of new standard links in the initial oscillatory structures, reflecting the new dynamic properties of the mechanisms that become the usual form of expansion and modification of the specific properties of modern machines.

In a number of scientific developments, taking into account the complication of the computational schemes is reflected in the concept of introducing so-called additional couplings, which have certain capabilities in terms of the mentioned circumstances. These approaches are evidenced in the elaboration and use of structural mathematical modeling techniques.

In these approaches, the mechanical linear oscillatory system is compared (on the basis of the principles of dynamic analogies) to the structural diagram of a dynamically equivalent automatic control system with the appropriate use of the capabilities of the analytical tools of automatic control theory.

The use of structural mathematical modeling techniques has the advantage over other approaches that makes it possible to distinguish various dynamic linkages between the elements of the system formed during the oscillatory motions of the object. Taking into account these factors allows one to significantly expand and detail the concepts of the peculiar features of dynamic properties and interactions of elements of technical objects, whose computational schemes can be considered as mechanical oscillatory motions with multiple degrees of freedom. At the same time, the proposed direction of development of the scientific and methodological basis requires revision and rethinking of a number of positions, approaches, ways and means of solving problems that arise in assessing the dynamic states of technical objects, in the evolvement of technologies for evaluating, correcting, and forming dynamic states of technical objects within certain limits.

Chapter 1
Problems of the Dynamics and Vibration Interactions of Elements of Machines and Mechanisms:
Dynamic Interactions Under Vibration Loads

ABSTRACT

The chapter examines dynamic problems and vibrational interactions in machine elements and mechanisms, providing a comprehensive overview of the design and technical characteristics of technological machines. It explores typical structural schemes of vehicles and lever connections that are often used in technical objects. The concept of computational chains is introduced, which takes into account the constraints that are when connecting moving elements in dynamic problems of machines. The chapter also presents the main concepts of structural mathematical modeling in the dynamics of mechanical oscillatory systems, which are necessary for solving vibration protection problems in machines and equipment.

INTRODUCTION

The problems of the dynamics of technological machines and transportation vehicles differ in complexity and variety, which has found its way into the formation and development of a theoretical basis, represented by an interdisciplinary scientific space. Modern machine science is based on the methods of theoretical mechanics, the theory of dynamical systems, the theory of oscillations, the theory of mechanisms and machines, the theory of automatic control, methods of mathematical modeling and information technologies.

DOI: 10.4018/978-1-6684-7237-8.ch001

Modern machines and equipment operate under conditions of intense dynamic loading, accompanied by the action of vibration factors of various physical nature. In general, for machine science, the problems of ensuring safety and reliability in conditions of vibration effects are topical (Antipov, 2006) and are developed in various directions of the general plan and in attention to the tasks of detailing the influence of dynamic factors on the functioning of power transmissions, aggregates, assemblies and machine parts (Zinoviev & Bessonov, 1964).

Attention to various forms of manifestation of vibrational interactions is typical for problems of vibration protection of machines and equipment, which was reflected in (Goverdovsky & Trofimov, 2009). Theoretical and applied developments related to the problems of vibration protection of machines, equipment and apparatus are presented in (Rakov & Sinev, 2011). The problems of controlling the oscillations of technical objects and developing methods and means for controlling the dynamic state of mechanical engineering objects under vibrations has become the topical direction of research in the problems of the dynamics of machines and mechanisms (Belokobylsky, Eliseev & Sitov, 2013; Belokobylsky, Eliseev & Kashuba, 2013; Wolfson, 2008). Attention to the problems of searching for and developing methods and means of protecting various technical objects from vibrations and shocks found its way in (Eliseev et. al., 2008), which considered the features of dynamic synthesis, vibration protection devices of various constructive and technical forms and purposes. Despite the variety of problems solved in the field of dynamic interactions of machine elements and mechanisms which determine the parameters of movement of units and the resulting loads specifying the possibilities for reliable operation and operational safety, the general basis for estimating the dynamic state is the methods of mathematical modeling. Within these representations, real engineering facilities and technical systems are displayed by computational schemes in the form of mechanical oscillatory systems containing composite elements with distributed and lumped parameters (Kamaev, 2005; Samarsky & Mikhailov 2002; Sovetov & Yakovlev, 2001; Belokobylsky & Eliseev, 2016).

Of course, while evaluating the formation of a theoretical basis in solving the problems of machine dynamics, such a position is quite a private one. This influences the features of the evaluation of vibrational dynamic interactions, characteristic, to a greater extent, to the tasks of vibration protection.

At the same time, considering mechanical engineering objects as the computational schemes whose dynamic state is associated with various forms of vibrational interactions, based on a detailed study of the dynamics of mechanical oscillatory systems consisting of elastic, dissipative and mass-and-inertia elements with lumped parameters, has become quite widespread in engineering practice (Vargunin et. al., 2006; Wolfson, 2013; Eliseev & Ermoshenko, 2010; Eliseev & Khomenko, 2014; Eliseev et. al., 2013).

One of the directions in the development of detailed approaches to assessing the dynamic state of technical objects has been the attention to taking into account the features of vibrational interactions in mechanical oscillatory systems, in the structure of which more complex devices are used than the traditional elementary units of vibrational systems. It was shown in (Eliseev et. al., 2016; Eliseev et. al., 2013; Eliseev, Reznik & Khomenko, 2011; Eliseev et. al., 2009; Eliseev, Khomenko & Barsukov, 2011) that in mechanical oscillation systems one can consider additional devices that implement the functions of motion transformation, and also have the ability to create a number of new effects characteristic of automatic control systems, i.e. dynamic managed systems. Essentially, in the solution of the problems of machine dynamics, the computational schemes of mechanical engineering objects became more complicated, as mechanisms of a different nature (pneumatic, hydraulic electrodynamic), in addition to elastic, dissipative and mass-and-inertia elements, were included in the structure of mechanical oscillatory systems, as well as motion transformation devices in the form of lever, screw, toothed and other mechanisms.

Expanding the notion of a set of typical elements, the use of external energy sources, as well as additional constraints that change the spectrum of the dynamic properties of oscillatory systems, initiates the development of new approaches and a methodological basis reflecting the properties of systems analogous to automatic control systems, which, in particular, found its way in the development of the structural theory of vibration protection systems.

Mechanisms in mechanical oscillatory systems are constituent elements of a more complex form than traditional elements of vibration protection systems. They have features that manifest themselves in certain relationships that are established between the parameters of the movement of individual units of the mechanism and cannot depend on the frequency of vibration effects, since the connections between the state parameters are kinematic in character. Such connections are visually implemented as the simplest lever mechanisms of the first and second genera. The evaluation of the dynamic states of mechanical oscillatory systems, having in their structure certain devices or formations from typical elements, requires the development of more detailed representations about the connections between the elements of the mechanical oscillatory system proposed as a computational scheme. It is also necessary to take into account the features of constructing mathematical models and methods for estimating the singularities of dynamic properties of real technical objects. In this respect, in the problems of vibrational interactions, there are certain established ideas associated with the use of linear models, the smallness of deviations from the position of static equilibrium, etc.

Structural and Technical Features of Technological Machines and Forms of Their Dynamic Loading

Technological machines, intended to implement vibration processes of transportation, separation and surface treatment of parts, find wide application in industry. These technical objects include vibration stands, vibrating feeders, molding machines, vibrating conveyors, etc. The theoretical foundations of technology processes are laid in the works of Russian scientists, reflecting the features of dynamic interaction of elements, the design of working elements, vibration excitation, ensuring the reliability of operation, protection of instruments and equipment, and also protection of human operator and the environment from harmful dynamic effects (Kopylov, 2011; Panovko, 2006). Many vibrating technological machines have complex structural forms and are designed to create certain vibrational fields of spatial and simplified species.

Table 1 presents calculations of vibrating machines. In the structure of these machines, various formations obtained by combining mass-and-inertia and elastic elements, as well as motion transformation devices and mechanisms of various types are widely used. From Table 1 it follows that vibrating technological machines are mechanical oscillatory systems, in the structure of which there are not only mass-and-inertia and elastic elements but also various mechanisms that are used to introduce certain constraints. Such relationships are implemented in lever forms and mechanisms that ensure the transformation of motions and create a certain spatial area for the interacting elements to be placed.

Table 1. Comparison of dynamic schemes for operational properties

Machine type	Drive type	Scheme	Balance state	Factor of amplification	Stability
Single-mass	Crank-and-rod mechanism with a rigid connecting rod (superresonance)		– –	–	+ +
			–	–	+
	Unbalanced (superresonance)		+	–	+
	Electromagnetic (resonance)		– –	+	– –
Two-mass	Crank-and-rod mechanism (resonance)		+	+	–
		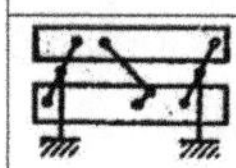	–	+	+
	Unbalanced (resonance)		+	+	– –
	Electromagnetic (resonance)	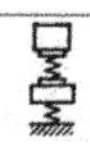	+ +	+	– –
Three-mass	Crank-and-rod mechanism (resonance)		+	+	–
	Unbalanced (interresonant)		+	+	+
	Electromagnetic (interresonance)	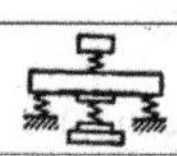	+ +	+	+

To obtain vibrational fields of the required type, it is important to design the spatial structure of the arrangement of the elements of the mechanical oscillatory system in order to obtain the desired shape of the amplitude distribution of the surface oscillations, as shown in Figure 1. A characteristic point is the node *A*, in which the amplitude of the oscillations is zero (Figure 1, *a*). The emergence of the node is due to the fact that the system has lever linkages created by the working element in the form of a solid body making angular oscillations. The node can be created by installing a dynamic absorber (Figure 1, *b*).

Figure 1. Schemes for providing a specific envelope of the effector displacement amplitudes: a – by means of an eccentric location of the vibration exciter; b – by means of installing a dynamic vibration absorber

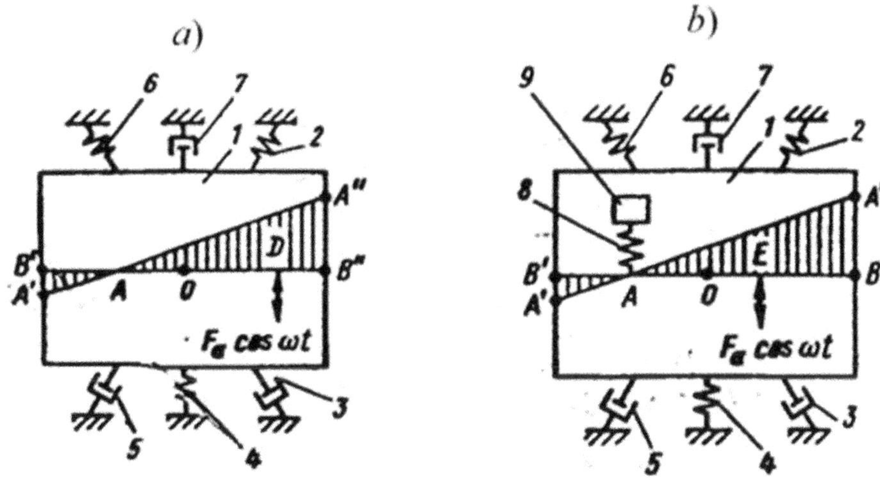

Elements of vibrating machines that form excitable elastic systems consist, as a rule, of several parts that can move relative to each other, creating the necessary relationships between the amplitudes and phases of oscillations of individual points of the vibrating machine.

Some Possibilities of Correction of Dynamic States of Technical Objects

The introduction of additional masses in the general case of mass M_1 and M_2 (Figure 2) perform spatial oscillations, and the elastic constraints k_1, k_2 hold the masses in all six coordinates, however, in the working (resonant) mode one can consider the unidirectional oscillation scheme described by two equations

$$M_1\ddot{x}_1 + (k + k_1)\cdot x_1 + kx_2 + c_1\dot{x}_1 + c(\dot{x}_1 + \dot{x}_2) = F(t); \tag{1}$$

$$M_2\ddot{x}_2 + (k + k_2)\cdot x_2 + kx_1 + c_2\dot{x}_2 + c(\dot{x}_1 + \dot{x}_2) = F(t); \tag{2}$$

where c_1, c_2 are the coefficients of conditional viscous resistances, including the losses between the medium and the masses; k, c are stiffness and resistance coefficient of the main constraint. Usually $k >> k_1$ (or k_2), and for $k_1 \approx k_2$, $c_1 \approx c_2$, we can assume that

$$\gamma_1 = \frac{M_1}{M_2} = \frac{A_2}{A_1} = \frac{c_1}{c_2} \tag{3}$$

that is, lever linkages find their way in the system.

Figure 2. Model of a two-mass vibrating machine

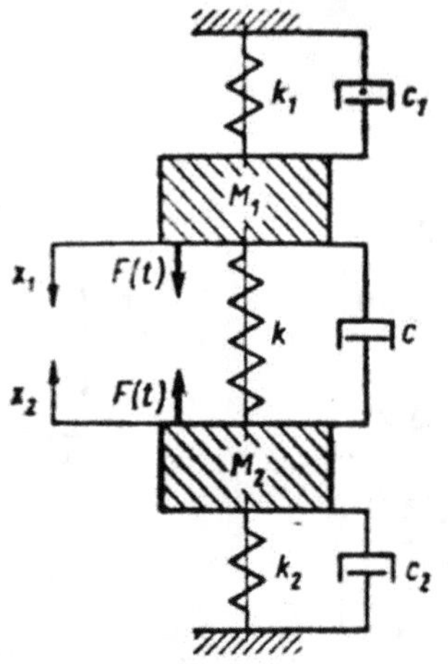

In this case, the mathematical model is simplified and takes the form

$$M_{red}\dot{x}+\left(A_1+A_2\right)\cdot\dot{x}+kx=F\left(t\right), \tag{4}$$

where $M_{red}=\dfrac{M_1}{1+\gamma_1}$; $c_{red}=\dfrac{c_1}{1+\gamma_1}$; $x = x_1 + x_2$; A_1, A_2 are the amplitudes; l_1, l_2 are coordinates of the center of the nodal point (center of inertia) (Figure 3).

Figure 3. Simplified model of a two-mass vibrating machine

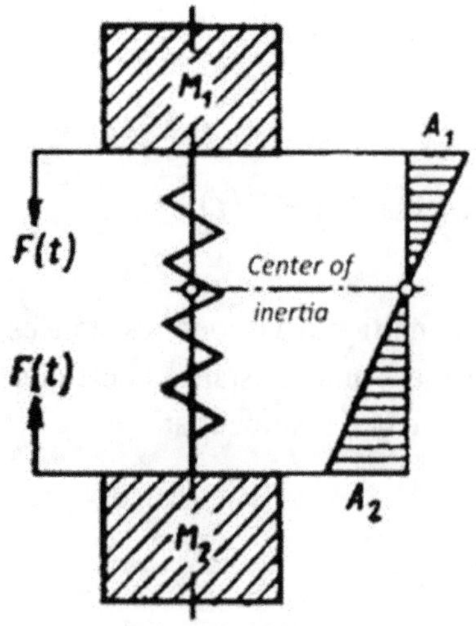

The frequency of free oscillations of the system is determined by

$$\omega_0 = \sqrt{\frac{k}{M_{\text{red}}}} \text{ rad / s} = \sqrt{\frac{5}{\Delta_{\text{red}}}}, \tag{5}$$

where $\Delta_{\text{sr}} = \frac{M_{\text{red}} g}{k}$ is the relative displacement of masses. For $\omega_0 = 50$ Hz, $\Delta_{st} = 0{,}01$ cm for given ω_B (for example, 314 or 50 Hz) and $\xi = \frac{\omega_B}{\omega_0}$ ($\xi = 0{,}85 \div 0{,}95$ is the pre-resonant adjustment; $\xi = 1{,}1 \div 1{,}15$ is the post-resonant adjustment). Required stiffness

$$k = \left(\frac{\omega_B}{\xi}\right)^2 M_{\text{red}}. \tag{6}$$

The amplitude of the oscillations according to equation (1.4) with $F(t) = F_0 \sin \omega t$ is:

$$A = \frac{F_0}{k\sqrt{\left(1-\xi^2\right)^2 + 4h^2\xi^2 / \omega_0^2}}; \tag{7}$$

where $h = \frac{c_{\text{red}} + c}{2M_{\text{red}}}$.

There is great variety of vibration exciters that create the necessary vibrational loads for technological purposes. The adjustment of such machines requires an evaluation of the dynamic properties of the working elements, the state of which is determined by the frequency characteristics.

Systems of the Formation of Dynamic States of Machines

Vibration systems with a centrifugal perturbation get widespread use. As an example, consider the planar linear scheme shown in Figure4, *a*. The body *1* of a single-shaft centrifugal vibration exciter whose unbalance *3* rotates around the *B* axis at a constant angular velocity ω is rigidly connected to the body *2* that is driven into the oscillatory motion. This body, together with the parts of the exciter that are rigidly attached to it, shall be called the effector. Its center of mass lies at point *A*, whose mid-position is taken as the origin of coordinates $xA_{\text{mid}}y$. Figures *4* and *5* denote the projections of the resultant elastic forces applied to the effector, characterized by the stiffness coefficients C_x and C_y, and the numerals *6* and *7* denote the projections of the resultant dissipative forces, characterized by the resistance coefficients *bx* and b_y. Elastic moment *8* and dissipative moment *9*, characterized by the coefficients of angular stiffness *s* and angular resistance *p*, are also applied to the actuator. The parameters of the dynamic state necessary for calculating the machine in the first stage are determined by the simplified scheme in Figure 4, *b*.

Figure 4. Single-mass computational schemes: a – with three degrees of freedom; b – with one degree of freedom

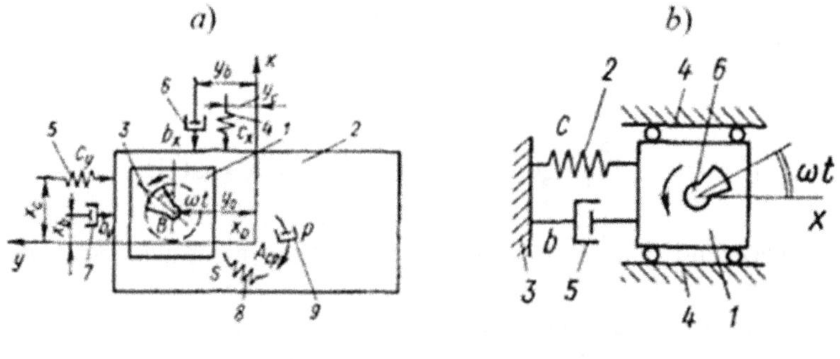

Vibration machines of special purpose can consist of mechanisms of electromagnetic, pneumatic and hydraulic nature. Figure 5 shows the dynamic schemes of the electromechanical stand.

Figure 5. Dynamic scheme of the vibration exciter

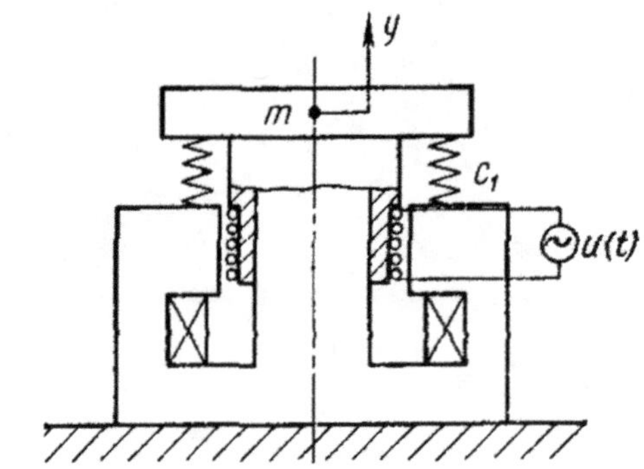

Structural diagrams of the vibration stand reflect an essential feature of a vibrating technological system: a vibration stand is an automatic control system.

Figure 6 shows the structural diagram of the system pictured in Figure 5.

Figure 6. Structural scheme of the vibration exciter

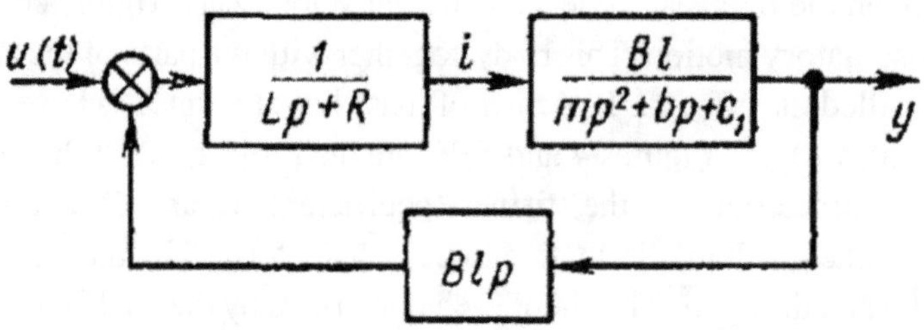

Even the simplest dynamic model of the electrodynamic oscillation exciter is a closed linear system of the third order with a negative speed feedback.

Figure 7. Computational schemes of the dynamics of the machine a flat motion of the working chamber

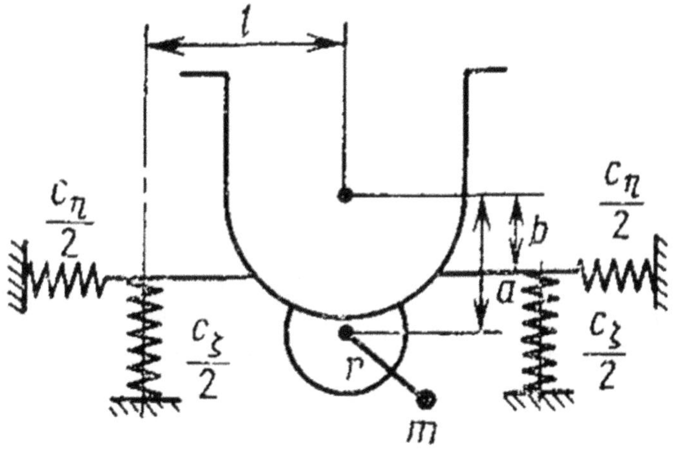

Vibration treatment of surfaces is based on the use of machines with complex kinematics of the motion of the units (Figure 8) (Harris & Crede, 2002; Selvinsky, 2009).

Figure 8. Computational schemes dynamics of the machine with a volumetric motion of the working chamber

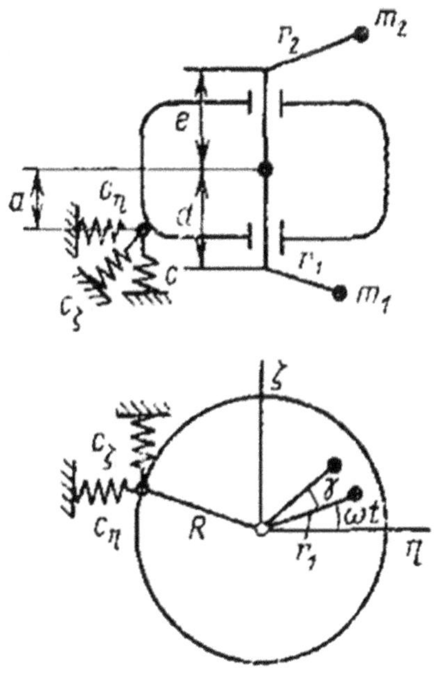

The aim of dynamic calculation is to select such installation parameters that provide the necessary amplitude and frequency (Figures 5-7).

Machines with a flat motion of the working chamber are calculated according to the scheme shown in Figure 7.

Sufficient accuracy for practical calculations is ensured by solving a linear system of equations:

$$\left.\begin{aligned} M\ddot{\zeta} + c_{\zeta}\zeta &= mr\omega^2 \sin\omega t; \\ M\ddot{\eta} + c_{\eta}\eta &= mr\omega^2 \cos\omega t; \\ J\ddot{\phi} + c_{\zeta}\phi l^2 + c_{\eta}\phi b &= mra\omega^2 \cos\omega t \end{aligned}\right\} \tag{8}$$

where M, J are the mass and moment of inertia relative to the central axis of the oscillating mass; c_{η}, c_{ζ} are the total stiffnesses in the directions of η and ζ.

From the analysis of the schematic diagrams of various vibrating technological machines, it can be concluded that for calculations it is necessary to consider mechanical oscillating systems with several degrees of freedom, in which the constituent elements are combined into specific structures whose dynamic properties are determined by spatial relationships, the possibilities of forming the lever interactions, motion transformation, transmission and dissipation of energy.

Vibrational feeders for moving and orienting parts in automatic assembly of products are distinguished by complex computational schemes. The computational schemes of such machines contain elastic elements and rod systems that implement small oscillations in various directions, using kinematic ones.

Transport Machines: Computational Schemes, Peculiarities of Suspension

Technical devices that ensure the reliability and safety of operation of transport machines and vehicles for various purposes, in particular, are elements of quite complex structural and technical solutions that have a multi-purpose orientation. Typically, such devices consist of several interconnected units, forming certain structures in which individual assemblies and mechanisms can be allocated. Suspension has elastic and dissipative properties, allowing one to read dynamic effects on protection objects. In terms of the formation of computational schemes, there is an orientation toward the use of mechanical oscillatory systems, the elements of which provide elastic and dissipative interactions, the transformation of motions, the creation of a spatial system for balancing the motion of elements, etc. (Harris & Crede, 2002; Selvinsky, 2009).

In solving the problems of ensuring the safety of the operation of the transport means, the suspension systems became widespread and, even in the initial structural and technical forms, represented mechanical oscillatory systems with an expanded set of composite elements (Figure 9) Domnin, 2005).

In modern cars, hydraulic and hydropneumatic shock absorbers have become widespread. The simplest computational schemes are shown in Figure 10.

A more detailed computational scheme of a car is shown in Figure 11 (Domnin, 2005). It reflects the dynamic properties of mechanical oscillatory systems with four degrees of freedom and the presence of lever linkages that form a spatial system of relationships between the coordinates of the motion of points in the system.

Figure 9. Car suspension with the shock absorber Truffaut: 1 – car frame; 2 – traction; 3 – lever

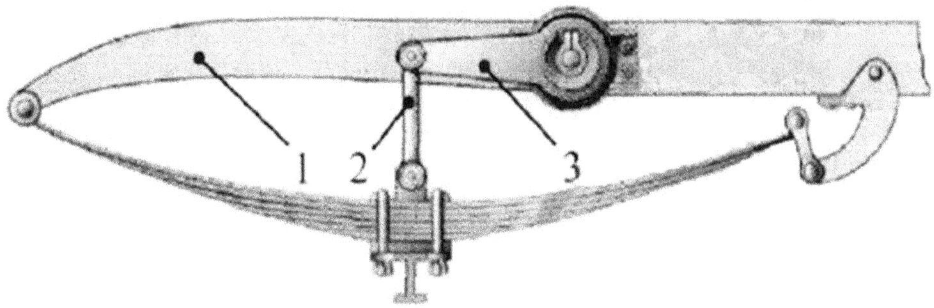

Figure 10. Single-mass (a) and two-mass (b) computational schemes of oscillations: M – sprung mass; m – unsprung mass; C_S and C_T – stiffness of the suspension and tires; K_S u K_T – damping in the suspension and tire; z_1 and z_2 – coordinates of the displacement of the centers of masses; q(t) – excitation from the road

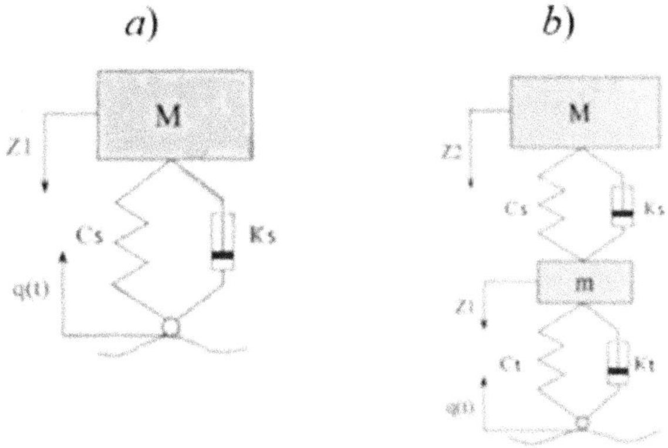

In such computational schemes the connectivity of motions between partial systems is implemented, it is possible to study specific regimes, various types of constraints, including inertial, dissipative-elastic and, including, lever ones, arising from the presence of an oscillating solid having mass M and moment of inertia J (see Figure 11).

As a result of the development of computer technology and software, spatial computation schemes for car oscillations started to be increasingly used. Specific complex schemes exist and are used in special developments, reflecting the possibilities of introducing additional constraints and mechanisms such as dynamic vibration dampers (Figure 12). For a mathematical model, whose computational scheme is shown in Figure 12, there are numerous inertial and elastic constraints between partial systems, as well as many constraints formed by the spatial arrangement of shock absorbers.

In this case, it is necessary to take into account the coordinates of the attachment of elastic-dissipative elements in the spatial system, which is due to the recognition of the emerging dynamic relationships that can be interpreted as the presence in the oscillatory structures of additional devices for the transformation of motion and mechanisms (Ryabov et. al., 2009; Tarasik, 2004). Computational schemes of cars

Figure 11. The computational scheme of oscillations with four degrees of freedom: C_{s1}, C_{s2}, C_{t1}, C_{t2} – damping in the front and rear suspension of the tires; K_{s1}, K_{s2}, k_{t1}, k_{t2} – damping in the front and rear suspension, in the tires; Ftr1, Frt2 – forces of dry friction; z_1, z_2, z_{m1}, z_{m2} – coordinates of the displacement of the centers of mass; α – the angle of rotation of the body relative to the transverse axis; q1(t) and q2(t) – perturbations from the road; V_{avt} – car speed

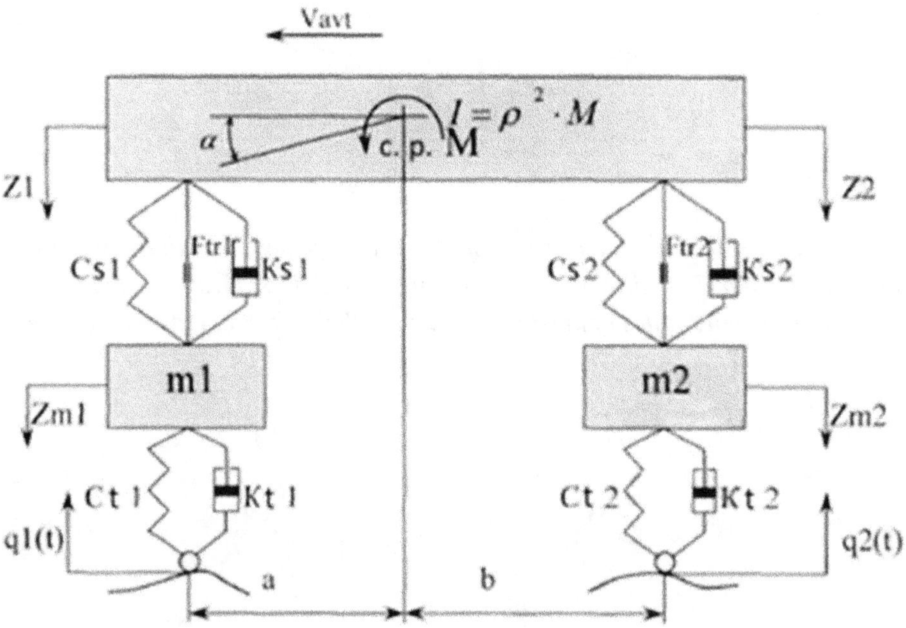

Figure 12. Spatial computational scheme of oscillations

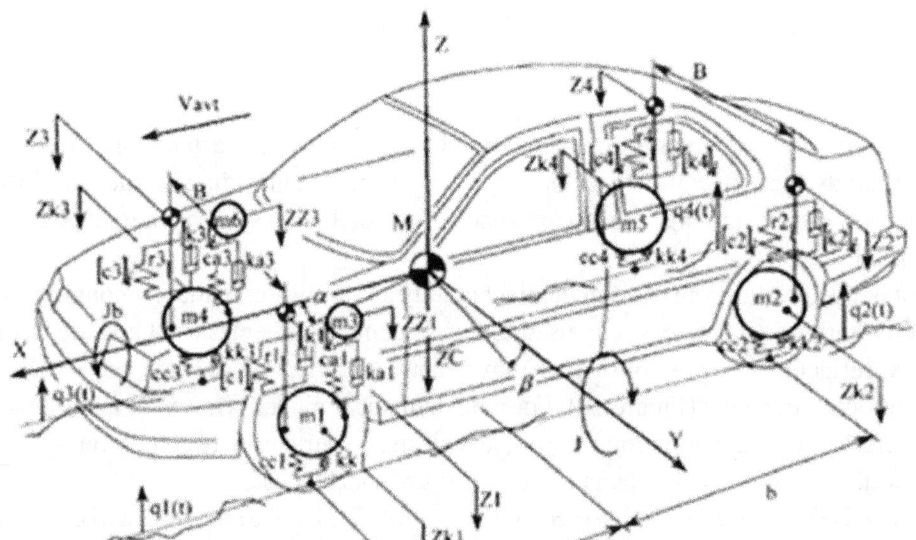

are very diverse. In particular, from Figure 12 it follows that the dynamic properties of a passenger car (smoothness of movement, stability under periodic disturbances, reduction of impacts from shock loads) are provided by a spatial mechanical oscillatory system with many degrees of freedom. Structural and technical relations form the interaction of partial systems, which are taken into account at the design stages by mathematical modeling methods (the location of shock absorbers and levers is meant), followed by experimental refinement of the dynamic state parameters to the required level of requirements (Lalanne, 2002; Khorychev, 2012).

The dynamics of trucks implies the inclusion of additional dynamic loads arising from the transported goods, including liquids in tanks.

The peculiarity of transport suspensions is the use of a fairly wide set of mechanisms that implement elastic interactions of elements in mechanical oscillatory systems, the dissipation of energy, the balancing of mutual motions of elements and assemblies, and the uniform distribution of loads. In Figure 13 is a general view of the rear axle of a truck. Suspension systems for such transport vehicles differ in structural and technical complexity.

Figure 13. Scheme of oscillating systems of a heavy-duty truck tractor (carrying capacity is up to 25 tons) I – wheel suspension system; II – carrier system (tractor frame and semi-trailer frame with platform); III – vibration protection system; 1 – tires; 2 – springs; 3 – vibration dampers; 4 – elastic suspension elements for the cabin and seat; 5 – cabin and seat suspension vibration dampers; 6 – bunk for the co-driver; response of the supporting surface $N_1 \leq 60$ *kN;* $N_2 \leq N_3 \leq N_4 \leq N_5 \leq 100$*kN*

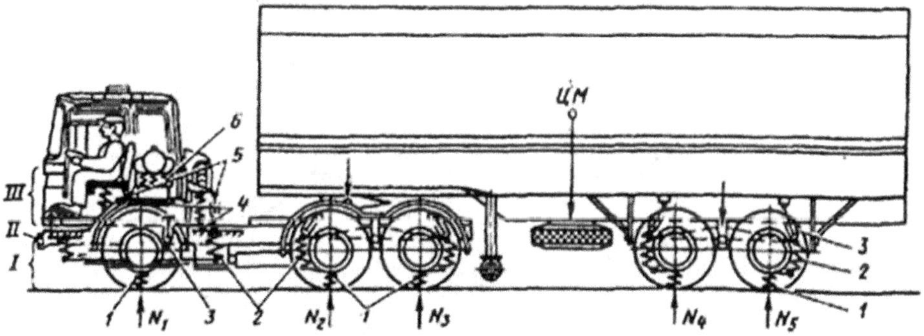

In structural and technical solutions related to car suspension systems, elastic and dissipative elements often represent the mechanisms of one or another type, combined by means of levers into mobile systems that move with several degrees of freedom, as reflected in (Wolfson, 2008; Eliseev, Khomenko & Barsukov, 2011; Bolshakov & Eliseev, 2013). In the problems of protection against periodic external disturbances, the computational schemes of transport vehicles are mechanical oscillatory systems in which components as a protection object (in the form of a solid body with one, two or more degrees of freedom) are identified clearly enough; elastic-dissipative elements in the form of various kinds of dynamic vibration dampers, as well as devices providing spatial interactions of the elements. These elements form kinematic schemes for the transformation of motion and forces, and create a system of definite links between the motions of protection objects when so-called partial systems are distinguished in them. The connectivity of motions between partial systems is ultimately reflected in the formation of a complex picture of vibrational motions, the estimation of the parameters of which, in itself, is a challenging task.

The majority of constructive and technical solutions in the schemes of protection from vibrations and impacts of technological and transport machines are mechanical oscillatory systems in the structure of which lever mechanisms are used, as well as mechanisms of other types such as toothed, screw, articulated, cam, etc. In essence, within these representations, the tasks of ensuring the necessary dynamic state of the object of vibration protection are reduced to solving problems of evaluation, monitoring and control of the oscillations of mechanisms (Wolfson, 2008).

Railway transport differs from automobile transport in that the railway track determines more stable forms of dynamic interactions of the rolling stock and the upper structure of the track. In the dynamics of the rolling stock, works of Russian scientists have created a theoretical basis for solving a wide class of dynamics problems. As for rolling stock of railway transport, it is represented by complex technical systems. As an example, Figure 14 shows a two-axle trolley of a freight car and a bolster.

Figure 14. Conventional biaxial trolley of the bolster

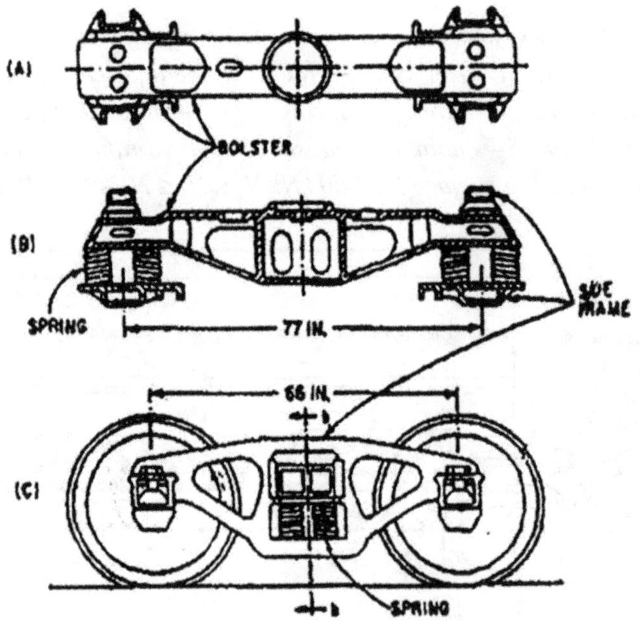

In general, even in the plane dynamics problem, the computational scheme will be a mechanical oscillatory system with several degrees of freedom. In this case, the trolley is only a fragment of a more complex system consisting of several parts. The trolleys, as can be seen from Figure 14, have elastic elements that can occupy different positions in the system, and mobile parts of the structure, also making movements that are characteristic for interactions with lever linkages (Belokobylsky, Eliseev & Sitov, 2013; Eliseev et. al., 2009). The computational scheme of the freight car is shown in Figure 15 and gives a representation of the features of the spatial oscillations of the car in the process of its motion.

Figure 15. The computational scheme of a four-axle freight car

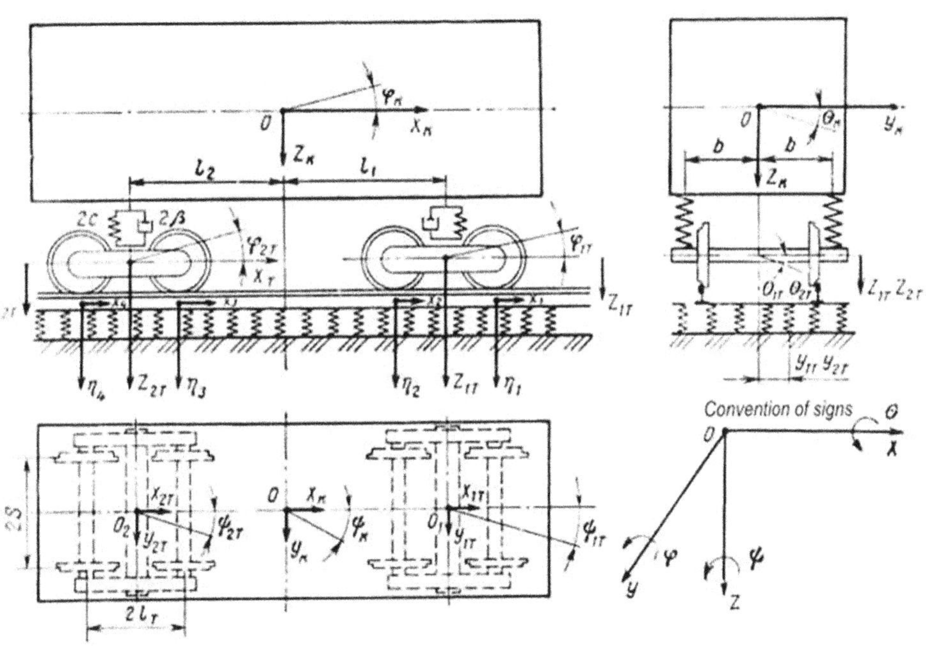

The standard range of natural frequencies of the suspension system in the vertical direction for different types of cars is between 2.5-7.5 Hz. Dynamic interactions of the elements of such a system provide diverse manifestations of lever linkages and the essential significance of such factors as the spatial distribution of the points of attachment of the elastic-dissipative and mass-and-inertia elements inside the object, as well as the ways of interconnecting the elements. Possible ways of using the lever couplings are given in Table 2.

4. An important task is to protect the human operator of transportation vehicles. In Figure 16 shows the computational schemes of vibration protection systems used in the development of the machinist's armchairs (Eliseev et. al., 2008; Eliseev & Ermoshenko, 2010).

Summarizing some results of a comparative survey of the dynamics of railway transport, we note that the providing a standard level of the dynamic state of complex technical systems requires work with complex computational schemes. The latter reflect a complex pattern of the dynamic interaction between individual blocks and devices of the original system. A detailed representation of the dynamic state requires mathematical models (even in a linear formulation) in the form of systems of differential equations of high order. The real needs of solving design and technical problems related to the development of individual blocks (trolleys, couplings, connection with the body of a locomotive or a car, etc.) or evaluation of dynamic properties in certain types of traffic are usually carried out on simplified models, which, in their basic forms, become mechanical oscillatory systems with one, two or several degrees of freedom.

Table 2. Computational schemes for the study of rolling stock vibrations taking into account linkages of the lever type

Features of models of technical objects	Structure and forms of interaction of elements
Schemes of suspension of a biaxial trolley	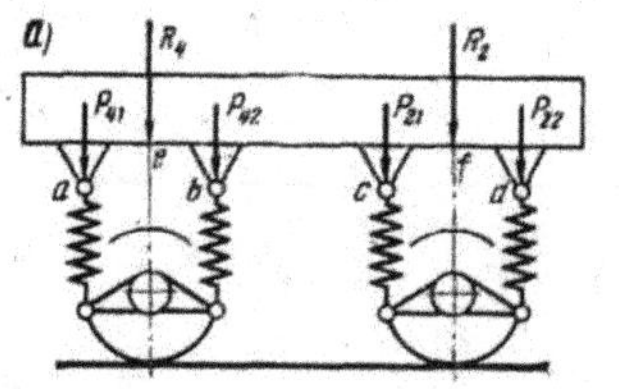 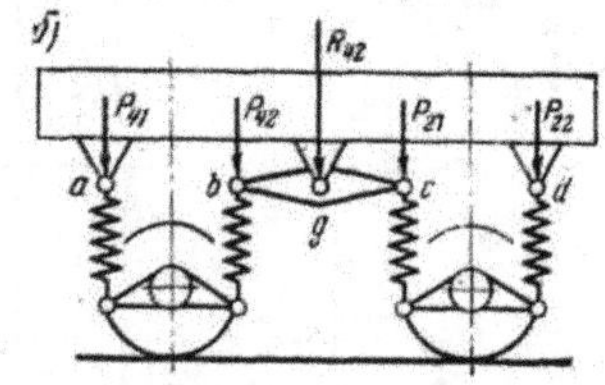
Schemes of suspension of a three-axis trolley	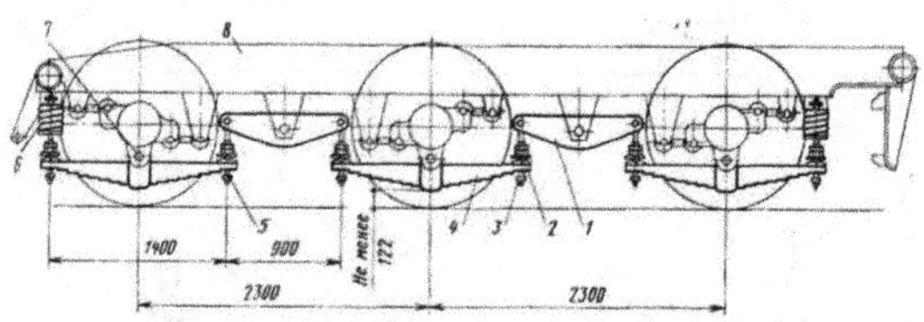
Scheme with traction links of two-way action	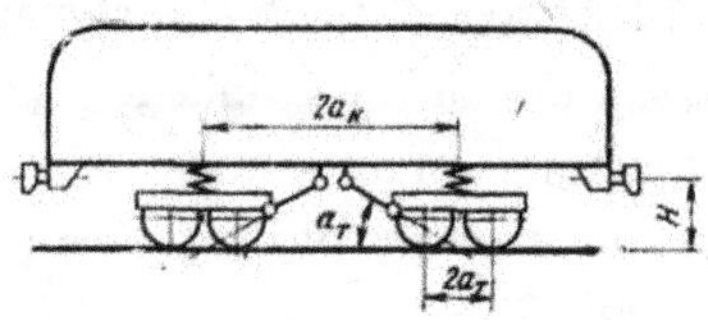
Scheme with traction links (behavior in tension)	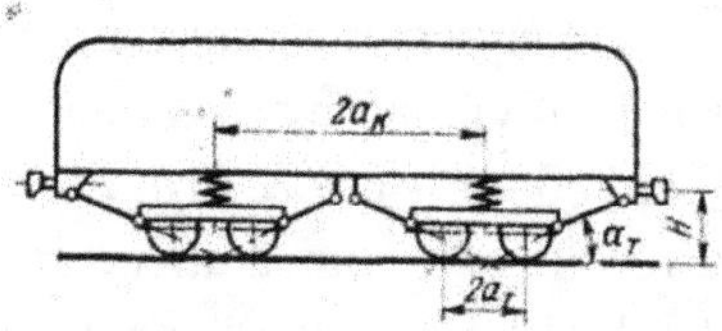

continued on following page

Table 2. continued

The scheme of transmission of longitudinal forces from the trolley to the body by inclined drags	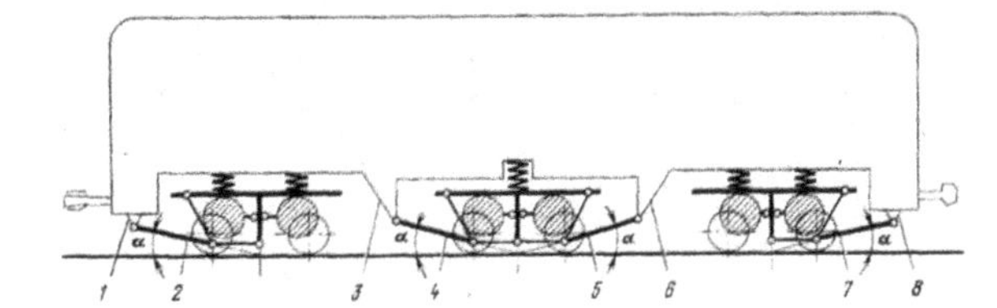
Balanced spring suspension schemes	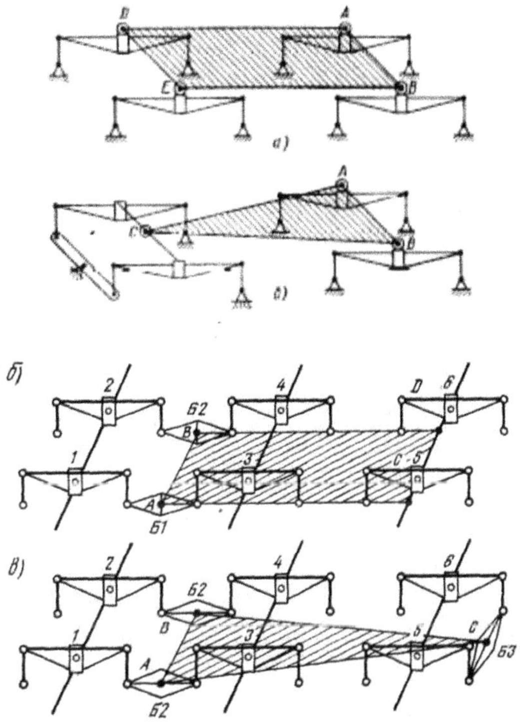

continued on following page

Table 2. continued

Schemes with transverse connection of the body and trolley	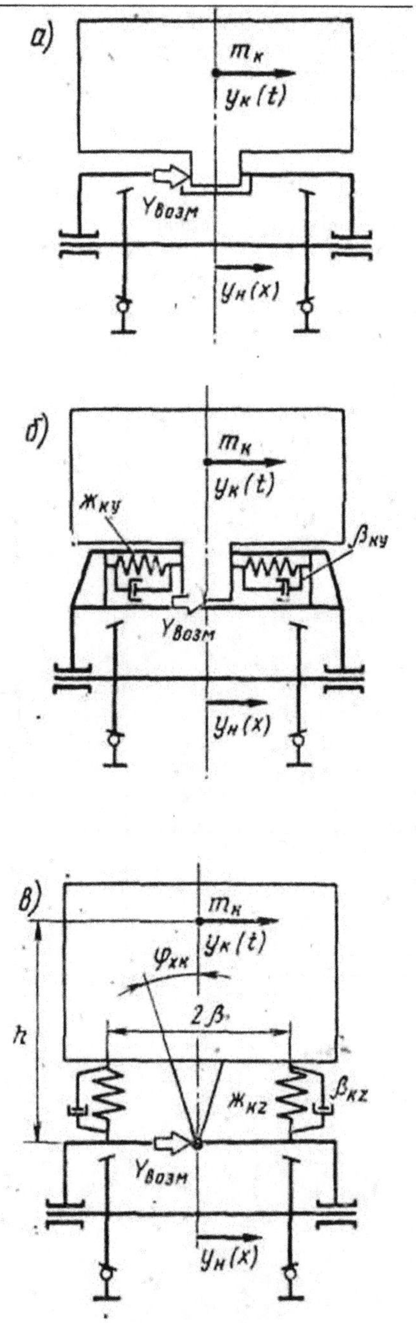

Elastic elements are used as the main elements of spring suspension systems at various levels. As a rule, these are coil springs or leaf springs, as well as devices for dissipating the vibration energy in the form of hydraulic dampers of friction dampers and, in some cases, pneumatic shock absorbers. Naturally, the inertia-and-mass units of the oscillating loop act as system-forming elements represented by solid

Figure 16. Schematic diagrams of the seat guide gears

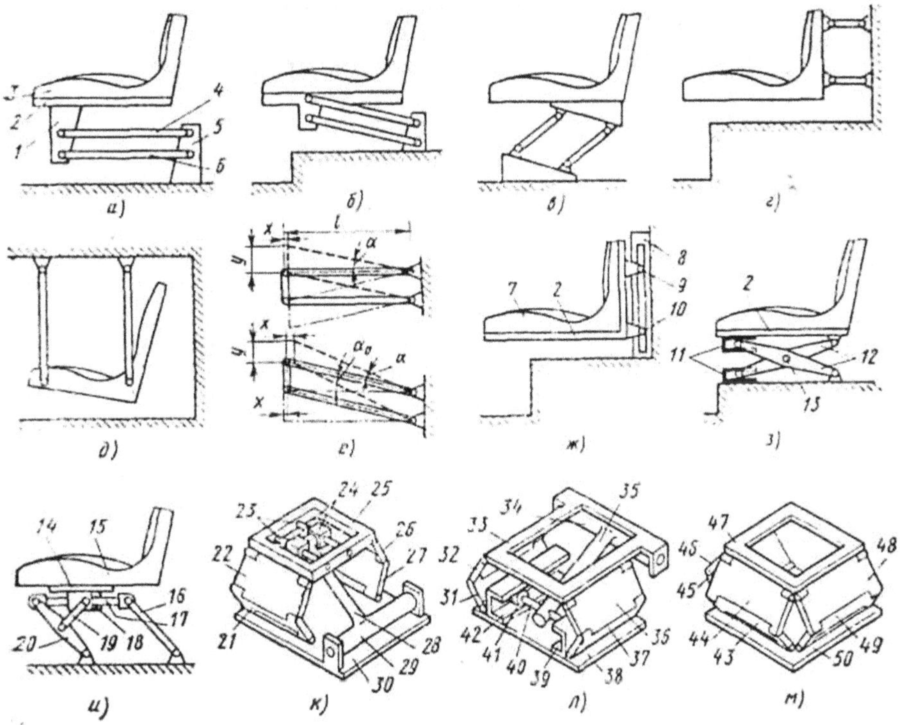

bodies having the shape of a material point; a solid of beam type; a solid body fixed at a point or having the ability to make more complex motions.

Lever Linkages and Mechanisms in Technical Objects

Lever mechanisms are widely used in solving problems of vibration protection of various machines and equipment, since they possess:

- a small propensity to self-oscillations and to the imposition of resonant vibration frequencies of elastic constraints on the vibrations of the elastically suspended elements of the effector (which, apparently, is far from being the case);
- the ability to work in a wide range of external loads with small strokes and forces of elastic units with self-adjustment of these parameters during the reorganization of operating modes;
- ease of maintenance and correction of output parameters under a varied amplitude-frequency spectrum of external loads;
- good mateability (ability to combine) with other types of vibration-isolating (dampening) devices with higher performance indicators of functional reliability, resource and economy of operation (consumed energy).

Figure 17 shows a classification diagram of articulated vibration protection devices.

Figure 17. Classification diagram of articulated vibration isolators of a passive type

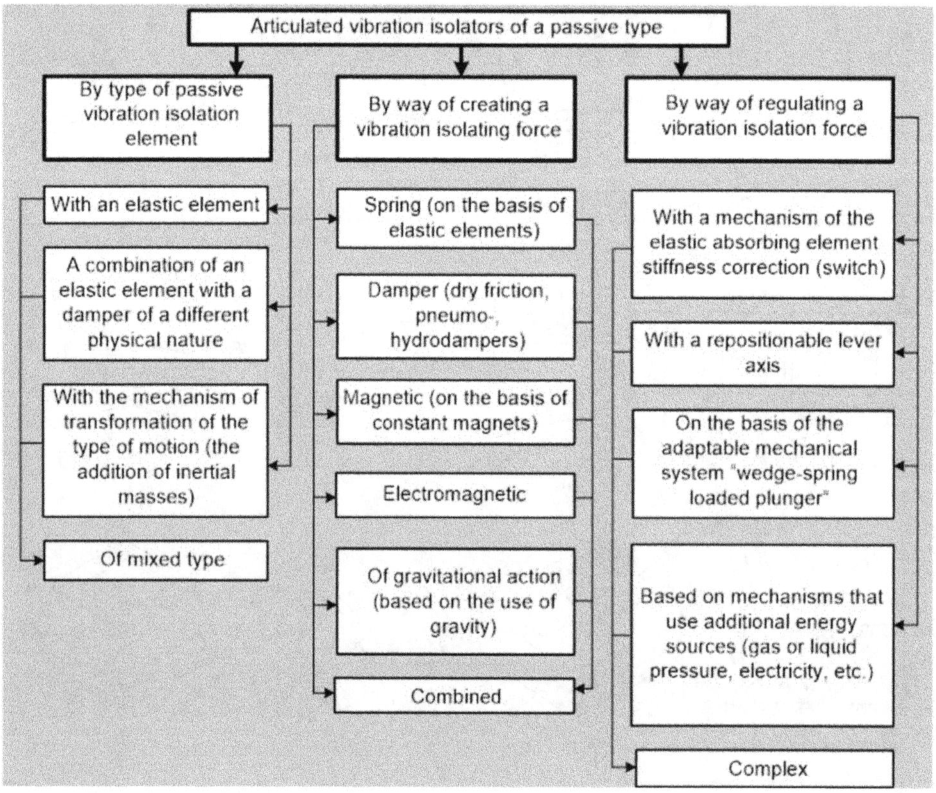

Figs. 18 and 19 describe classification diagrams and design and technological methods on the construction of mechanisms and elements that form the basis of articulated vibration isolators of a passive type. These structures have a constructive simplicity, manufacturability in production and assembly and are used in vibration protection mechanisms to transfer high traction forces when subjected to significant inertia loads. A variety of constructive and engineering forms of lever mechanisms is also presented in (Khomenko et. al., 2013).

Lever mechanisms can be effectively used in combination with pneumatic devices, as shown in Figure 20 (Vargunin et. al., 2006).

Lever mechanisms are often used as a basis for constructing vibration isolators. A spring articulated vibration isolator of general form is known (Figure 1.21).

In this structure, an equal-arm lever *8, 13* with an axis of rotation *12* is used. With hinges *6, 14*, the lever arms *5, 7, 15* are connected to the protection object *A* and the block of counter-acting springs *3* and *10*, providing vibration isolation of the object protection under the disturbing effect of the vibratory base *B*. The lever mechanisms in the problems of adjusting the vibration protection systems can act as the correctors of the stiffness of the elastic elements of the vibration energy dissipation, as shown in Figure 22 (Vargunin et. al., 2006).

Lever mechanisms are used in other constructive and technical options, for example, in rotary motions, power transmissions. Known works in this direction have been performed on elastic couplings. In a number of works, the introduction of lever mechanisms, more precisely, the use of effects created by

Figure 18. Classification diagram of articulated mechanisms of vibration protection devices by the form of kinematic engagement of arms of the swinging type lever with an effector

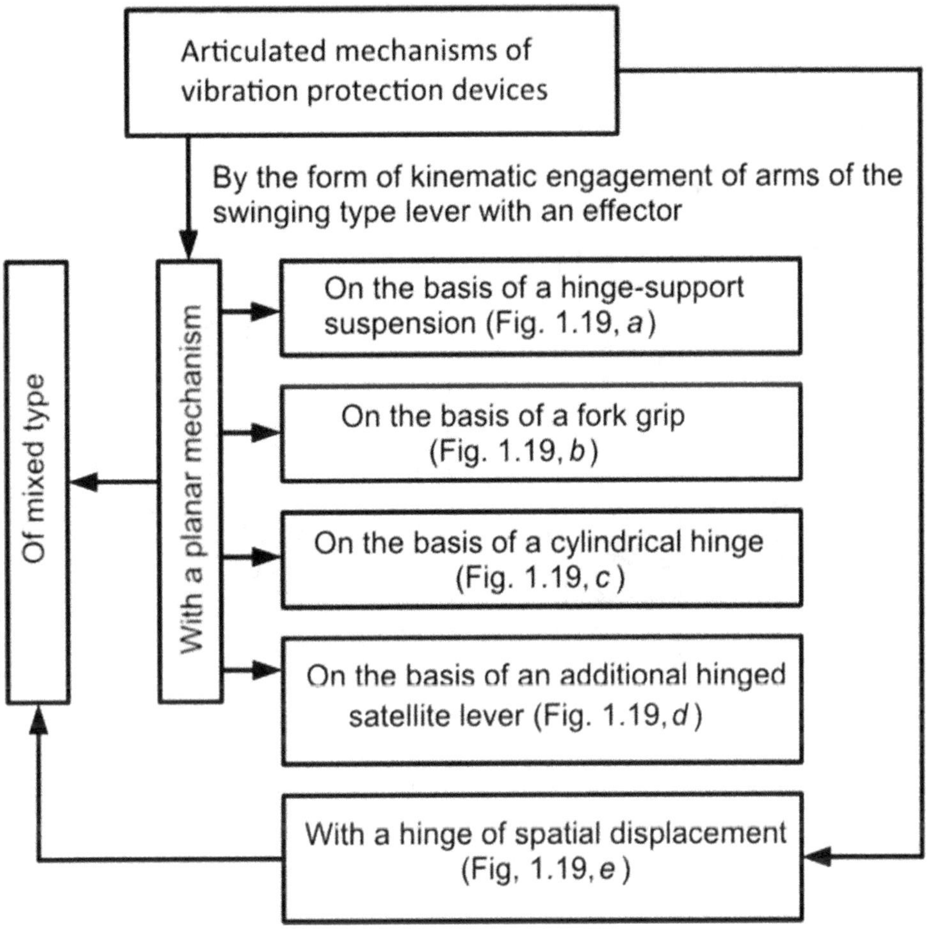

Figure 19. Types of kinematic engagement of the swinging type lever with an effector of reciprocating action: 1 – lever; 2 – effector; 3 – hinge of spatial displacement; 4 – removable ring; 5 – groove in the fork end of the effector; 6 – planar cylindrical hinge rigidly fixed at the end of the lever's effector; 7 – reference through groove for the hinge 6 in the lever; 8 – fork grip of the effector; 9 – groove in the fork end of the lever; 10 – satellite lever; 11 – articulated support

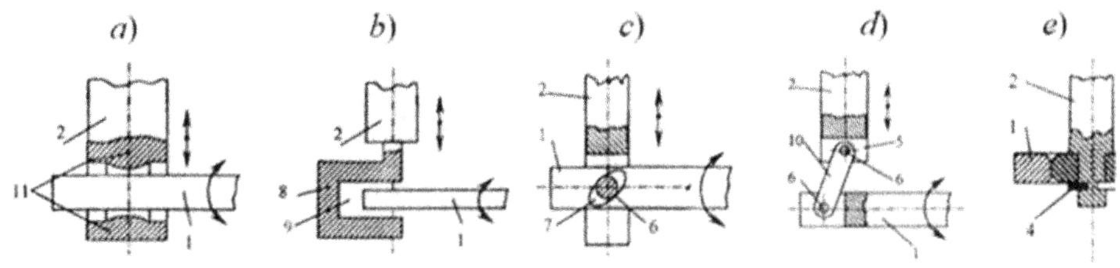

Figure 20. Structural-computational scheme of pneumatic vibration isolator with a lever-piston control mechanism: 1, 2 – working piston (effector or power actuator); 3 – damper-plunger (a sensing element); 4 – the axis of rotation of the lever (movable support); 5 – lever; 6 – case; A – receiver; B_1 – charging unit

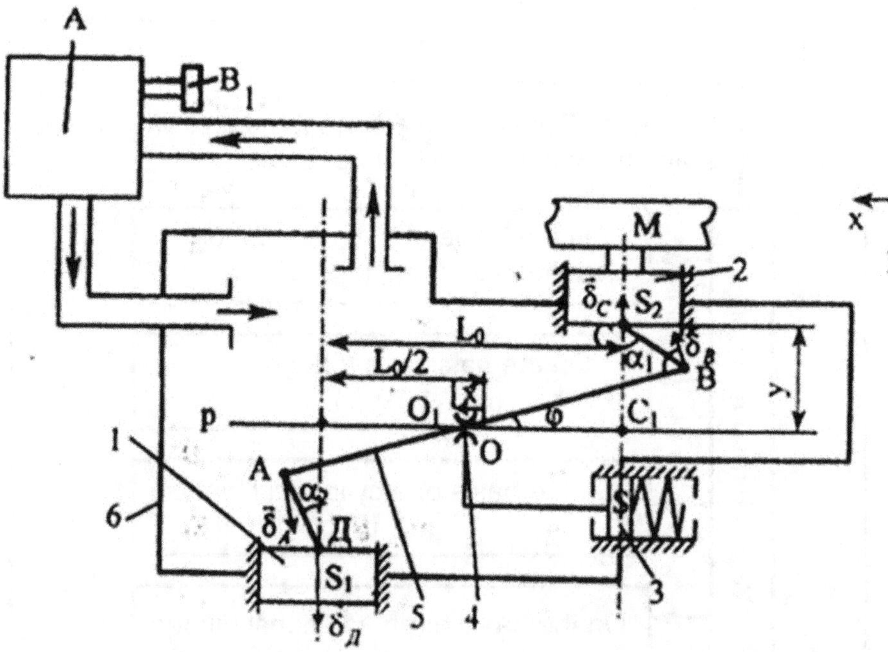

Figure 21. Typical construction of a spring articulated vibration isolator: 1 – housing; 2, 11 – thread trimmer; 3, 10 – spring; 4, 9 – support of the spring; 5, 7, 15 – stock; 6, 14 – hinge; 8, 13 – lever arm; 12 –axis of rotation of the lever; A – object of protection; B_1– vibroactive base

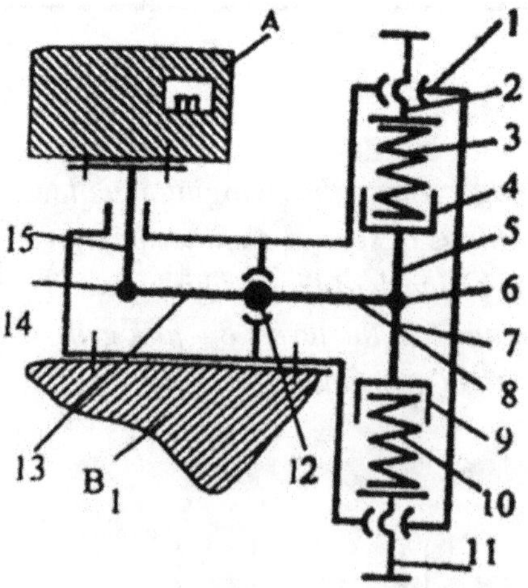

Figure 22. Spring vibration isolator with lever articulated-piston mechanism of energy dissipation: 1 – housing; 2 – power piston; 3 – threaded shank; 4 – spring; 5 – piston rod; 6 – groove; 7, 9, 13 – hinge; 8, 12 – lever; 10, 14 – plunger; 11, 15 – jet; A – object of protection; – vibroactive base

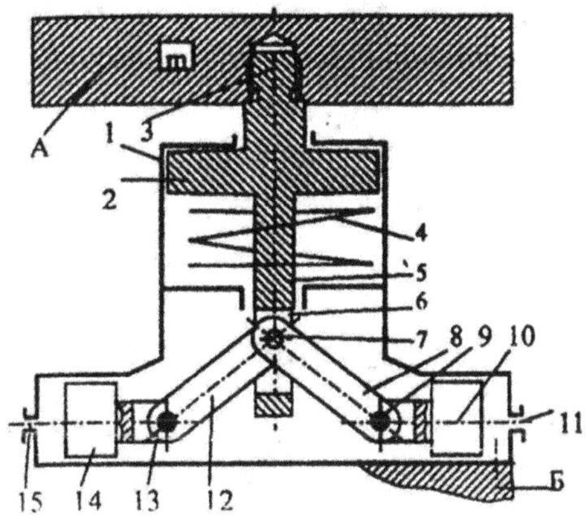

lever mechanisms, but on the basis of other constructive and technical solutions, is presented in (Eliseev & Ermoshenko, 2012), where the notion of additional constraints is used. An interesting variant of the interpretation of the lever interactions is the helical non-locking mechanism (Eliseev et. al., 2008), in which the relationships between the parameters of the translational and rotational motions are implemented, as well as between the power factors "force–torque". In Figure 23 some schemes of vibration protection systems with mechanisms for the transformation of motion are presented.

Figure 23. Variants of computational schemes of vibration protection systems with additional constraints: a – external engagement; b – internal engagement; c – epicyclic transmission; g – transmission through a toothed rack

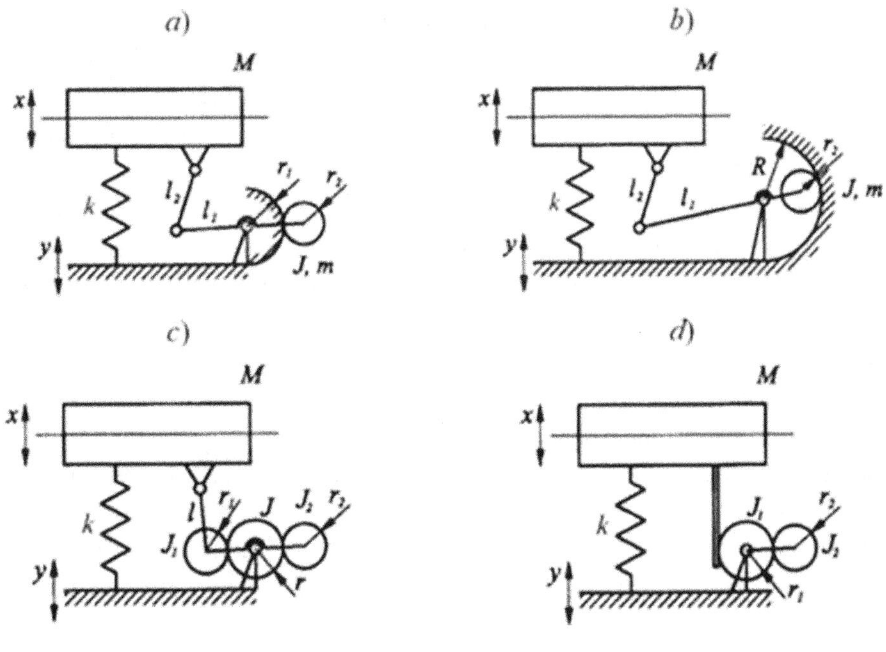

Lever mechanisms in their simplest constructive and technical forms in the system of electromechanical analogies are interpreted in the form of transformers, which makes it possible to note the essential significance and role of lever linkages in the dynamic interactions. The latter ones manifest themselves through kinematic and force relations of the parameters of the static and dynamic states of mechanical oscillatory systems. In those cases when external perturbations are of a harmonic nature, the usual characteristics of the system can be regarded as dynamic in the sense that the stiffnesses of the elastic elements, the mass-and-inertia properties become dependent on the frequency of the external action. Such approaches are also connected with the use of the notions of reduced stiffnesses and masses (Belokobylsky, Eliseev & Sitov, 2013; Belokobylsky, Eliseev & Kashuba, 2013; Eliseev et. al., 2008; Belokobylsky & Eliseev, 2016; Eliseev, Reznik & Khomenko, 2011). The general properties and manifestations of lever linkages are connected with the consideration of the oscillations of solids on elastic supports when a system of generalized coordinates is used, connected with the center of mass and rotations around it.

Features of Computational Schemes in the Problems of Machine Dynamics: Approaches and Methods

Technical objects in the form of technological and transport machines are distinguished by a complex dynamic structure and a variety of motions of the constituent elements. Specific problems of dynamics determined by the need to assess the dynamic state of individual aggregates, assemblies and parts, as a rule, represent a local distribution of interactions of elements, which is associated with the construction of appropriate mathematical models. Usually such an abstraction and formalized description of the research task at the level of use of a certain analytical tools are preceded by a preliminary assessment of the properties of the technical object and the construction of a computational scheme corresponding to the problem under investigation. The papers (Harris & Crede, 2002; Lalanne, 2002) deal with the methodological issues, focused on the need to implement systemic approaches in the construction of mathematical models, the need to take into account the laws of physical nature. The main problem in mathematical modeling is related to the rational choice of the dimension of the model, that is, to the evaluation of the possibilities of introducing simplifications that make it possible to obtain mathematical models allowing the possibility of analytical methods. These methods give a representation of the existence of certain regularities, general dynamic properties, the possibility of manifesting special dynamic regimes such as resonance phenomena, dynamic absorbing of vibrations, loss of static or dynamic stability. General recommendations in the implementation of such approaches are reflected in (Rakov & Sinev, 2011).

The computational scheme of many mechanical engineering objects as machines that traditionally consist of components such as engine, power transmission or transmission mechanisms, working elements in the form of mechanisms interacting with the working medium in formalized representations are displayed in the form of mechanical oscillation systems of varying complexity (Wolfson, 2013). Although mechanical systems consist of a limited set of typical elements (mass-and-inertia, elastic and dissipative), the problem of investigating dynamic properties can be very complex because of the presence of nonlinear properties in real objects. Such properties are manifested in connection with the mass variability, nonlinear properties of elastic and dissipative elements. Gaps, backlashes, preliminary deformations (tensions) of elastic elements, forces of dry friction, etc., arising in the work of compounds of individual elements, affect fundamentally the dynamic properties of machine engineering objects (Belokobylsky, 2002). In technical calculations conducted at the stages of pre-project research and development, methods of linearization of non-linear systems are widely used, which, in many cases, provides the needs of frequency analysis and dynamic synthesis of vibration protection systems (Harris & Crede, 2002).

Linearization of mathematical models can be performed at various stages of the formation of representations about the properties of the object being involved.

The problems of the vibration protection of objects, in the case of the dynamics of technological and transport machines, in a certain sense, can be regarded as "milestone" problems of the dynamics of machines, since in setting the aims of vibration protection, choosing the object of protection, evaluating external influences, determining the parameters of the dynamic state, choice of methods and means of state management, there are all stages of calculating the system parameters necessary to provide operating of parts and machine components.

In recent years, in connection with the development of methods of automation, design and research of the dynamics of various technical systems, applied software packages of specialized and general purpose have become widespread (ADAMS, Maple, Matlab, Mathcad). The basis of most software packages used to assess the state of technical objects is the finite element methods in their various modifications (Pykhalov & Milov, 2007; Sobolev, 2003). With regard to the evaluation of dynamic properties of systems, the packages of applied programs, due to the specific nature of their methodological basis, have a certain range of capabilities. To evaluate the dynamic properties of technical objects in general, and engineering objects in particular, the most commonly used computational schemes are mechanical oscillatory systems with one, two and three degrees of freedom.

The most widely used method is based on the application of the Lagrange equations of the second kind (Luric, 1961). In this case, the motion of the system is considered in generalized coordinates with the determination of the kinetic and potential energies of the system and the corresponding generalized forces. Details of the application of this method are presented in many works, including applications to vibration protection systems (Belokobylsky, Eliseev & Kashuba, 2013; Eliseev, Reznik & Khomenko, 2011).

Taking Account of Constraints in Connections of Mobile Elements

When constructing mathematical models, the constructions of mechanical oscillation systems are taken into account. Such systems are flat or spatial formations of mass-and-inertia elements (material points, solids), elastic constraints and dissipative devices for the dispersion of oscillation energy. When connecting such elements, closed mechanical chains and more complex structures are formed, in which different mechanisms can be located. The latter ones can be considered (Druzhinsky, 1977) as closed mechanical chains with one fixed unit. In all cases of connection, a mobile formation is regarded, which is obtained from two units that can move relative to each other. This mobile formation of two units is called a kinematic pair. There are different types of compounds and their classifications. Table 3 shows the classification of kinematic pairs, reflecting the possibility of relative motions of the units.

Constraints are an important concept in mechanics, and their complexity determines, to a large extent, the complexity of mathematical models of technical objects.

In constraints are defined as certain restrictions that are imposed on the motion of a system consisting of material points. Formally, the constraints can take the form of equations connecting the coordinates of the motion of individual points. The constraints may depend on time in explicit form or indirectly. Depending on the singularities of the equations of constraints, they are divided into holonomic and non-

Table 3. Conventional notations of kinematic pairs

Pair class	Number of conditions of constraint	Number of degrees of freedom	Pair name	Figure	Conventional notation
I	1	5	Ball-and-plane	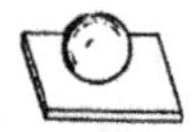	
II	2	4	Ball-and-cylinder		
III	3	3	Spherical		
III	3	3	Planar		
IV	4	2	Cylindrical		
IV	4	2	Spherical with a finger	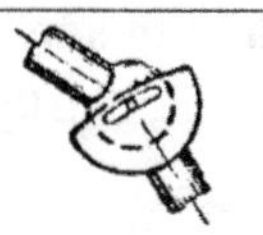	
V	5	1	Translational	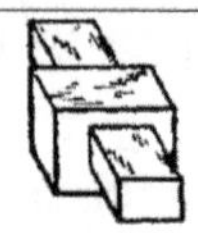	
V	5	1	Rotational		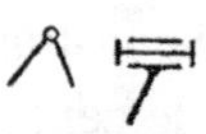
V	5	1	Screw		

holonomic. Existing constraints are divided into bilateral and unilateral, depending on whether they can change the sign of the application of loads, there may (and may not) be contact failures.

Such constraints are often found in transport vehicles ("contact – wheel – road", "wheel pair – rail", etc.) and in vibrating technological machines (Panovko, 2006; Kogan, 1997).

In mechanical oscillatory systems, elastic elements or springs are regarded as units in a computational scheme or a mechanical system. In this sense, attention should be paid to the shapes and connections of the mass-and-inertia elements with elastic elements, and also to the interconnections between the elastic and dissipative units. The features of such representations about the connections of elements of oscillatory systems are characteristic for the applied theory of vibration protection systems. Mass-and-inertia elements in mechanical oscillation systems can also be interconnected, including using the lever linkages implemented by rods having fixed or movable support points. Partly such opportunities were considered in the above sections.

The notion of a constraint has a broad conceptual field, and therefore specific constructive and technical forms of interaction of elements should be displayed with appropriate comments on the nature of the formation of interactions of elements in order to identify the generality of this concept. Kinematic pairs are constraints. The elastic member may be associated with another member or a support surface.

At the same time, the technical literature often uses the concepts of elastic and dissipative constraints implemented by springs and dampers. In addition, the notion of mass-and-inertia constraints that can arise between individual elements, as well as between blocks of oscillatory systems, is applied. Interpartial systems can also be considered as such blocks.

A number of works dealing with the dynamics of vibration protection systems develop ideas about possible forms of joining two elements with the help of intermediate units, whose properties take the limiting values (tend to zero or infinity) (Eliseev, Gordeeva & Fomina, 2011). In such cases, the constraint between the elements disappears, or is transformed into a compound of a certain type. Such compounds are called joints, which broadens the notion of possible forms of joining elements of mechanical oscillatory systems and their constraints (Eliseev & Ermoshenko, 2012).

Mathematical models of mechanical oscillatory systems with distributed parameters are based on the use of partial differential equations. Elements with distributed parameters have the form of elastic beams, plates, membranes, which is widely used in technical systems. The mathematical tools for the study of elastic vibrations are reflected in works on the dynamics of oscillatory systems, the dynamics of structures and construction mechanics.

In the first approximation, elastic vibrations of elements are considered as oscillations of systems with lumped parameters on the basis of simplification methods developed in the theory of oscillations, where the approximate model with one degree of freedom reflects the properties of motions in the lower forms of oscillations. In general, the dynamics of systems with distributed parameters have their own peculiarities, the consideration of which requires detailed ideas about the conditions for fixing elastic elements and their combination with elements having lumped properties.

A preliminary review of the methods of mathematical modeling of various mechanical systems shows the trends in the formation of fairly general approaches, in which the original mechanical systems are represented by dynamic systems that have developed forms of various constraints and influences from the external environment. In this case, force disturbances can be considered in certain situations as controlling forces.

In such representations, the analytical tools for studying the dynamic properties of systems become, in essence, the application of general methods of the theory of systems, and in special cases, the theory of automatic control (Galperin, 1964).

Structural Mathematical Modeling in the Dynamics of Mechanical Oscillation Systems

1. The ideas of structural mathematical modeling in the dynamics of mechanical oscillation systems are based on the generality of the representations of the properties of automatic control systems and mechanical oscillatory systems represented by systems of ordinary differential equations with constant coefficients. The objects under consideration have linear properties. The use of controlled forces in mechanical oscillatory systems turns such systems, including vibration protection systems, into automatic control systems (Eliseev et. al., 2008; Eliseev, Reznik & Khomenko, 2011). The basis for the formalization of ideas about possible types of constraints between elements is the use of Laplace transforms (Lurie, 1961), which allow linear differential equations to be transformed into algebraic ones. The subsequent transformations of structural interpretations of a system of algebraic equations are associated with the development of the formalization tools proposed in papers on technical cybernetics and the theory of automatic control (Eliseev & Khomenko, 2014; Kim, 2003; Olson, 1947).

In the transition to structural mathematical models, the system of initial linear differential equations is interpreted by a dynamically relevant structural diagram of the automatic control system. In this case the dynamic properties of the original system are correlated with the transfer function, which depends on the composition of the elements, their interconnections and the support surfaces, and also on the type and places of application of external disturbances. In such a formulation of the problems of evaluation and control of the dynamic state, the study of the problems of the dynamics of machines in vibrational disturbances reduces to studying the properties of the corresponding transfer functions. In this case, the response of the mechanical system is considered at the input harmonic signal, and the response is estimated from the frequency characteristics reflecting the ratio of the modulus of the dynamic parameter of interest to the modulus of the input action (or signal).

This approach makes the inverse Laplace transform to determine the response in the time domain unnecessary and allows obtaining information using the features of the resulting transfer function. In this case, a certain part of the useful information is lost, since only linear forms of the characteristics of the elements are assumed, transient processes are not taken into account when external disturbances are applied, etc. The shortcomings of such preliminary estimates of dynamical systems can be eliminated to a certain extent by using special methods for taking into account nonlinear factors and complicating the relationships inside the system.

2. Figure 24 shows the main stages of the transformation of a system with one degree of freedom while determining the dynamic properties of constraint reactions.

It is shown that each element of the mechanical oscillatory system (mass-and-inertia element, spring, damper) in the structural diagram corresponds to an element of the structural scheme with the corresponding transfer function of the second-order integrating unit, the first-order amplifying and differentiating units. Note that the structural diagram reflects the features of the defining problems of the dynamics of machines associated with vibration protection and vibration isolation from external influences (Belokobylsky, Eliseev & Kashuba, 2013; Eliseev et. al., 2008; Eliseev, Reznik & Khomenko, 2011).

Figure 24. Principal positions of the method of determination dynamic responses

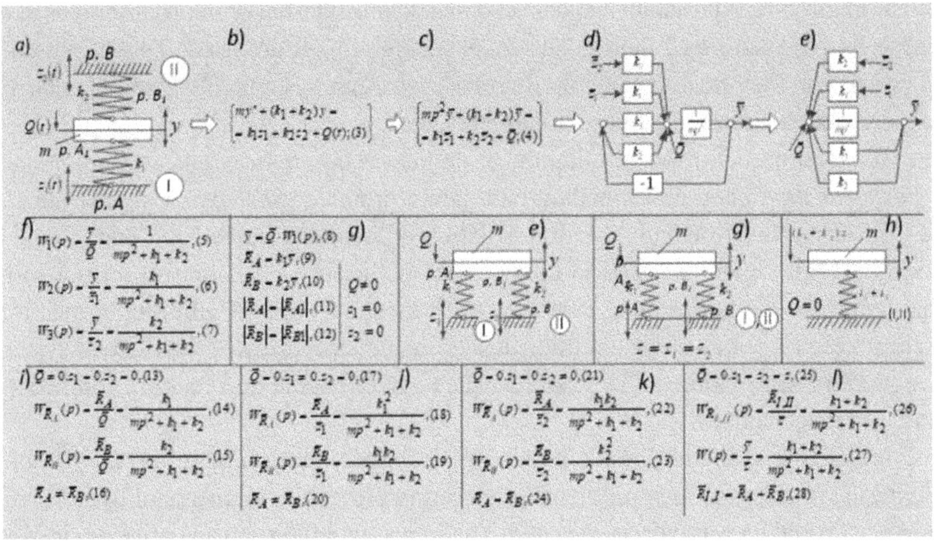

The object of protection in vibration protection systems is the basic element and it has the transfer function of an integrating unit of the second order. Concerning this unit, direct and reverse constraints (feedbacks) are formed. When introducing additional components into the computational scheme, as shown in Figure 24, corresponding changes occur in the structural diagram. When introducing a complex element, for example, a motion transformation device implemented by introducing a screw non-locking mechanism, the structural diagram acquires an additional constraint of the corresponding type. Details of the evaluation of the introduction of additional constraints in the form of various mechanisms are considered in (Eliseev et. al., 2008; Eliseev et. al., 2013). For further applications and the development of the method of structural mathematical modeling, it is important to refine representations about constraints in systems with several degrees of freedom and expand ideas about the possibilities of searching for and developing new forms of constructive and technical linkages and their influence on the dynamic properties of vibration protection systems.

Vibration Protection of Machines and Equipment

Several directions can be identified in solving the problems of protecting technical objects from external and internal periodic disturbances. In each of them the specificity of the arising physical processes is taken into account.

These approaches assume the elimination of such features of parts and assemblies as a mismatch in the rotational motions of the position of the center of mass and the center or rotation axis. Sources of such disturbances have a different nature, including inaccuracy in the manufacture of parts. Methods and means for balancing rotating parts and assemblies have been developed, which was presented in (Belokobylsky, Eliseev & Sitov, 2013; Wolfson, 2008; Eliseev et. al., 2008; Khomenko & Barsukov, 2011). A more complex problem in this direction is to balance the mechanisms in which the reduced masses and stiffnesses change when implementing periodic movements of individual units. This requires the use of special balancing means by introducing additional masses and their corresponding location on moving parts of mechanisms and devices.

In a more general formulation of this vibration protection problem, one can consider a task of reducing the limitation or modifying external influences. In cases where external or internal sources of excitations cannot be limited by the nature of their occurrence, special means are used. Together with the object of vibration protection, they form so-called vibration protection system. Elements of such systems are springs, dampers, shock absorbers, vibration dampers, etc.

Vibration isolators, used for the installation of engines, compressors and technological machines for various purposes have been developed and are being produced serially. A significant place in the problems of vibration protection and vibration isolation of technical objects has been found by motion transformation devices, which are used in connections with elastic, dissipative and mass-and-inertia elements that belong to the traditional set of typical elementary units of vibration protection systems. Motion transformation devices are most often levers in the structures of oscillatory systems. At the same time, motion transformation devices can take the form of helical non-locking, toothed, pneumatic, hydraulic and other mechanisms.

In some cases, it is possible to directly reduce the level of vibration based on the use of active devices using external sources. In such problems of vibration protection elements of hydro-, pneumo- and electroautomatics forming a servo drive are used. The input signal to such an active device is formed by a certain measuring circuit and is then implemented by a servo that creates a control force. In essence, the vibration protection system in advanced form can be implemented as a specialized system of automatic control of the dynamic state. Such systems have found application in transport and technological complexes for special purposes.

A generalized approach to the problems of dynamic synthesis of vibration protection systems predetermines the development of ideas of feedback and structural mathematical modeling within the framework of the analytical tools of the theory of automatic control. Structural mathematical modeling in the problems of the dynamics of mechanical oscillatory systems has the potential to create original solutions, which use the notion of expanding a set of typical elementary units, new types of constraints (including lever linkages) between elements, and new rules for structural transformations of mathematical models of vibration protection systems.

CONCLUSION

Thus, for mechanical engineering objects, mechanical oscillating systems with lumped parameters are widely used as computational schemes. These computational schemes are used for vibrating technological and transport machines. In the practice of preliminary studies and estimations, the computational schemes of machine building objects are simplified, primarily by introducing assumptions about the smallness of the oscillations and the linearity of the properties of typical elements (mass-and-inertia, elastic and dissipative ones). As mathematical models, systems of ordinary linear second-order differential equations with constant coefficients are used. When constructing mathematical models, methods of reducing the number of degrees of freedom of the system are used; expressions for kinetic energy are simplified, representations are used about the possibilities of expanding a set of typical elements and systems, including increasing the various forms of their connection.

Essentially, in real objects, the typical units (springs and dampers) are, in a certain sense, mechanical devices and mechanisms, that is, they consist of several parts or elements, including movable ones, which in a number of cases allows us to use the notions of different forms connecting elements to blocks and structures.

In the formation of spatial forms of mechanical oscillation systems, lever mechanisms and lever linkages play a key role. In this regard, the problems of oscillation of solids on elastic supports are particularly indicative, since in rotational movements relative to the center of mass, the relationships characteristic of lever mechanisms of various types are established between the coordinates of the attachment points of elastic elements. Stable relationships between the parameters of the state of mechanical oscillatory systems can be considered as lever linkages, implemented in a variety of constructive and technical forms.

In structures of computational schemes of machine building objects, mechanisms can be used as certain elements of a structure, but of a more complex kind. Due to the preference of linear mathematical models in mechanical oscillation systems, structural mathematical modeling methods based on Laplace transforms have certain advantages. Structural models provide taking account of features of formation of dynamic constraints and possible transformations of a system for obtaining necessary transfer functions.

REFERENCES

Antipov, V. A. (2006). *Suppression of vibration of Gordeeva and units of transport systems*. Marshrut Publ.

Belokobylsky, S. V. (2002). *Dynamics of systems with dry friction and its application to the problems of mining mechanics*. Mashinostroenie Publ.

Belokobylsky, S. V., & Eliseev, S. V. (2016). Dynamic analogies in problems of the theory of circuits and automatic control: Structural interpretations. *Systems. Methods. Technologies*, *32*, 7–16.

Belokobylsky, S. V., Eliseev, S. V., & Kashuba, V. B. (2013). *Applied problems of the structural theory of vibro-protective systems*. Polytekhnika Publ.

Belokobylsky, S. V., Eliseev, S. V., & Sitov, I. S. (2013). *Dynamics of mechanical systems. Lever and inertial-elastic constraints*. Polytekhnika Publ.

Bolshakov, R. S., & Eliseev, S. V. (2013). *Methodological foundations of problems of determining the reactions of constraints in mechanical oscillatory systems*. Dep. in the VINITI RAS.

Domnin, D. A. (2005). *A method for improving the vibration-damping parameters of an automobile suspension by selecting rational parameters for dynamical wheel vibration dampers*. Research Gate.

Druzhinsky, I. A. (1977). *Mechanical chains*. Mashinostroenie Publ.

Eliseev, S. V. Belokobylsky, S.V., Upyr', R.Yu. & Gozbenko, V.E. (2009). Lever linkages in the problems of the dynamics of mechanical oscillatory systems. Theoretical aspects. Irkutsk: Dep. in the VINITI RAS.

Eliseev, S. V., Artyunin, A. I., Ermoshenko, Yu. V., Moskovskikh, A. O., & Kaimov, E. V. (2013). *The method of transformation of mechanical circuits based on the introduction of intermediate devices in the compounds*. Dep. in the VINITI RAS.

Eliseev, S. V., & Ermoshenko, Yu. V. (2010). Dynamic properties of vibration protection systems. Standing States. *Modern technologies. System analysis. Modeling.*, *28*, 24–30.

Eliseev, S.V. & Ermoshenko, Yu.V. (2012). *Joints of units in the dynamics of mechanical oscillation systems*. Irkutsk: IrGUPS Publ.

Eliseev, S. V., Gordeeva, A. A., & Fomina, I. V. (2011). Dynamic properties of vibrational systems at limiting values of the parameters. *Herald of the VSSUTU*, *35*, 12.

Eliseev, S. V., Kargapoltsev, S. K., & Kashuba, V. B. (2016). Structural Transformations Mapping of Dynamic Interactions in Oscillatory Systems. *Bulletin of the SSUPS*, *33*, 27–41.

Eliseev, S. V., & Khomenko, A. P. (2014). *Dynamic damping of oscillations: the concept of feedback and structural methods of mathematical modeling*. Nauka Publ.

Eliseev, S. V., Khomenko, A. P., & Barsukov, S. V. (2011). *Modern problems of machine dynamics. Protection from vibrations and shocks*. Dep. in the VINITI RAS.

Eliseev, S. V., Moskovskikh, A. O., & Kaimov, E. V. (2013). *Dynamics of mechanical oscillatory systems: structural analogies, mechanical circuits*. Dep. in the VINITI RAS.

Eliseev, S. V., Reznik, Yu. N., & Khomenko, A. P. (2011). *Mechatronic approaches in the dynamics of mechanical oscillation systems*. Nauka Publ.

Eliseev, S. V., Reznik, Yu. N., Khomenko, A. P., & Zasyadko, A. A. (2008). *Dynamic synthesis in generalized problems of vibration protection and vibration isolation of technical objects*. IGU Publ.

Galperin, I. I. (1964). *Automation as one-sided mechanics*. Energiya Publ.

Goverdovsky, V. N., & Trofimov, A. N. (2009). Problems and perspectives of creation of vibration protection for high-speed railway transport. *Modern technologies. System analysis. Modeling*, *24*, 86–92.

Harris, C. M., & Crede, C. E. (2002). *Shock and Vibration Handbook*. McGraw — Hill Book Co.

Kamaev, V. A. (2005). *Conceptual design. Development and improvement of methods*. Mashinostroenie Publ.

Khomenko, A. P., Eliseev, S. V., Artyunin, A. I., Parshuta, E. A., & Kaimov, E. V. (2013). *Mechanisms in elastic oscillatory systems: peculiarities of dynamic properties, tasks of vibration protection of machines, devices and equipment*. Dep. in the VINITI RAS.

Khorychev, A. A. (2012). *Conceptual design of functional subsystems of cars based on heuristic classifications*. NIH.

Kim, D. P. (2003). Theory of automatic control.: Vol. 1. *Linear systems*. Fizmatlit Publ.

Kogan, A. Ya. (1997). *The dynamics of the track and its interaction with the rolling stock*. Transport Publ.

Kopylov, Yu. R. (2011). *Dynamics of the processes of vibro-impact hardening*. Nauchnaya Kniga Publ.

Lalanne, C. (2002). *Mechanical vibrations shock. Specification Development* (Vol. V). Hermes Penton Ltd.

Lurie, A. I. (1961). *Operational calculus and application in technical applications*. GITTL Publ.

Lurie, A. I. (1961). *Analytical Mechanics*. GIFML Publ.

Olson, G. (1947). *Dynamic analogies*. Politizdat Publ.

Panovko, G. Ya. (2006). *Dynamics of vibrational technological processes*. Institute for Computer Research.

Pykhalov, A.A. & Milov, A.E. (2007). *The contact problem of the static and dynamic analysis of the assembly rotors of turbomachines*. Irkutsk: IrSTU Publ.

Rakov, D.L. & Sinev, A.V. (2011). Structural analysis and synthesis of new technical systems based on the morphological approach. *Problems of machine-building and automation, 2,* 73-80.

Ryabov, I. M., Chernyshev, K. V., Vorobyov, V. V., & Urukov, I. N. (2009). A mathematical model of a vehicle suspension with an inertia-frictional shock-absorber. *Modern technologies. System analysis. Modeling*, *21*, 29–31.

Samarsky, A. A., & Mikhailov, A. P. (2002). *Mathematical Modeling: Ideas. Methods. Examples.* Fizmatlit Publ.

Selvinsky, V. V. (2009). *Dynamics of the contact interaction of solids.*

Sobolev, V. I. (2003). *Discrete-continual mathematical models in algoritmic and programmatic resolution of problems of vibration suppression of construction and equipment.* Research Gate.

Sovetov, B. Ya., & Yakovlev, S. A. (2001). *Modeling of systems.* Vysshaya Shkola Publ.

Tarasik, V. P. (2004). *Mathematical modeling of technical systems.* Dizayn PRO Publ.

Vargunin, V. I., Gusarov, V. N., Ivanov, B. G., & Levchenko, A. S. (2006). *Construction and calculation of artlculated means and assemblies.* Samara State Technical University Publ.

Wolfson, I.I. (2008). *Oscillations in the machines.* St. Petersburg: SPbGUTID Publ.

Wolfson, I. I. (2013). *Dynamics of cyclic machines.* Polytekhnika Publ.

Zinoviev, V. A., & Bessonov, A. P. (1964). *The fundamentals of the dynamics of machine aggregates.* Mashinostroenie Publ.

ADDITIONAL READING

Eliseev, A. V. (2017). Dyads in mechanical systems: Features of dynamic properties. Part I. *Vestnik Irkutskogo Gosudarstvennogo Tehniceskogo Universiteta*, *21*(7), 26–38. doi:10.21285/1814-3520-2017-7-26-38

Eliseev, A. V., & Eliseev, S. V. (2016). Approaches to developing motion sensors in problems of granular media dynamics. *Bulletin of Irkutsk State Technical University*, *8*(115), 20–35. doi:10.21285/1814-3520-2016-8-20-35

Eliseev, A. V., Kashuba, V. B., Kinash, N. Zh., & Eliseev, S. V. (2016). Features of the dynamic properties of working elements of technological machines under inertial excitation of vibrations. *Bulletin of Don State Technical University*, *16*(4), 63–71.

Eliseev, A. V., Kuznetsov, N. K., & Mironov, A. S. (2023). Maps of dynamic invariants in assessing motion modes of mechanical vibrational systems. *Trudy*, (MAI), 128.

Eliseev, A. V., Kuznetsov, N. K., & Mironov, A. S. (2023). System approach to assessing dynamic states of technical objects: elements of the concept of dynamic invariants. *Vestnik Astrakhanskogo Gosudarstvennogo Tekhnicheskogo Universiteta. Seriya: Upravlenie. Vychislitelnaya Tekhnika i Informatika, 1*, 16–24.

Eliseev, A. V., Kuznetsov, N. K., & Nikolaev, A. V. (2022). The concept of dynamic invariants in assessing the structural features of mechanical oscillatory systems. *Transport. Mining and Construction Engineering: Science and Production, 15*, 18–30.

Eliseev, A. V., Mamaev, L. A., & Sitov, I. S. (2015). Some approaches to substantiating the scheme of inertial excitation in technological vibration machines. *Systems. Methods. Technologies, 4*(28), 15–24.

Eliseev, A. V., Sitov, I. S., & Kuznetsov, N. K. (2022). Systemic approaches to assessing dynamic states of technical objects under vibrational loads: Frequency functions, dynamic invariants, regularization methods. *Systems. Methods. Technologies, 4*(56), 7–15.

Eliseev, A. V., & Vyong, K. Ch. (2016). Some possibilities of controlling a one-dimensional vibration field of a technological machine. *Modern technologies. System analysis. Modeling, 1*(49), 33–41.

Eliseev, S. V., & Eliseev, A. V. (2013). General approaches in problems of determining contact reactions in solids under static loads with account for non-holding connections. Modern technologies. *System analysis. Modeling, 4*(40), 51–60.

Eliseev, S. V., Kargapol'tsev, S. K., Bol'shakov, R. S., & Eliseev, A. V. (2020). Lever connections: Opportunities for forming dynamic states in mechanical oscillatory systems. *Transport of the Urals, 3*(66), 17–23. doi:10.20291/1815-9400-2020-3-17-23

Jakóbczak, D. (2007). 2D and 3D Image Modeling Using Hurwitz-Radon Matrices. *Polish Journal of Environmental Studies, 4A*(16), 104–107.

Jakóbczak, D. (2010). *Shape Representation and Shape Coefficients via Method of Hurwitz-Radon Matrices* (Vol. 6374). Lecture Notes in Computer Science. Springer-Verlag Berlin Heidelberg. doi:10.1007/978-3-642-15910-7_47

Chapter 2
Vibration Impacts and Dynamic Conditions of Elements of Technical Objects:
Vibrational Effects in Machine Dynamics

ABSTRACT

The chapter describes the features of vibration protection systems, the elemental basis, and structural and technical forms of additional connections, lever and gyroscopic mechanisms, special-purpose mechanisms, additional connections in the structures of vibration protection systems, systems with mechanisms providing quasi-zero rigidity, some means of solving problems of vibration protection in transport, as well as ways and means of creating low-frequency vibrations. Various methods and means for reducing vibrations are considered, including the use of additional connections, mechanisms, and devices. The features of lever mechanisms and their consideration in assessing the dynamic properties of technical systems are described. Methods of dynamic synthesis of vibration protection and features of calculation and selection of system parameters are considered.

INTRODUCTION

In the dynamics of machines, vibration effects play a key role and manifest themselves in various forms, reflecting the processes of interaction of the elements of machines, as well as the peculiarities of the contacts of the working elements with the external environment. The causes of vibrations are quite varied: to reduce or limit vibrations, a sufficiently large arsenal of methods and means of changing the dynamic state has been developed. These methods and means are defined in several directions related to balancing mechanisms, balancing rotating parts, vibration protection and vibration isolation, etc. Tasks to be solved, for all their diversity, also have some commonness, which is manifested in the sequence of certain stages associated with the selection of the object of protection, the state of which requires assessment, control or change. Most often, such approaches are based on detailed consideration of the

DOI: 10.4018/978-1-6684-7237-8.ch002

functional features of machines, the distinguishing of mechanisms in them, the refinement of the formation of a system of forces that arise from the transfer of energy from the engine to the working element, and taking into account dynamic factors of interaction with the working environment (Andreychikov et al., 2011; Ganiev, 2010; Lontsikh & Eliseev, 2014). Taking into account the elastic properties of the units of the mechanisms, as well as the elasticities of the hinged joints, results in computational schemes, which are mechanical oscillatory systems with one, two or several degrees of freedom (Birger, 1988; Galiev et al., 2010).

The traditional problems of the theory of machines and mechanisms associated with the determination of the laws of motion of units, forces acting in kinematic pairs, static and dynamic balancing of motion, are complemented by the need to take into account the oscillations occurring in a mechanical system. In the simplest form, the problems of vibration interactions are considered as problems of vibration protection and vibration isolation, reflecting the characteristics of the emerging vibrations and requirements for their normalization at selected observation points (Lakanne, 2002; Averyanov et al., 2008). In this regard, the problems of vibration protection and vibration isolation of machines and mechanisms can be regarded as significant, since the statement of assessment, control and change of the dynamic state involves identifying the object to be protected, determining its mass-and-inertial, elastic and dissipative characteristics that form a certain systemic representation about the properties of the systems displayed by the above parameters (Eliseev et al., 2008). Work (Makhutov et al., 2008) reflected the importance and significance of solving the problems of vibration strength, vibration protection of technical systems and the need to increase the reliability and service life of machines and equipment under intensive dynamic loading within the scope of development programs for power and transport engineering, as well as within the framework of similar sectoral programs in other areas, in particular, shipbuilding and defense industry. In the section of scientific, technical and forecasting developments one can indicate the need for further evolvement of machine science and the theory of mechanisms and machines with respect to creating new vibration technological machines and equipment. The main problems that need to be solved include: mathematical modeling of machines as oscillatory systems, for which compliance with certain acoustic and vibration criteria is important. Many issues of the dynamics of machines and mechanisms using the above and variable parameters are presented in (Andreevsky & Blekhman, 2001). Developing ideas about the possible forms of vibration effects in the mechanisms, one could identify the following research directions:

1. *Oscillations of mechanisms* on a vibrating basis, which implies the study of the possibilities for the occurrence of certain motions of the mechanism, i.e. the motion of the initial unit, with the vibrations of the rack or the bases of the mechanism. In such cases, most often, the elasticity of units and hinges is not taken into account, and the resulting motions and their features correlate with the phenomena of nonlinear mechanics, self-synchronization, etc. (Blekhman, 1994).
2. *The imbalance (static and dynamic) of the units of the mechanism.* This results in the uneven motion, which manifests itself in the form of periodic oscillations of the angular velocity of the drive unit of the mechanism, as well as other units. Such problems are close in their formulation to those of vibration protection, however, they are specific and are represented by dynamic balancing technologies (Frolov et al., 2004) and dynamic balancing of rotating shafts. Compensation of dynamic imbalance in some cases can also be considered in the class of vibration protection tasks, for example, as a problem solved by installing dynamic oscillation dampers (Genkin & Ryaboy, 1988; Belokobylsky, Eliseev & Sitov, 2013; Eliseev et al., 2012).

3. Many *means* of changing the dynamic state of the object of protection take (or have) the form of mechanisms (lever, toothed) used in oscillatory systems as additional units for the transformation of motion. Mechanisms, in this case, have the ability to provide the flexibility of adjustment of parameters or improving the adaptability of systems.

The traditional formulation of the problem of vibration protection (or vibration isolation) is usually associated with an object having one degree of freedom, whereas the vibration protection system can be more complex and have a large number of degrees of freedom. Less common are spatial problems of vibration protection, due to the complexity of mathematical models. Representations of such problems can be obtained in a number of works by domestic and foreign authors (Belokobylsky, Eliseev & Kashuba, 2013; Myamlin, 2002; Khokhlov, 2002).

The above does not so much confirm the functional complexity of the vibration protection systems by the mechanisms in the composition of the technological machine, as the complex nature of the dynamic loading of all parts of the machine and some commonness in the formulation of problems of assessing the dynamic state inherent in the problems of vibration protection. It should be noted here that vibration loads require additional consideration when considering mechanisms in the structures of vibration protection systems, since the reliability of the machine's operation must be assessed as a whole, i.e. also in the modes of vibration dynamic effects. This implies the need to develop methods for assessing the dynamic state, determining the emerging forces and capabilities of vibration protection systems as mechanical oscillatory systems, including mechanisms.

METHODS AND DEVICES FOR CHANGING THE VIBRATION STATE: USE OF ADDITIONAL CONSTRAINTS, MECHANISMS, AND DEVICES

Although a large number of means of vibration protection have been developed and used so far, the requirements for vibration isolation of machines and mechanisms continue to grow; not only the vibration protection systems become more complex, but also the criteria for evaluating their effectiveness. Analysis of the causes of vibration disturbances of various objects shows that they are caused by many factors. It is practically impossible to determine the magnitude and nature of all impacts at the design stage, therefore, the product needs to be refined after manufacturing. These circumstances result in the need to create special means of vibration protection, capable of ensuring high-quality and reliable performance of objects in various operating modes, which necessitates such devices being included in the composition of objects and products at the design stage (Brysin & Sinev, 2008; Goverdovsky, 2006; Rakov & Sinev, 2011). During operation, the characteristics of a technical object or product may change, for example, as a result of wear, change in the mode or conditions of work.

The degree of complexity of the vibration protection system depends on the frequency change, as well as on the vibration activity of the machines within the frequency spectrum range. Vibration isolation and dynamic damping are the main ways of abatement of vibration on the paths of its propagation, however, dynamic damping is effective in a narrow frequency band (Sobolev, 2002; Fomichev & Fomichev, 2004).

The means of vibration protection from variables according to the frequency of impacts can be divided into two groups. The first includes tools and methods based on the use of passive devices, i.e. not using the energy of an external source to create forces that reduce vibration, but providing only a targeted change in the dynamic parameters of the systems (inertial, stiffness parameters and damping characteristics)

(Sobolev, 2002; Fomichev & Fomichev, 2004; Burkov & Efremov, 2004; Vargunin et al., 2006; Eliseev et al., 2013; Eliseev, Khomenko, & Logunov, 2010). The second group includes the means of vibration protection using active working elements (pumps, stepper motors, various drives). Methods of protection against influences with a determinate change in the frequency of the disturbance are divided according to the type of parameter settings to unregulated, self-adjusting and adjustable settings. Unregulated ones include non-linear dynamic torsional vibration dampers, non-linear bearings and couplings, non-linear vibration dampers. Assuming that high-frequency vibrations are quite effectively absorbed by the internal resistance forces of various structural elements of machines and equipment, dissipative properties of gaskets and moving joints, we note that the detuning from low-frequency vibrations in the range of 25-50 Hz is a fairly autonomous problem. This range also includes the protection of a human operator, due to the fact that the human body responds most strongly to the frequencies of 8-10 Hz. Therefore, the principles of ensuring effective vibration protection and vibration isolation of the human operator can be extended to a wide class of tasks related to the vibration protection of machines, equipment, instruments and equipment interacting in one way or another with the human operator. The focus is on the low frequency range: the features of the dynamic interactions of mechanisms that are used in vibration protection schemes are studied, as well as the features of the dynamic interaction of the protected object in the structure of the vibration protection system, when the object itself is not a finite solid (or "spatial metric") (Eliseev & Yermoshenko, 2012; Eliseev & Zasyadko, 2004).

Existing approaches for effective vibration isolation in the frequency range of 2–20 Hz suggest that the natural frequency of a passive vibration isolation system should be about 1 Hz. There are various schemes of systems of passive vibration isolation of objects, differing by combinations of elastic and damping elements, guiding mechanisms and mechanisms of motion transformation. The most widely used vibration isolation systems with guiding mechanisms ensure the reduction of vibration transmission in one, as a rule, vertical direction (the motion of a mechanism is described by one generalized coordinate). In (Eliseev, Kovyrshin & Parshuta, 2013) the possibilities of vibration protection systems within a wide range of frequencies of external influences are considered in detail.

In practical problems of vibration protection, passive type systems are most widespread and widely used. Figure 1 presents the classification of the elements of passive vibration isolation systems, which provides an idea that the vibration protection system in itself is already a rather complex system that includes various mechanisms. In this regard, one can add that the object of protection can also represent not only a material point (or a solid body that performs a vertical motion), but can also act as a complex object with a sufficiently developed system of dynamic interactions with elements of the vibration protection system.

The capabilities of vibration protection systems depend on many factors, therefore the properties of vibration protection systems are controlled by the corresponding state standards and regulatory documents (Lakanne, 2002). For example, the source data for the calculation of the security system of a human operator is GOST 12.4.025–76, namely:

- vibration characteristics of the seat base, established experimentally for a particular type of machine when operating in the characteristic mode;
- the mass of the operator acting on the seat, equal to $^5/_7$ of his entire mass;
- the mass of the sprung seat;
- the computational scheme of the oscillating system "operator – seat", determined by the design of the seat and the selected dynamic model of the human body;

Figure 1. Classification of the components of passive vibration isolation systems

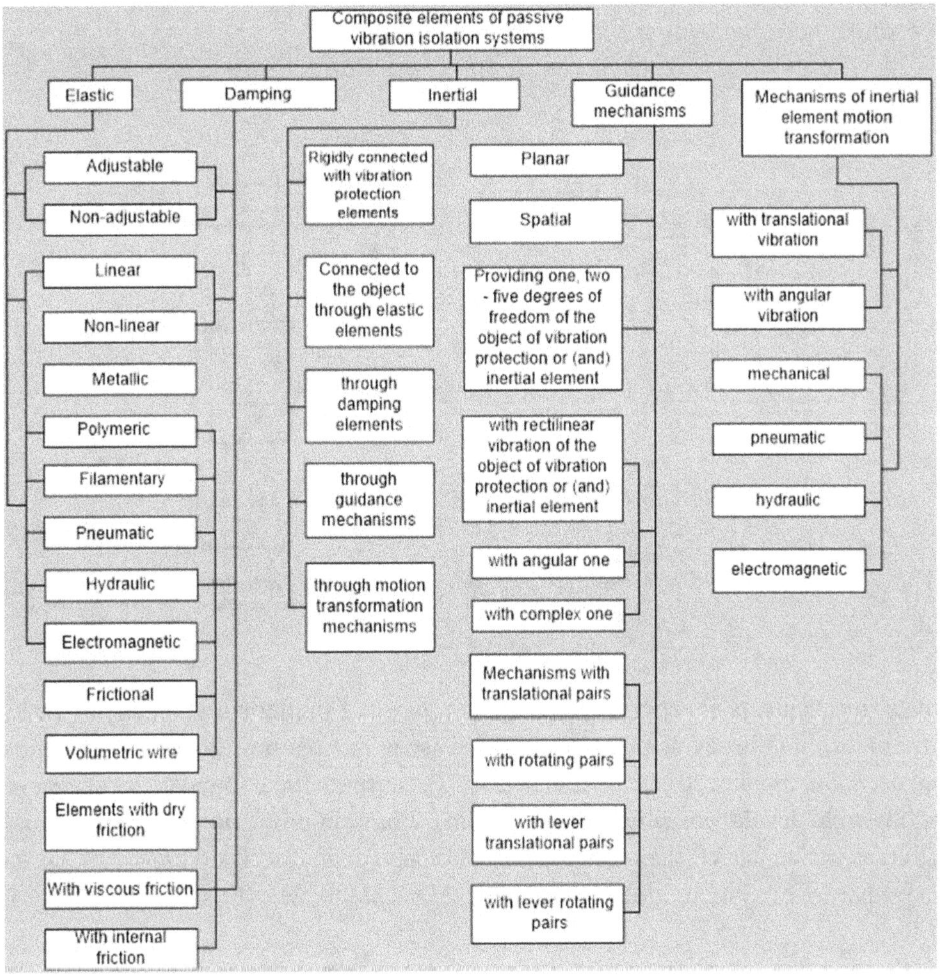

- characteristics of forces arising in the oscillatory system (stiffness and friction coefficients for restoring and dissipative forces);
- free wheeling of the seat, which should not exceed the permissible values established for machines of a particular type. Without going into the details of the calculation, we give the schematic diagrams (Fig. 2) which are developed and recommended for calibration estimates.

Special-purpose vibration protection systems (see Figure 2, *b*, *c*) are mechanical oscillatory systems using typical elements in the form of mass-and-inertia, elastic, and dissipative units. The introduction of additional masses (or the formation of two cascades) creates conditions for the implementation of the effects of dynamic oscillation damping (see Figure 2, *d*). The properties of such systems are well studied and are presented in (Belokobylsky, Eliseev & Kashuba, 2013).

Introduction to the basic mechanical systems of additional constraints implemented by various mechanisms creates the conditions for changing the dynamic state due to the effects of the transformation of motions of the elements.

Figure 2. Computational schemes of typical problems of vibration protection: a) linear system with stops; b) linear system with dry friction; c) nonlinear system with dry friction; g) linear vibration isolation system with a single-mass model of the human body

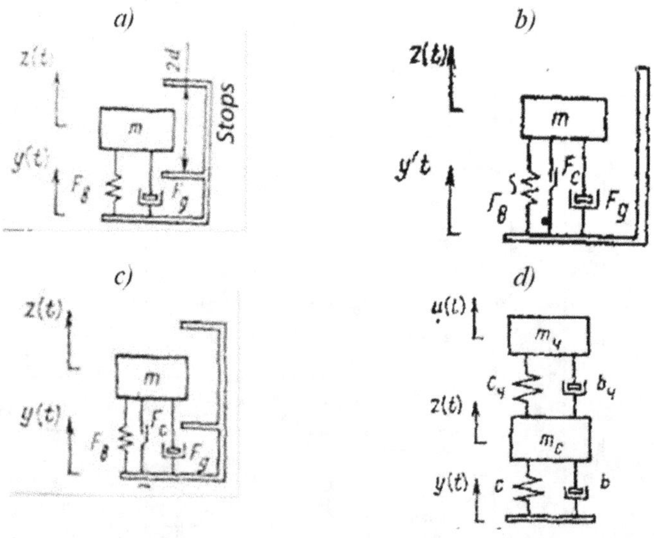

5. Pneumatic mechanisms are characterized by simplicity of implementation and availability of energy resources. In (Eliseev & Logunov, 2009; Eliseev & Logunov, 2010; Eliseev, Moskovskikh, Bolshakov, & Savchenko, 2012; Druzhinsky, 1977), the theoretical foundations of equivalent transformations were developed, which allow building vibration protection systems that use common typical elements, however, these approaches also have features. They are based on the fact that certain properties of systems are provided in mechanical linkages (Eliseev, Parshuta & Bolshakov, 2013).
6. A typical diagram of vibration protection systems using pneumatic mechanisms is shown in Figure 3.

   Pneumatic mechanisms in the structures of vibration protection systems are widely spread, which is reflected in (Atabekov, 2006; Eliseev, Parshuta, & Bolshakov, 2013). In recent years, a number of curious elaborations are presented in (Eliseev, Parshuta & Bolshakov, 2012; Eliseev, Parshuta, & Kaimov, 2013), where methods for constructing mathematical models of pneumomechanical systems have been proposed. Pneumatic mechanisms in the structures of vibration protection systems are widely spread, which is shown in (Eliseev, Kovyrshin & Parshuta, 2013; Druzhinsky, 1977; Atabekov, 2006; Eliseev, Parshuta, & Bolshakov, 2013). In recent years, a number of interesting extensions are presented in (Eliseev, Parshuta & Bolshakov, 2012; Eliseev, Parshuta & Kaimov, 2013), where methods for constructing mathematical models of pneumomechanical systems have been proposed.
7. Procedures for computer-aided design of vibration protection systems are well developed if we keep in mind their traditional forms (Eliseev & Yermoshenko, 2012; Eliseev, Parshuta & Bolshakov, 2013).

   There are search engines that allow forming series of predictable design solutions in relation to automobiles; such an example is shown in Figure 4.

Figure 3. Computational scheme of the pneumatic system with internal damping: 1 – additional (damper) camera; 2 – intercameral throttle; 3 – pneumatic cylinder (working chamber); V_w, V_d – the volume of working and damping chambers

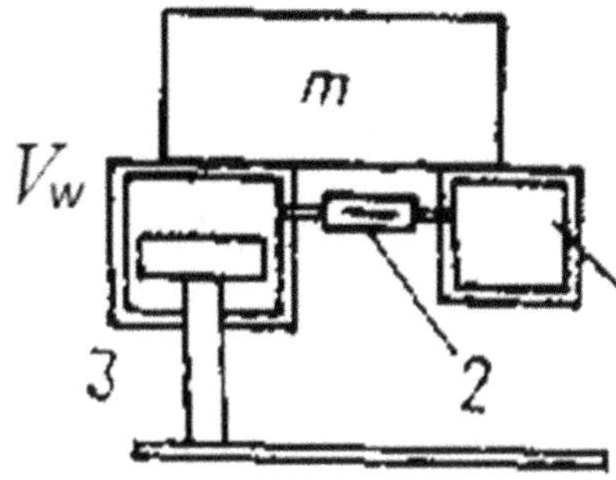

A comparative analysis of the capabilities of systems with pneumatic mechanisms shows that in the low-frequency region they have the same disadvantages as systems with ordinary elastic elements. Opportunities for improving the properties of systems are associated with the transition and construction of combined structures in which elastic pneumatic elements work in motion transformation schemes created by articulated mechanisms. Certain possibilities when expanding the range of useful properties of vibration protection systems for various purposes are opened when using active vibration isolation systems. The schematic diagrams of such systems are presented in Figure 5.

8. Details of obtaining the dynamic characteristics of the given computational schemes of vibration protection systems in Figure 5, are considered using the concepts of the transmission coefficients of the amplitude of oscillations (from the base to the object), as well as the dynamic coefficients of the transmission of forces, taking into account the distribution of characteristics over certain frequency zones. As an example, Figure 6 shows the typical characteristics of the system vibration isolation with external harmonic perturbations. At the same time, the analytical tools used in the calculations are based on mathematical models in the form of differential equations with the necessary characteristics being subsequently determined without the use of structural interpretations. Although those would be quite appropriate due to the need to assess the level of dynamic interactions, which is usually used for strength calculations.

Detailed information on the development of research in the field of the theory and practice of active vibration protection systems is given in (Averyanov et al., 2008; Reznik, 2001; Brysin et al., 2012). To assess the potential capabilities of adjustable systems, Figure 7 shows the frequency response of a vibration protection system with pneumomechanical devices that extend the capabilities of protection systems, as compared with conventional approaches.

The effect of vibrations on technical objects is quite different. This is considered in various aspects in the dynamics of machines. The interaction of the working elements of machines with the external environment is often accompanied by the occurrence of vibrations and is detrimental to normal operation and functioning. At the same time, a complex problem arises, requiring attention to the details of dynamic processes and the systemic perception of the possibilities for the manifestation of various forms of interaction between the elements of machines.

Figure 4. Fragment of the functional structural table to predict new areas of innovation. F_1 – asymmetric characteristic; F_2 – adjustable quasi-zero; F_3 – multistage; F_4 - jumping; F_5 - piecewise linear; F_6 - negative stiffness; F_7 - frequency dependent; F_8 - amplitude "notch"; F_9 – "overshoot"

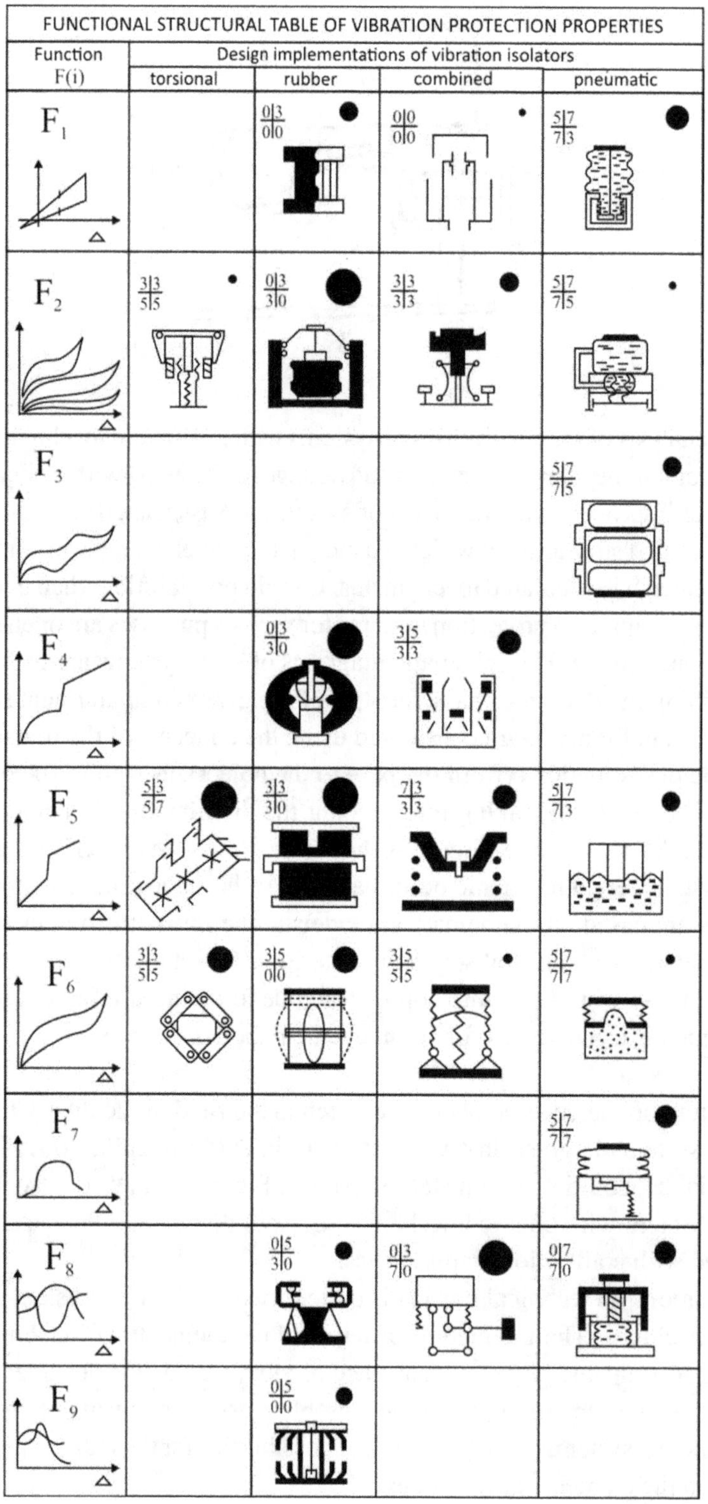

Figure 5. Computational schemes of active vibration protection systems: a – hydromechanical with relative position regulator; b – hydropneumatic with relative position regulator; c – pneumatic with external damper and relative position regulator; d – pneumatic with internal throttle and relative position regulator, the regulator is connected to the working chamber; e – the regulator is connected to the damper chamber; f – electrohydraulic vibration insulation

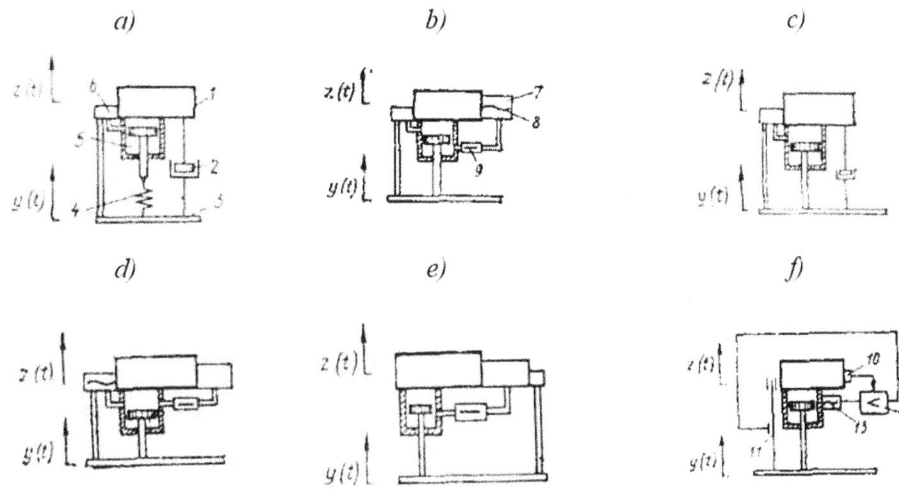

Figure 6. Relative transmission gain $T_x(v)$ of the linear single-mass vibration isolation system (according to Figure 2, a)

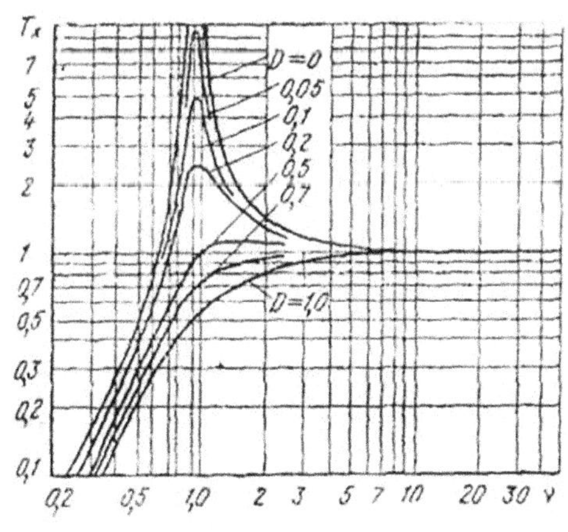

Vibration protection options are determined by the specifics of technological machines, although, in essence, the problems of vibration protection and vibration isolation can be considered in a generalized setting (Sobolev, 2002; Eliseev, Reznik & Khomenko, 2011).

In connection with the characteristics of industry specialization, there were works on the dynamics of road and rail transport, equipment in the mining industry, and in shipbuilding that came to prominence. This stimulated the development of a theoretical basis in the problems of vibration engineering and vibration protection in particular (Ganiev, 2010; Khomenko, Eliseev & Yermoshenko, 2012). The need to

Figure 7. The absolute transmission gain $T_z(w)$ of the pneumatic vibration isolation system with internal throttling (Figure 3): 1 – with the throttle resistance equal to 0; 2 – with throttle resistance, equal to infinity; 3 – with optimal throttle resistance.

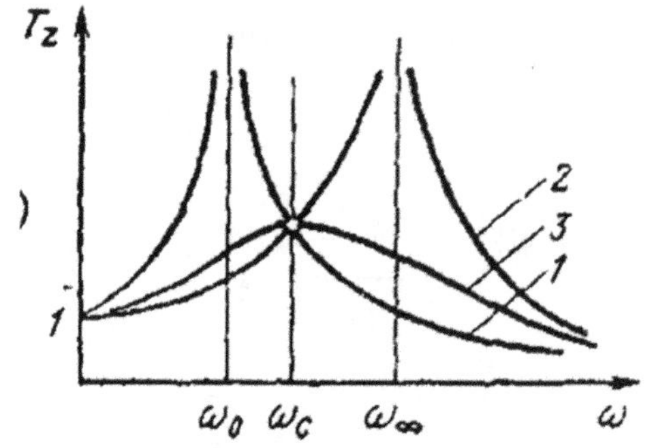

ensure the dynamic accuracy and reliability of robotic systems has initiated interest in the dynamics of mechanisms in the form of closed and open mechanical circuits and the theory of mechanisms in general.

It can be noted that the variety of problems in the dynamics of mechanical oscillatory systems (that, in many cases, became computational schemes for various technical objects) greatly contributed to the development of an analytical tools based on structural methods (Eliseev & Yermoshenko, 2012; Eliseev, Moskovskikh, Bolshakov & Savchenko, 2012; Eliseev et al., 2009), which were developed in applications of automatic control theory. These methods have proven to be useful for modern approaches in the development of vibration protection systems that, in essence, are specialized automatic control systems. We should also note that the vibration protection systems themselves, due to their complexity, started to be considered at the level of extended notions about the structure of the vibration protection systems, as well as the ideas about the inclusion of various mechanisms in their composition. This is partly due to the fact that the real interpretations of elastic and dissipative elements in the computational schemes of technical objects de facto are different mechanisms, which is clearly seen when considering the features and details of shock absorbers, dampers, suppressors, etc. (Eliseev et al., 2012; Vargunin et al., 2006; Kiryukhin et al., 2012).

FEATURES OF VIBRATION PROTECTION SYSTEMS, ELEMENTAL BASIS, AND CONSTRUCTIVE AND ENGINEERING FORMS OF ADDITIONAL CONSTRAINTS

Linear mechanical oscillatory systems have the property of vibration isolation. Under certain conditions, they work in such a way that dynamic vibration absorbers (or dampers) are considered to be one of the advantages of the effective operation of passive vibration protection means capable of suppressing the established forced vibrations of mechanisms and structures under monoharmonic perturbations.

Lever and Gyroscopic Mechanisms

In (Belokobylsky, Eliseev & Kashuba, 2013), the properties of manipulation systems are considered, which are manifested in the possibilities of implementing dynamic oscillation absorbing modes. Figure

Figure 8. Computational schemes of technical objects (manipulators) in modes of elastic oscillations under the action of external disturbances

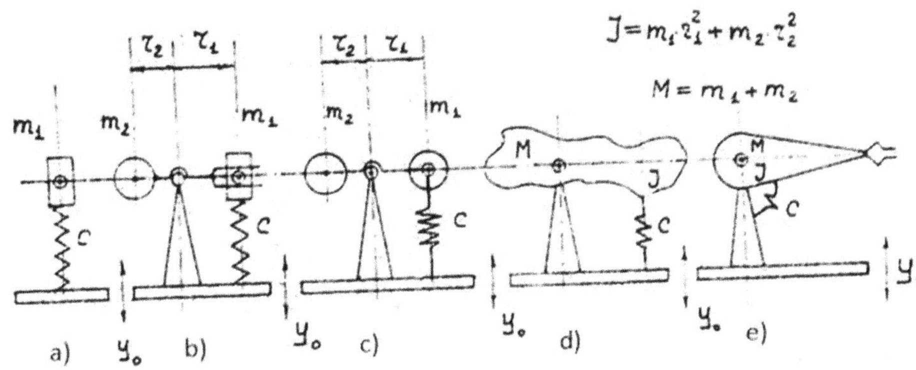

8 shows the computational schemes of industrial manipulators, which are based on the use of simple mechanisms in combination with typical inertial and elastic elements. If these schemes lead to the simplest form, then the object of protection can be represented as an oscillating solid on elastic supports. The counterbalances shown in the diagrams (Figure 8, *a*, *b*) create additional control possibilities, which, in turn, simultaneously solves the problem of static and dynamic balancing for the flat mechanism.

Mechanisms of inertial damping are described in (Lee, 2008), where gyroscopic devices are used as vibration absorbers in instrument structures. In the context of conducted research, a gyroscopic device is an example of the use of an additional constraint in the form of a spatial mechanism in a vibration protection system, which distinguishes the gyroscopic oscillation absorber (GOA) from other previously considered schemes (Eliseev et al., 2008).

In some problems of oscillation absorption, it becomes necessary to weaken the motion excited by a force with a variable frequency within a given range. For such applications, the usual oscillation absorber is often inappropriate. Broadband absorption is possible due to the fact that the antiresonant frequency is a function of an adjustable absorber parameter. Continuous reduction of unwanted motion to zero is achieved by adjusting the setting of the antiresonant frequency to a variable excitation frequency.

An idealized gyroscopic oscillation absorber is shown in Figure 9. It consists of a rotor, a casing and a frame. The GOA is depicted associated with the object of protection. The angular motion of the rotor around its own axis is described by the coordinate ψ, and the oscillations of the casing and frame around their axes are determined by the coordinates φ and θ. Vibrational motion of the structure is given by the coordinate δ. The device is a standard gyroscope with a displaced external axis of the cardan suspension, which is associated with a vibrating structure. The displacement of the external axis of the cardan suspension allows the linear motion of the structure to be converted into an inertial moment, acting relative to the rotor axis, parallel to the θ axis. This moment, in turn, causes the gyroscopic precessional moment of the rotor around the φ axis. As this gyroscopic moment is reduced to zero, changes sign and then repeats, the resulting gyroscopic precession moment causes alternating oscillations of the rotor axis in angle φ at the frequency of harmonic excitation.

Figure 9. Idealized scheme of the gyroscopic oscillation absorber

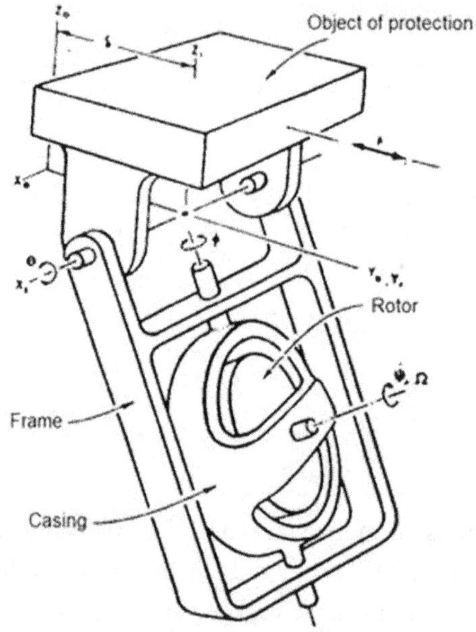

The harmonic inertial force of the response with a frequency, equal to the frequency of excitation, simultaneously has the opposite effect on the structure. For small motions, the antiresonant frequency ω_A is a linear function of the constant speed of rotation of the rotor Ω:

$$\omega_A = \frac{I_S}{\sqrt{I_\theta I_S}} \Omega, \tag{1}$$

where I_S is the moment of inertia of the rotor relative to the axis of its own rotation;

I_θ is the moment of inertia of the GOA with respect to the θ axis;
I_Φ is the moment of inertia of the rotor-casing system about the ψ axis; therefore, the maximum weakening of the oscillatory structure occurs at the synchronization of the rotor speed and the excitation frequency.

A number of issues related to the evaluation of the possibility of using a GOA to control the oscillation processes in mechanical systems are presented in [48]. Mechanisms with rotating units (shafts, rotors, disks) can be attributed to a special group of mechanisms whose properties are determined by gyroscopic interactions, which requires the development of special approaches. At the same time, this area has a great potential for creating vibration protection systems and, in general, controlled oscillatory systems for various purposes.

Mechanisms for Special Purposes

The detailed elaboration of ideas about dynamic interactions in the structures of vibration protection systems, having in its composition elastic-inertial units, is reflected in the works (Eliseev & Yermoshenko, 20121; Eliseev et al., 2009). In recent decades, a lot of new elastic-inertial vibration protection mechanisms have been proposed and investigated. One could name multi-stage systems, motion transformation mechanisms of the "screw-nut" type, two-stage vibration protection systems with a dynamic damper on an intermediate mass, lever, symmetrical cantilever and rod suspensions with vibration compensation, balanced systems and others, including multi-mass systems. In these systems, an essential role is performed by mechanisms that provide not only the transformation of motions, but also the functions of combining elements into certain structures. The latter can reflect the generalized properties of vibration protection systems using the concepts of dynamic stiffness and compliance of systems, impedances, etc. (Eliseev et al., 2008; Eliseev et al., 2013; Druzhinsky, 1977; Eliseev, Volkov & Kukharenko, 1990).

With all the variety of existing schemes of elastic-inertial vibration isolation, it is necessary to note one common feature that limits the range of applicability of this class of systems. To obtain sufficiently effective vibration isolation in the low-frequency region, it is impossible to infinitely reduce the stiffness of the vibration isolation. The dependency between the required stiffness and vibration isolation characteristics is valid for linear conservative systems with any number of degrees of freedom.

Additional Constraints in The Structures of Vibration Protection Systems

The methods and means of changing the dynamic state of oscillatory mechanical systems are quite diverse and are determined by the specificity of the problems being solved. These can be problems of drive dynamics, in which elastic elements are widely used in the form of various couplings, limiting the parameters of dynamic interactions in acceleration and deceleration modes, as well as when the motion parameters fluctuate with changes in external loading conditions or significant changes in the internal state. The features of vibration isolation of the drive mechanisms are described in detail in (Lilov, 1993).

Figure 10. Oil switch in closed position with power piston in rest position

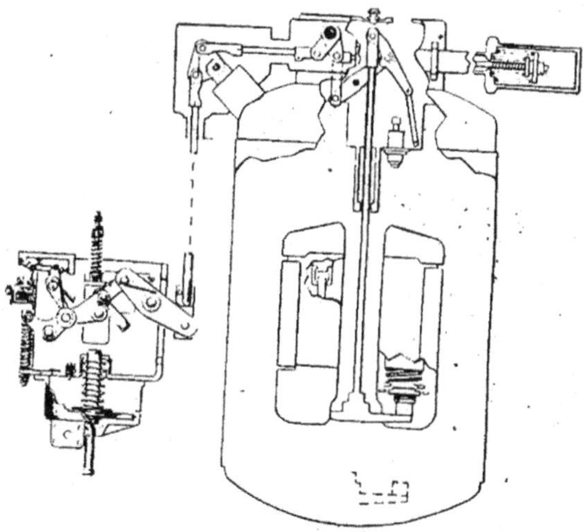

Systems that implement the control of the dynamic state of an object under the conditions of the need for consistency between several elements can include several mechanisms. A typical example of this kind is a high-voltage oil switch, shown in Figure 10.

The lever mechanism connects the contacts with the drive, and also serves to increase the stroke of the power piston in order to obtain the required large stroke of the contacts. The elasticity of the system is taken into account in a simplified model that includes two equivalent masses, thus forming one equivalent spring (Figure 11).

Figure 11. A simplified model of a system with two degrees of freedom, in which three branches are combined into one, and the distributed stiffness is replaced by one equivalent spring: y_c and z_c – position of the traverse in the system with absolutely rigid units and its actual position

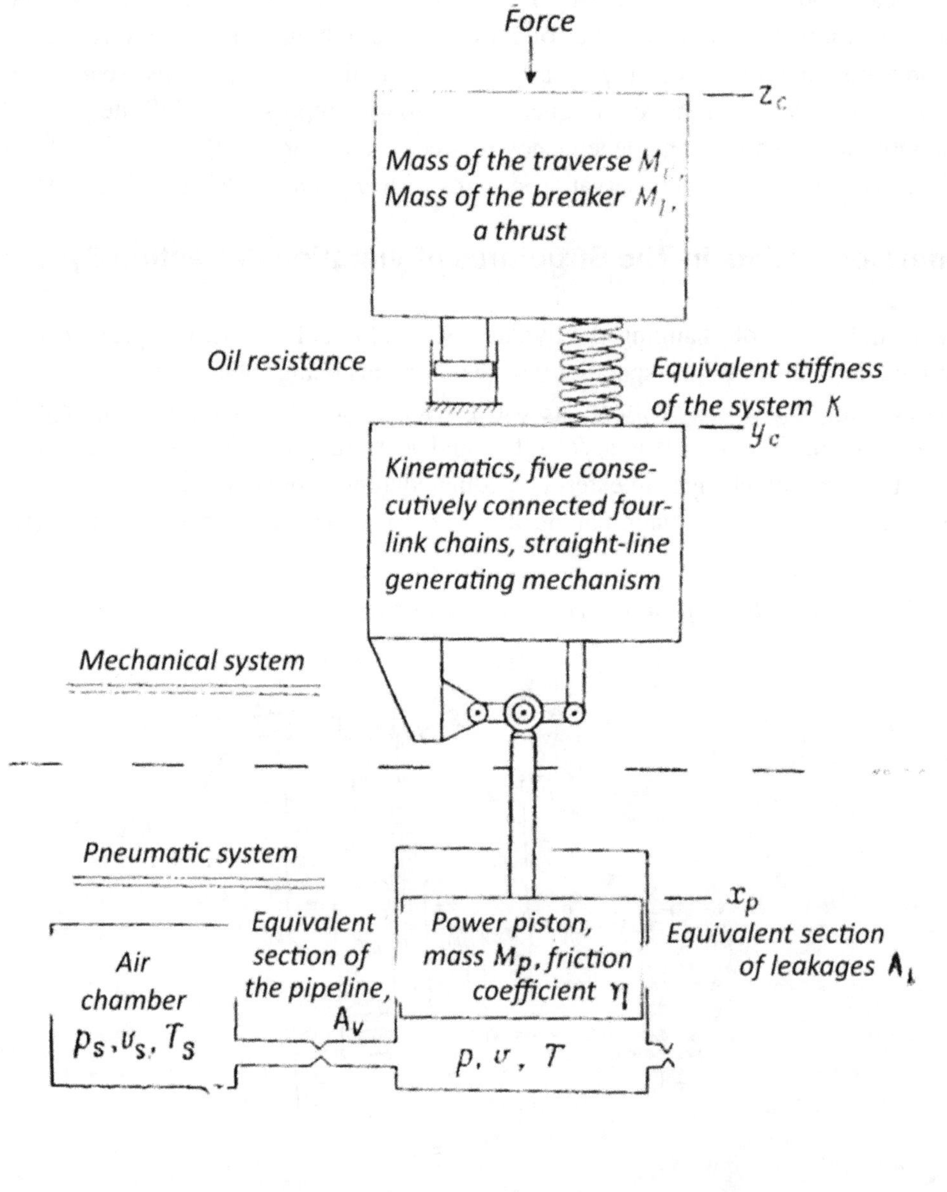

In this paper, the very idea of the concept of an equivalent spring is of interest. This spring consists of several elements connected together, combined in several branches, but constituting one whole. Such an approach is often used in solving problems of the dynamics of mechanical systems considered in (Burkov & Efremov, 2004).

Systems With Mechanisms That Provide Quasi-Zero Stiffness

The use of mechanisms is typical for almost all means of vibration protection, from vibration insulators to dynamic oscillation dampers.

The work (Burkov & Efremov, 2004) describes the so-called stiffness compensators, which are installed parallel to the main elastic element. The mechanism of zero stiffness can be inferred by Figure 12.

Figure 12. Zero stiffness vibration isolation mechanism

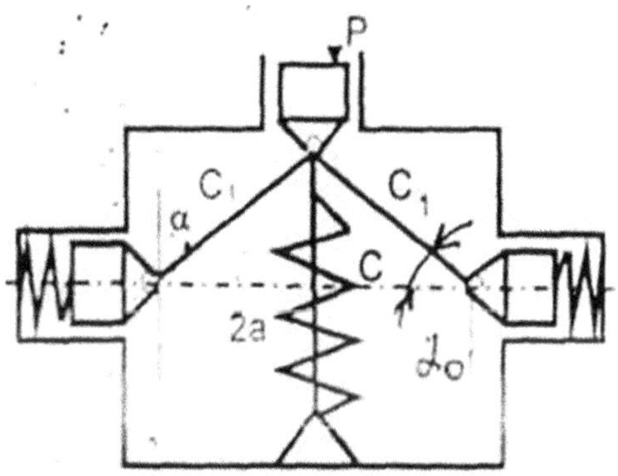

The relationship between the load P and the vertical displacement x of the compensator assembly is expressed by:

$$P_k(x) = 2C_1\left(1-\frac{x}{a}tg\alpha_0\right)\left[\frac{1}{\sqrt{tg^2\alpha_0+\left(1-\frac{x}{a}tg\alpha_0\right)^2}}-\cos\alpha_0\right], \tag{2}$$

where C_1 is the stiffness of the guides of the compensator.

Works (Saveliev, 2003; Salpin et al., 2002) describe some elastic systems with the so-called non-adjacent equilibrium form (or "overshoot" systems). In (Saveliev, 2003) studies were carried out on the synthesis of surface parameters of moving segments that form the optimal trajectory of the motion of roller pairs of elastic elements. The essence of improving the vibration protection properties of the system (in this case, the seat suspension) is explained by the analysis of the force relationship determined by the structure of the mechanism (Figure 13).

Figure 13. The force interconnection scheme in the mechanism of the vibration protection assembly

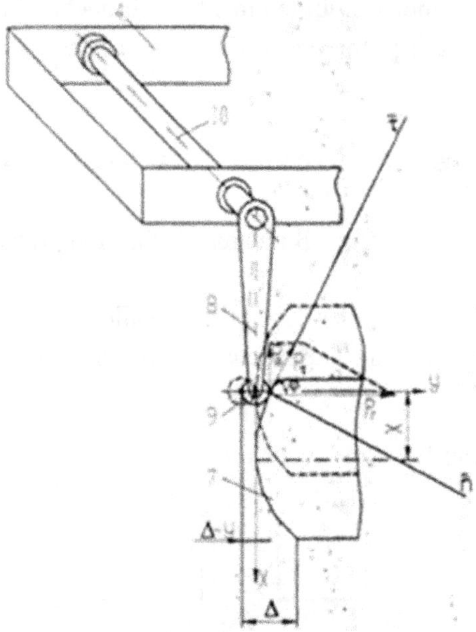

Thus, the properties of the vibration protection system are determined by the peculiarities of the dynamic interactions of the elements of the vibration protection systems combined into mechanisms forming the structures. Under certain conditions, such mechanisms can be reduced to typical elements or be considered as new formations of the quasi-spring type (Burkov & Efremov, 2004).

Some Tools for Solving Problems of Vibration Protection in Transport

In a developed form, motion transformation mechanisms have found application in vibration protection suspension schemes for transport vehicle seats. The main structural schemes of such a suspension are systems with a parallelogram guide mechanism and a scissor-type guide mechanism (Figure 14).

Figure 14. Typical seat schemes for truck drivers: a – parallelogram; b – scissors

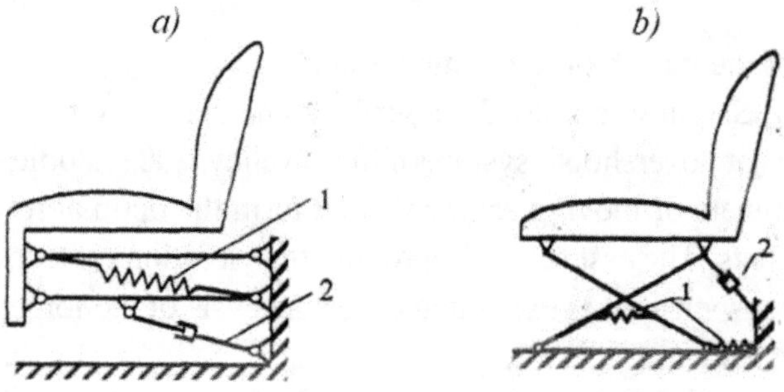

There are hydraulic telescopic shock absorbers installed in the suspensions of modern trucks. Figure 15 shows the schematic diagrams of the two options for the car driver seats: the design solutions are characterized by the use of articulated mechanisms (Belokobylsky, Eliseev & Kashuba, 2013).

Figure 15. Car driver seats: a – KamAZ; b – MAZ

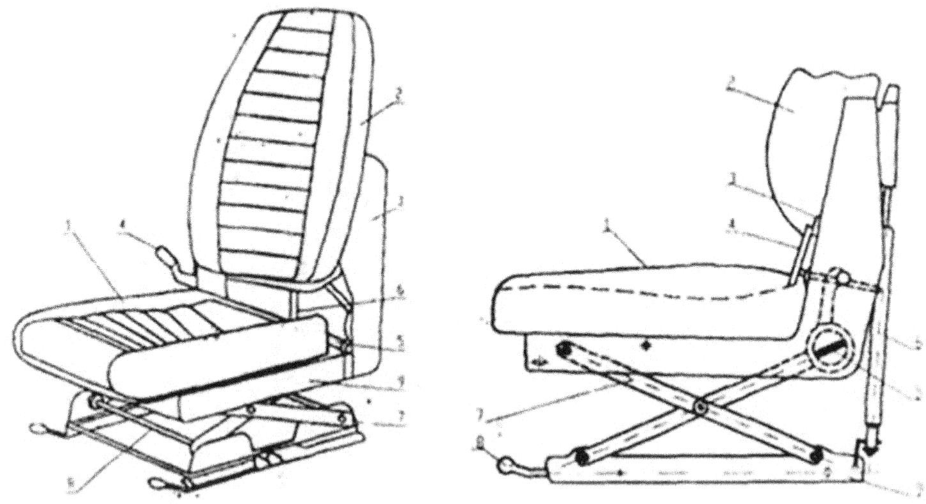

In some cases, passive vibration protection systems work better in the high-frequency region, and active ones – in the low-frequency region. On this basis, combined ("hybrid") vibration protection systems are developed. Recently, in the automotive industry, elastic systems with non-linear power characteristics, having a reduced stiffness in the working area of motions, are used as vibration protection devices. Multiwire cable segments are used as elastic elements in these systems. One ends of these segments are fastened on a fixed base, and the latter ones – on the seat frame. Works (Galiev et al., 2010; Belokobylsky, Eliseev & Kashuba, 20131; Eliseev, Parshuta & Kaimov, 2013) describe vibration protection systems with elastic elements in the form of curved rods enclosed in a sheath of vibration-damping materials (rubber-metal tapes, or a pneumatic element). The given examples show that in the practice of solutions of vibration protection problems there are tendencies to use not only elastic and dissipative elements, but also the mechanisms of transformation of relative motions of elements. They introduce additional possibilities into the systems while implementing certain dynamic interactions. Further use of the ideas of development of vibration protection systems for transport devices (suspensions) is associated with the increasing complexity of the mechanisms used.

Attempts to improve the properties of vibration protection systems essentially result in the formation (Figure 16) of specialized systems for the automatic regulation of dynamic properties (Salpin et al., 2002).

When designing special additional vibration protection devices, depending on the mass and operating conditions of the protected object, various pneumatic systems are often used.

Figure 16. Schematic diagram of the operator of transport vehicles with lever-elastic regulation system

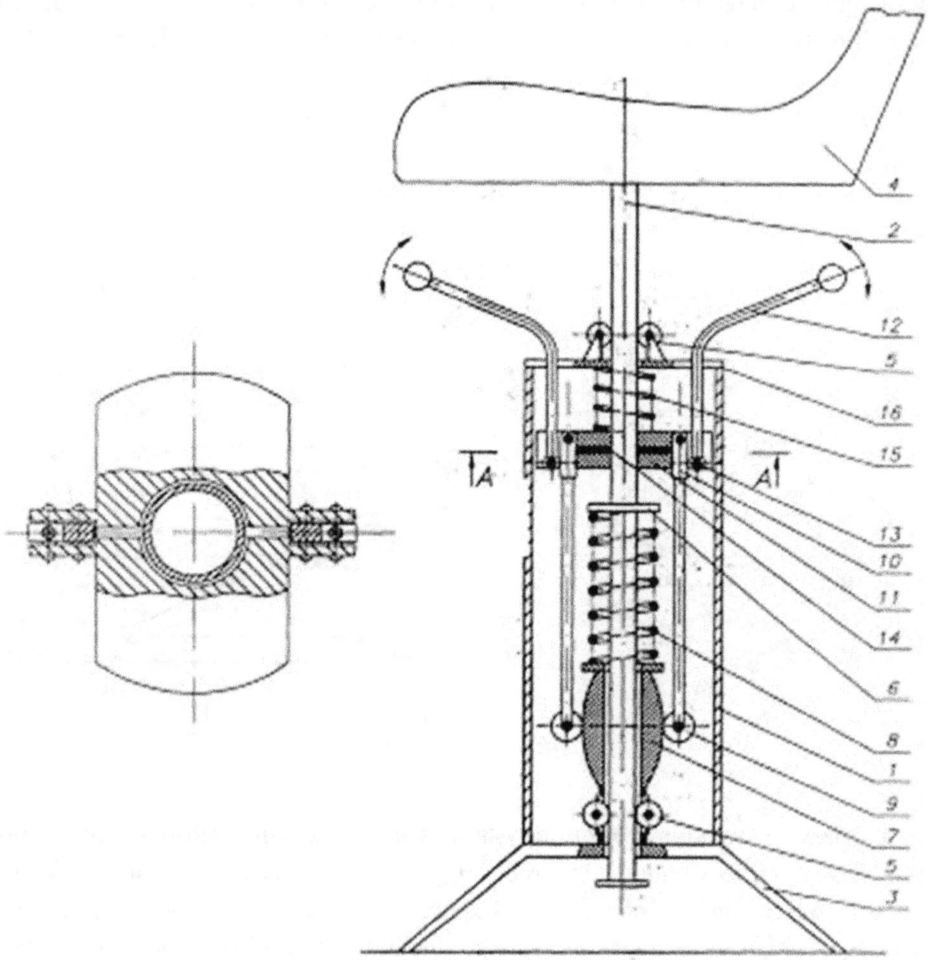

The scheme of a circular device is shown in Figure 17. Elastic elements here can be torsion bars, coil springs or brackets. The resulting total nonlinear characteristics depend on the characteristics of the main and additional suspensions.

The formation of the power characteristics occurs due to the motion of roller hinges in circular segments and the corresponding deformation of elastic elements. The power characteristic can be represented by a nonlinear equation (Shedlos', 2005):

$$P_n = 2k_y \left(1 - \frac{\sqrt{R^2 - b^2}}{\sqrt{R^2 - x^2}}\right) x, \tag{3}$$

where k_y is the stiffness of the elastic element;

R is the total radius of the trajectory of the center of the roller joint;

Figure 17. Circuit of force couplings in a circular device

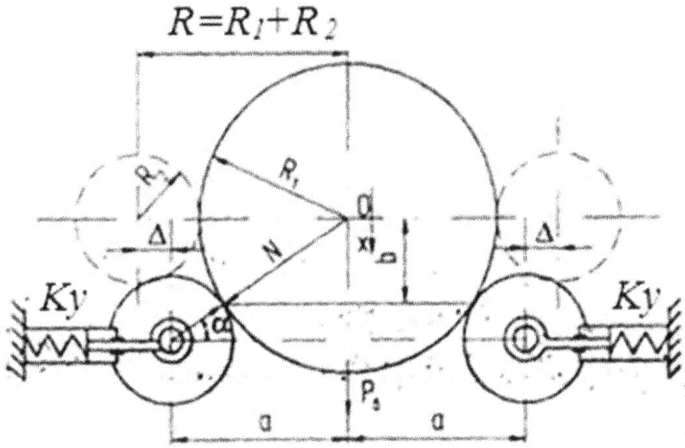

b is the stroke of the device from the middle position to the extreme.

From (2.3) it follows that, in particular, the main parameters affecting the force characteristic of the device are k_y, R, b.

If we take k_y and b as constant values and construct dependencies $P_n f(x)$, then we can be convinced that, with increasing of R, the waveform of the characteristic is distorted. In the first approximation, the evaluation of the dynamic properties of an oscillating system with mechanisms can be made on the basis of the computational scheme shown in Figure 18.

Figure 18. Computational scheme of single-mass oscillatory system

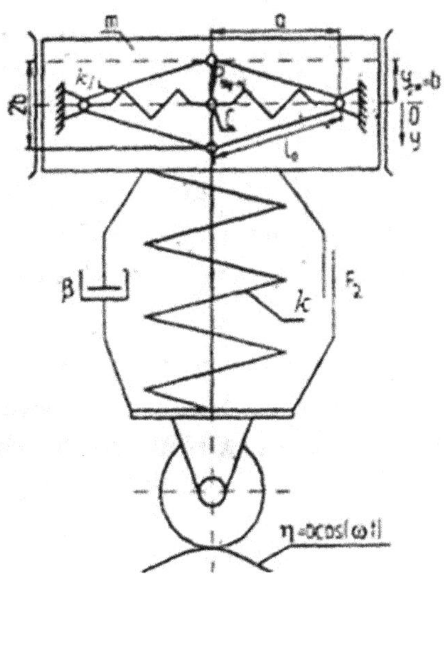

Natural oscillations of such a system are described by a differential equation.

$$m\ddot{y} + ky + P_B = 0, \tag{2.4}$$

where m is the mass of the oscillating body.

Figure 19 presents the computational scheme that makes it possible to simulate the vertical oscillations of an electric locomotive as a mechanical system with four degrees of freedom (Galiev et al., 2010).

Figure 19. The computational scheme of the linear four-degree mathematical model of the VL10 electric locomotive

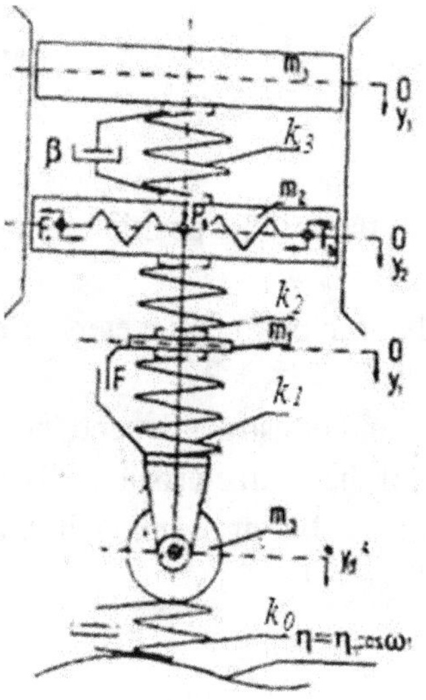

The following notation is used in this scheme:

m_2,m_3- mass of the body and trolley per one locomotive wheel;
m_1- conditionally introduced mass (equivalent to the mass of screw springs);
m_0=(m_{stat}+m_{dyn})- the sum of the unsrpung mass of the locomotive and the reduced mass of the track; hard-bone paths;
k_1,k_2,k_3- the stiffness of the elastic elements of the first and second steps of the suspension;
P_b – nonlinear vertical force of the “overshoot” device;
F_B- friction in the hinge assemblies of the device;
β- coefficient of viscous friction;
F- leaf spring friction force;
y_0,y_1,y_2,y_3- linear displacement of the masses;
η=B$\cos\omega t$- *t*rack irregularity.

The mathematical model of the system is:

$$\begin{cases} m_3\ddot{y}_3 + \beta\dot{y}_3 + k_3y_3 - \beta\dot{y}_2 - k_3y_2 = 0 \\ m_2\ddot{y}_2 + (k_2 + k_3)y_2 + \beta\dot{y}_2 - k_2y_1 + P_B + F_B\mathrm{sign}(\dot{y}_2 - \dot{y}_0) = 0 \\ m_1\ddot{y}_1 + \mathrm{Fsign}(\dot{y}_1 - \dot{y}_0) + (k_1 + k_2)y_1 - k_2y_2 - k_1y_0 = 0 \\ m_0\ddot{y}_0 - \mathrm{Fsign}(\dot{y}_1 - \dot{y}_0) + (k_0 + k_1)y_0 + \beta_0y_0 - k_1y_1 - k_0\eta - F_B\mathrm{sign}(\dot{y}_2 - \dot{y}_1) = 0 \end{cases} \quad (2.5)$$

The solution of a system of nonlinear differential equations is made by the Runge – Kutta method according to a standard program and analyzed in (Galiev et al., 2010).

Ways and Means of Creating Low-Frequency Oscillations

Figure 20 shows the calculation scheme of a low-frequency vibration stand designed for testing instruments in the conditions of ship's motions (Shedlos', 2005).

The IS-M test bench allows producing simultaneous alternating horizontal accelerations and inclinations that coincide in phase. It should be noted that the presence of various mechanisms in oscillatory structures is characteristic of vibration technological and testing machines.

Figure 20. Kinematic scheme of the IS-M installation

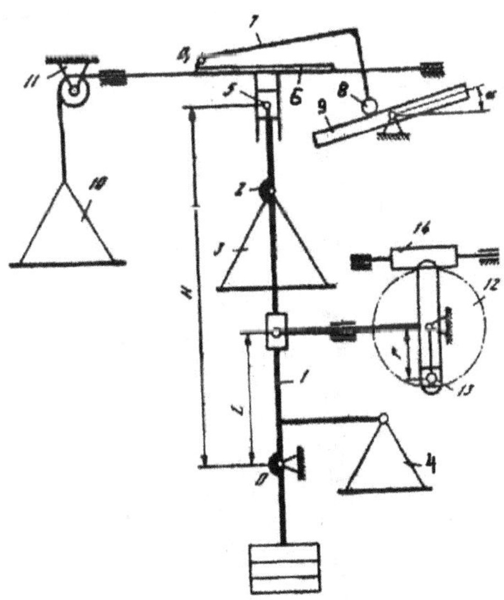

Figure 20 provides the theoretically precise rectilinear oscillatory motion of the *EF* unit, with which the work bench is connected.

The motion of the work bench according to the harmonic law is provided with a sinus toothed mechanism. In the considered constructive solution as part of the mechanical oscillatory system, both lever and toothed mechanisms are used.

The oscillatory system for linear horizontal acceleration HAS-M is built according to the scheme "drive – driver mechanism – work bench guide mechanism" (Shedlos', 2005). The horizontal trajectory of the work bench is provided by the Roberts articulated mechanism (Figure 22).

The test vibration machine has a complex structure, which is ensured by solving static and dynamic equation problems of equilibration due to the need to comply with regulations on the quality of reproducible oscillatory processes.

Figure 21. Kinematic scheme of the horizontal displacement stand

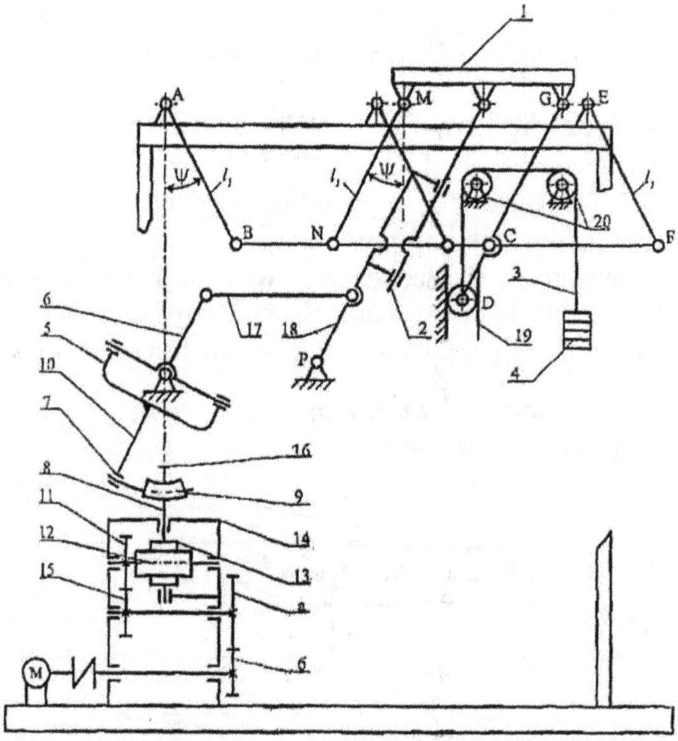

A characteristic computational scheme for vibration machines is the scheme (Figure 23) of an infra-low-frequency stand, in which the work bench performs angular oscillatory motions. It can be noted that in terms of meeting certain requirements for the tasks of the dynamic synthesis of mechanical oscillatory systems, vibration protection problems formally have a lot in common with the problems of reproducing certain signals.

Figure 22. Infra-low frequency stand scheme

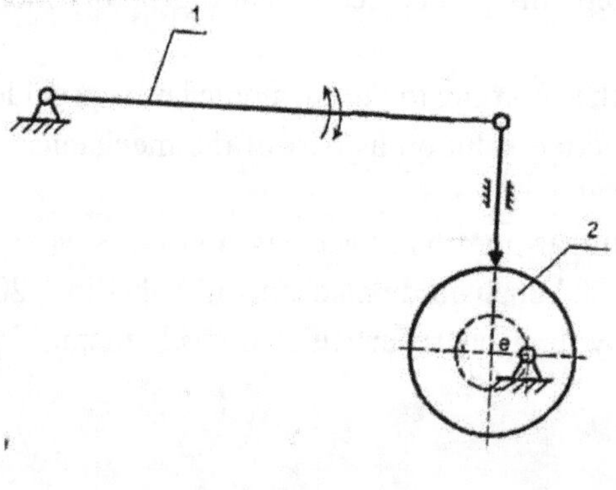

In both cases, it is of interest to develop methods and means of controlling the dynamic state of certain technical objects.

The computational scheme of a three-dimensional stand with a double articulated parallelogram is typical in this respect (Figure 23). Vibration testing machines are characterized by a sufficiently large number of elements of mechanical oscillatory systems that make angular rotating and oscillating motions.

Figure 23. The scheme of the articulated parallelogram

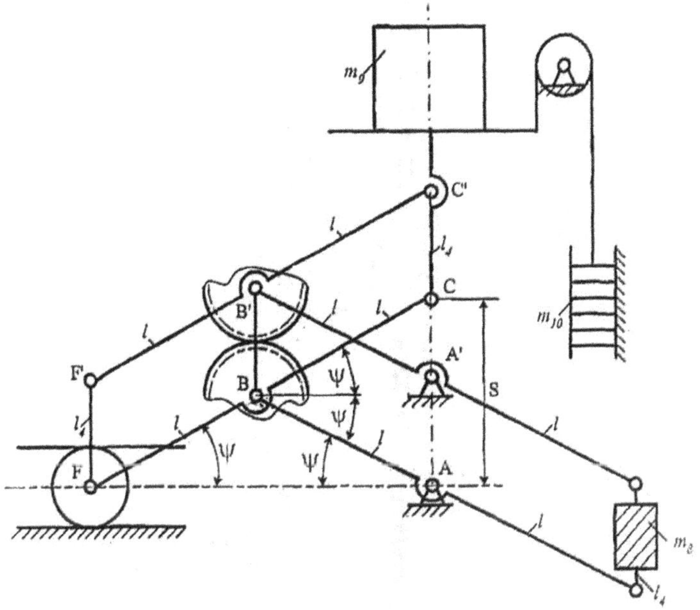

The natural development of design solutions is the use of pendulum circuits, as shown in Figure 25.

Figure 24. The scheme of the pendulum stand

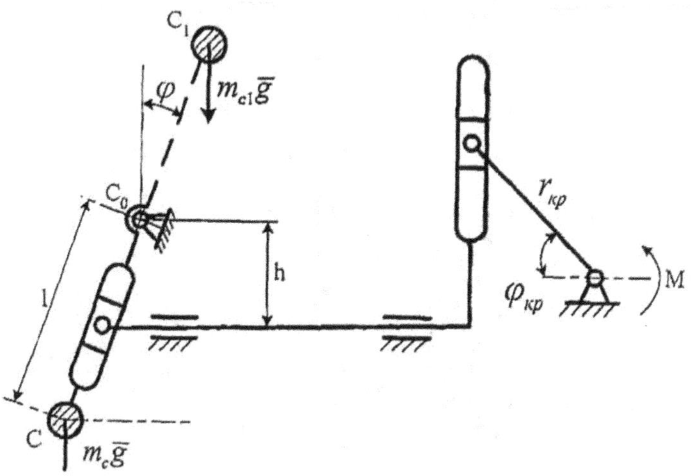

Interest in pendulum devices is also characteristic of vibration protection systems, in which pendulum oscillation dampers become common (Makhutov et al., 2008). In essence, the pendulum systems in their simplest versions can be considered as variants of the use of levers of the first and second kinds, which provide the ability to perform rotating and oscillating motions under regulated conditions.

Figure 25. System of precise angular orientation and stabilization of a rotating antenna

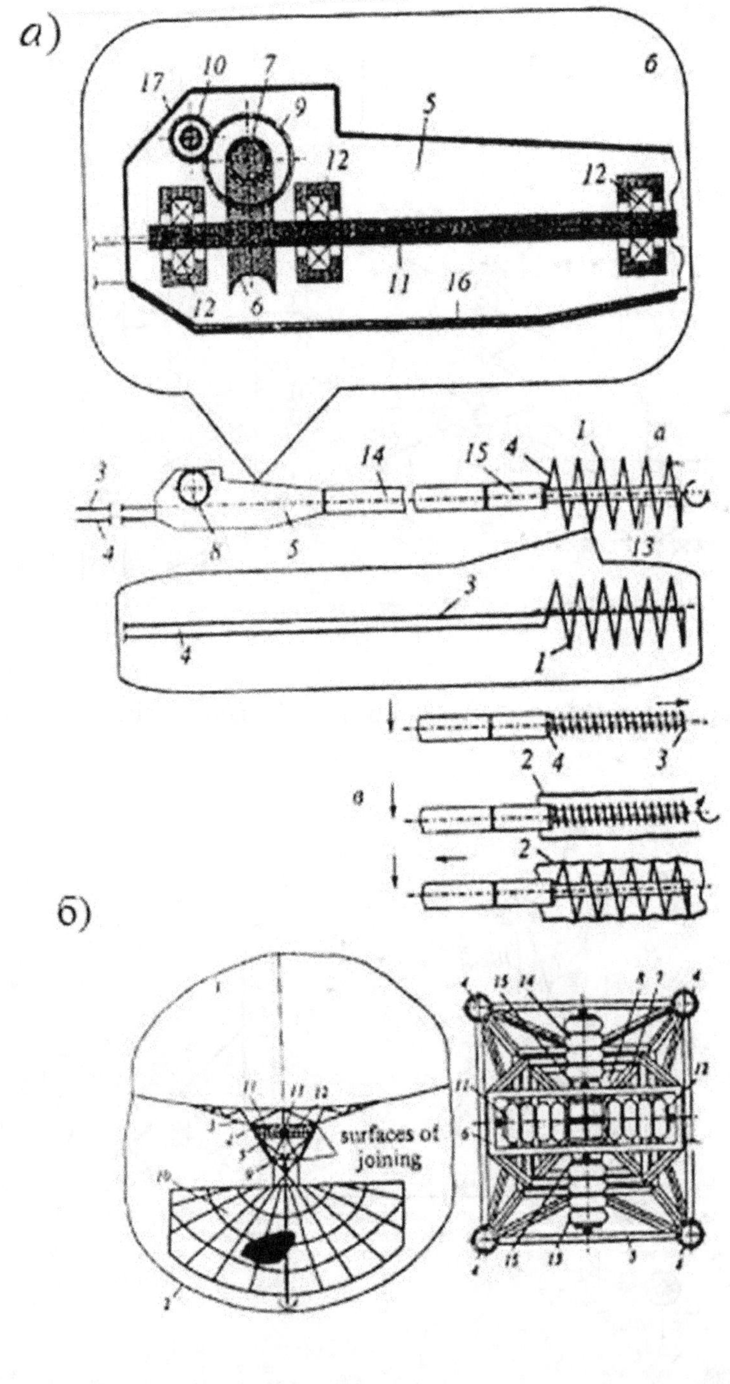

The range of use of systems with the tasks of reproducing oscillating motions is quite wide. Figure 25 shows the computational scheme of a modern radar system.

FEATURES OF LEVER MECHANISMS AND TAKING THEM INTO ACCOUNT IN THE ASSESSMENT OF DYNAMIC PROPERTIES OF TECHNICAL SYSTEMS

Ensuring reliable operation of machines in an ever-increasing energy, power and speed characteristics requires the improvement of methods for assessing dynamic characteristics. The various assumptions accepted in traditional methods (absence of gaps in the mechanism of friction conjugations, absence of deviations during the assembly of modules, absence of inaccuracies of the geometric shape of the surfaces of rubbing parts, etc.) do not quite correspond to the actual motions of the machine, which significantly reduces the accuracy of the computational methods.

Improving computer equipment makes it possible to create more and more complex models that contribute to a more accurate description of the motion and are close to the actual behavior of the machine. Creating a calculation model of the mechanism, taking into account the interaction of parts throughout the life cycle, is critical. The solution of this task allows us to increase the reliability of technical diagnostics, to improve the quality characteristics of reliability, safety, and efficiency. The gaps in the conjugations, despite their smallness in comparison with the nominal dimensions of the units, have a significant impact on the dynamics of the vibration protection system and its components, including mechanisms. In the process, there is a multiple increase in responses, an increase in the stress-strain state of parts and the emergence of vibro-impact modes of motion of the mechanism. Currently, various approaches are used to describe the motion of the mechanism and determine responses in kinematic pairs (KP), from which two basic methods can be distinguished: the kinetostatic method and the description of the motion of the mechanism in generalized coordinates using the Lagrange equations of the second kind (Frolov et al., 2004; Korolkov, 2010; Burkov & Efremov, 2003). The development of computing technology has provided great opportunities for the application of these methods.

Classical methods are reduced to the solution of the problem by the method of kinetics. In order to find the responses (R_x, R_y, R_z) in each kinematic pair (KP) of the mechanism, where there is a gap and descriptions of the motion of the node points of the units (with x, y, z coordinates and their derivatives) in the gap field, according to the d'Alembert principle, differential equations of motion are written for the corresponding units of the mechanism. The positions of these units, taking into account the errors caused by the presence of gaps, are determined by the analytical method of vector contours.

$$\begin{cases} \sum_{i=1}^{n} m_i \ddot{x}_i = \sum_{i=1}^{n} F_{ix} + \sum_{i=1}^{n} R_{ix}, \\ \sum_{i=1}^{n} m_i \ddot{y}_i = \sum_{i=1}^{n} F_{iy} + \sum_{i=1}^{n} R_{iy}, \\ \sum_{i=1}^{n} m_i \ddot{z}_i = \sum_{i=1}^{n} F_{iz} + \sum_{i=1}^{n} R_{iz}, \end{cases} \tag{6}$$

where n – the number of points of the mechanical system;

m_i- the mass of the i-th point of the system;

F_i – the projection of the active force $\overline{F}$ on the coordinate axes *x, y, z* respectively.

To determine the unknown functions (x_i, y_i, z_i) we integrate system (6) under the given initial conditions.

Equations (6) describe the motion of a mechanism adequately only in the case of contact motion of units, that is, with a closed kinematic chain. However, if there are gaps in the kinematic pairs, the KP elements will be detached. This is observed when the response of the corresponding KP vanishes. The difficulties of determining the inertia forces of the units of the mechanism arising from this, due to the fact that the position of the points of the units in the gap field is no longer a strictly defined parameter, do not allow the system (6) to be used to calculate the parameters of collisions.

In (Korolkov, 2010), the points in time at which the displacements occur are calculated based on the equality of the response force to zero. The number of such sections determines the number of collisions per work cycle of the mechanism. At the same time, is not possible to describe of free motion of KP components in the gap field. Therefore, it is impossible to determine the speed, impact force and, moreover, the mutual position of parts in the gap field when contact has occurred without using other additional methods.

The calculation methods for infinitesimal displacements are based on the use of the Lagrange equation of the second kind:

$$\frac{d}{dt}\left(\frac{\partial T}{\partial \dot{q}_j}\right) - \frac{\partial T}{\partial q_j} = Q_j, (j = 1, 2, ..., k), \qquad (7)$$

where *T* is the kinetic energy of the system;

k- the number of independent generalized coordinates;
q_j- generalized coordinates;
$\dot{q}_j$ – generalized speeds;
Q_j- generalized forces (Burkov & Efremov, 2003; Bakalov, 2006).

Integrating (7), they determine the law of motion of a mechanical system $q_j(t)$. The solution of this problem is often carried out by numerical methods, that is, by successive consideration of the static state of the system after a small period of time Δt or a small change in the generalized coordinate Δq_j.

METHODS OF DYNAMIC SYNTHESIS OF VIBRATION PROTECTION. SPECIAL FEATURES OF THE CALCULATION AND SELECTION OF SYSTEM PARAMETERS

The choice of research methods for vibration protection systems is determined by the required accuracy of the results and depends on the properties of the object of vibration protection. The methods of the linear theory of oscillations (Atabekov, 2006; Bakalov, 2006; Khomenko et al., 2012), which are used if the disturbing influence is a deterministic function and the system makes small oscillations, are widely known. In this case, the practical calculation is reduced to the selection of parameters that provide the natural frequency lower than the frequencies contained in the spectrum of external influences. The transmission coefficient determines the ratio of the amplitude of the steady-state response to the amplitude of the external influence. At the same time, there are systems whose behavior cannot be studied using linear mathematical models due to nonlinearities of dynamic characteristics, which requires the use of nonlinear mechanics methods for research. To solve the problem in a nonlinear formulation, the method of a small parameter is used. Its disadvantage is the computational complexity, which makes it difficult to use it in practical calculations. Ultimately, the calculations of nonlinear systems are guided by linearization methods, which allows us to detect the basic properties inherent in nonlinear vibration protection systems. The disadvantage of the linearization method is the impossibility of obtaining exact solutions to the differential equations of motion.

Matrix calculus methods allow simplifying certain parameters of oscillatory processes using computers, however, they require quite time-consuming preliminary transformations (Kiryukhin et al., 2012).

Approaches based on methods for determining optimal transfer functions are based on solving the Wiener-Hopf equation (Kiryukhin et al., 2012). The problem of synthesizing optimal vibration protection systems is reduced to the problem of synthesizing an optimal linear filter. The widespread use of the Wiener-Hopf method of optimal filtering is limited to systems for which the perturbation is described by the correlation function. As a rule, these models describe a stationary random process, and the object is represented by a concentrated mass for a one-dimensional case. A criterion of the total mass of the vibration protection system is of great importance for the formulation of optimization problems. Methods for the synthesis of electrical circuits are described in (Druzhinsky, 1977; Atabekov, 2006; Lee, 2008).

The structural method, as already mentioned in the first chapter of the monograph, consists in reducing the model to a structural diagram, which allows using the methods of the theory of automatic control for its analysis and synthesis. The use of the mathematical tools of the theory of automatic control makes it possible to represent the computational scheme in the form of an equation with Laplace operators and implement the methods of dynamic synthesis that are characteristic of the theory of automatic control.

Numerical methods and applied in the evaluation of deformation of complex objects under the action of forces. They are based on splitting the surface into many simplest elements. The greatest accuracy is achieved when the object is divided into triangular elements. The disadvantage of numerical methods is the problem of resources, growing non-linearly with an increase in the dimension of the problem and the number of nodes of the finite-element grid. Difficulties arise when taking into account the dynamic nature of the loads.

In general, due to the complexity of vibration protection systems, the orientation towards the methods of the theory of automatic control using frequency research techniques and concepts about transfer functions, their responses to typical external influences and dynamic synthesis methods has become quite characteristic. This has become a particularly noticeable direction due to the growing interest in the capabilities of active vibration protection systems.

To study the elastic-inertial vibration isolation, the problem of the limiting possibilities of vibration protection systems is often formed, which is reflected in the development of a number of areas:

- the problem of determining the optimal frequency dependence of dynamic coefficients;
- problems on systems of minimum mass with a given transfer function or with a given frequency spectrum;
- the problem of the uniqueness of a mechanical system having specified frequency characteristics;

- the problem of the actual construction of dynamic matrices and schemes with given or optimal characteristics.

Determining the optimal parameters of control interactions and the corresponding value of the extremum makes it possible to find an estimate above the limiting possibilities of vibration protection under the given conditions. However, this does not solve the synthesis problem, since a system implementing the control law that has been found may not be feasible. The estimation of the limiting possibilities of vibration protection systems for certain classes of external disturbances and the properties of the connected elements of the systems is considered in (Eliseev & Zasyadko, 2004; Eliseev & Logunov, 2010).

Controlled mechanical systems in their various manifestations, connected, in particular, with robotics, mechatronics, protection against noise and vibrations, can be attributed to such complex objects for which the construction of devices for controlling the state (or motion) requires consideration of many constructive and engineering features. This requires an assessment not only of the specifics of the motion of inertial objects, but also of the processes and implementation of forces, taking into account the responses of elements to impacts, as well as the use of certain technologies for collecting and processing information. In this multi-sided integration of ideas, the problems of machine science are solved when constructing walking machines, active vibration protection systems, and modern robotic and flexible production complexes.

CONCLUSION

Thus, at a certain stage of assessing the dynamic properties of technical objects, it is fully justified to use system approaches based on generalized ideas about the interaction of a mechanical system with the external environment, the interaction of elements and the conditions for passing signals through a dynamic system. The choice of computational schemes becomes, in turn, the initial stage of the formation of mathematical models. This, to a certain extent, predetermines the possibility of analytical approaches in assessing the range of dynamic properties of the systems under consideration and the possibilities of developing and applying methods and means of controlling the dynamic state.

The problems of the dynamics of modern machines, in particular, the vibration protection of machines, equipment, instruments and devices are quite diverse and are expanding as technical systems become more complex. A transition is being formed from the traditional formulation of the problems of vibration protection and vibration isolation to the problems of assessing, monitoring and controlling the dynamic state of the object of protection from external influences. In a developed form, vibration protection systems can be automatic control systems specialized in maintaining certain parameters of the response of the object of protection to the action of a disturbance.

REFERENCES

Andreevsky, B. R., & Blekhman, I. I. (2001). *Management of mechatronic vibration installations*. Nauka Publ.

Andreychikov, A. V., Khorichev, A. A., & Andreychikov, O. N. (2011). Expert system for forecasting strategic innovations. Quality. Innovation. *Education, 12*, 26.

Atabekov, G.I. (2006). *Fundamentals of the theory of chains: a textbook.* St. Petersburg: Lan' Publ.

Averyanov, G. S., Zubarev, A. V. & Khamitov, R. N. (2008). Oscillating system dynamics with controlled vibration protection devices. *Bulletin of mechanical engineering, 6*, 27-30.

Bakalov, A. N. (2006). *Circuit Theory*. Nauka Publ.

Belokobylsky, S. V., Eliseev, S. V., & Kashuba, V. B. (2013). *Applied problems of the structural theory of vibration protection systems*. Politekhnika Publ.

Belokobylsky, S. V., Eliseev, S. V., & Sitov, I. S. (2013). *Dynamics of mechanical systems: lever and inertial-elastic constraints*. Polytekhnika Publ.

Birger, I. A. (1988). Basics of resource design. In Mechanics and scientific and technical progress: in 4 vols. (Vol. 4, pp. 174-180). Moscow: Nauka Publ.

Blekhman, I. I. (1994). *Vibration mechanics.* Fizmatlit Publ.

Brysin, A. N., Shokhin, A. E., Sinev, A. V., Solovyov, V. S., Gordeev, A. B., Erokhina, T. V. & Stepanova, L. A. (2012). Improving the method of testing vibration protection systems with inertia transducers. *Problems of mechanical engineering and automation*, (4), 80-83.

Brysin, A. N. & Sinev, A. V. (2008). New approaches in solving problems of vibration isolation. *Problems of mechanical engineering and automation, 3,* 74-83.

Burkov, S. N. & Efremov, A. M. (2003). Theoretical study of the dynamic stability of pneumohydraulic bearings. *Scientific problems of transport in Siberia and the Far East, 2,* 124-131.

Burkov, S. N., & Efremov, A. M. (2004). Vibration strength and reliability of autonomous dielectric power plants. *Proceedings of the International conference Energy saving Technologies ISTU*, (pp. 68-74). IEA.

Druzhinsky, I. A. (1977). *Mechanical chains.* Mashinostroenie Publ.

Eliseev, S. V., Artyunin, A. I., Logunov, A. S., Nasnikov, D. N., Bolshakov, R. S., Kaimov, E. V., Mironov, A. S. & Parshuta, E. A. (2013). *Dynamic interactions of machine elements: computational schemes and mathematical models of vibration states.* Irkutsk: IrGUPS Publ.

Eliseev, S. V., Khomenko, A. P., & Logunov, A. S. (2010). Use in the schemes of pneumomechanical protection of mechanisms and motion transformation devices. Modern technologies. System analysis. *Modeling*, *1*, 8–13.

Eliseev, S. V., Kovyrshin, S. V., & Parshuta, E. A. (2013). Some questions of the theory of vibration isolation: Justification of structural approaches. *News of IHEA*, *3*, 121–127.

Eliseev, S. V., & Logunov, A. S. (2009). Features of transient processes in pneumatic vibration protection systems. Modern technologies. System analysis. *Modeling*, *1*, 8–15.

Eliseev, S. V., & Logunov, A. S. (2010). Mechatronics of pneumatic vibration protection systems. Systems. Methods. *Technologies*, *7*, 9–21.

Eliseev, S. V., Moskovskikh, A. O., Bolshakov, R. S., & Savchenko, A. A. (2012). The possibilities of integration of the methods of the theory of chains and the theory of automatic control in problems of the dynamics of machines. *Science and Education, 5*, 25–26.

Eliseev, S.V., Parshuta, E.A. & Bolshakov, R.S. (2012). On the possibilities of mechatronic approaches to the problems of vibration protection of technical objects. Vibration of machines: measurement, reduction, protection. *scientific and technical and manuf. magazine, 4*(31), 46-50.

Eliseev, S. V., Parshuta, E. A., & Bolshakov, R. S. (2013). Features of vibration protection system with the object of protection in the form of a solid. Dynamic responses. *Science in the modern information society: Materials of the Intern. scientific and practical conf. Moscow. Academic Publ.*

Eliseev, S. V., Parshuta, E. A., & Bolshakov, R. S. (2013). Transfer functions of mechanical oscillatory systems. Possibilities for estimating unit stiffness. Intern. *Journal of Applied and Fundamental Research, 1*, 11–18.

Eliseev, S. V., Parshuta, E. A., & Kaimov, E. V. (2013). Vibration protection systems. Dynamic responses and methods for their determination. Engines 2013: materials of the Intern. scientific-practical conf. Omsk.

Eliseev, S. V., Reznik, Y. N., & Khomenko, A. P. (2011). *Mechatronic approaches in the dynamics of mechanical oscillatory systems*. Nauka Publ.

Eliseev, S. V., Reznik, Yu. N., Khomenko, A. P., & Zasyadko, A. A. (2008). *Dynamic synthesis in generalized tasks of vibration protection and vibration isolation of technical objects*. ISU Publ.

Eliseev, S. V., Reznik, Y. N., Upir' R. Yu., et al. (2009). *Mechatronics of vibration protection systems. Elements of the theory*. Dep. in VINITI.

Eliseev, S. V., Trofimov, A. N., Bolshakov, R. S., & Savchenko, A. A. (2012). The concept of feedback in the dynamics of mechanical systems and dynamic oscillation damping. *Science and Education, 5*, 25.

Eliseev, S. V., Volkov, L. N., & Kukharenko, V. P. (1990). *Dynamics of mechanical systems with additional constraints*. Nauka Publ.

Eliseev, S. V. & Yermoshenko, Yu. V. (2012). *Links in the dynamics of mechanical oscillatory systems*. Irkutsk: IrGUPS Publ.

Eliseev, S. V. & Zasyadko, A. A. (2004). Vibration protection and vibration insulation as a control of object vibrations. Modern technologies. *System analysis Modeling, 1*, 24-31.

Fomichev, P. A., & Fomichev, E. V. (2004). *Automation of ship engines vibration protection*. Nauka Publ.

Frolov, K. V., Kraynev, A. F., & Kreinin, G. V. (2004). *Designing machines: reference book* (Vol. 1-2). Mashinostroenie Publ.

Galiev, I. I., Nekhaev, V. A., & Nikolaev, V. A. (2010). Methods and means of vibration protection of railway carriages. Moscow: GOU Publ.

Ganiev, R. F. (2010). Problems of mechanics of machines and technologies. *Problems of mechanical engineering and machine reliability, 3*, 3-20. Blagonravov Institute of Engineering.

Genkin, M. D., & Ryaboy, V. M. (1988). *Elastic-inertial vibration isolation systems: ultimate opportunities, optimal structures*. Nauka Publ.

Goverdovsky, V. N. (2006). *Development of the theory and methods of designing machines with infra-low-frequency vibration protection systems*. Research Gate.

Khokhlov, A. A. (2002). *Dynamics of complex mechanical systems*. MIIT Publ.

Khomenko, A. P., Eliseev, S. V. & Yermoshenko, Yu. V. (2012). *System analysis and mathematical modeling in the mechatronics of vibration protection systems.* Irkutsk: IrGUPS Publ.

Kiryukhin, A. V., Tikhonov, V. A., Chistyakov, A. G. & Yablonsky, V. V. (2012). Active vibration protection – purpose, principles, state. Vibration protection and noise insulation of pipelines and experimental research. *Problems of mechanical engineering and automation, 4*, 102-110.

Korolkov, M. V. (2010). *Development and research of analytical models of the dynamics of mechanisms with gaps in the interfaces of parts*. Research Gate.

Lakanne, C. (2002). *Mechanical vibrations & shock. Specification development*. HPS Publ.

Lee, M. (2008). *Damping of resonant oscillations of gyroscopic systems with an active dynamic damper*. Research Gate.

Lilov, L. K. (1993). *Modeling of related bodies*. Nauka Publ.

Lontsikh, P. A., & Eliseev, S. V. (2014). *Dynamic quality of machinery and equipment as a tool to ensure the reliability of production and competitiveness of processes.* NI ISTU Publ.

Makhutov, N. A., Petrov, V. P., Kuksova, V. I. & Moskvitin, G. V. (2008). Modern trends in the development of scientific research on the problems of mechanical engineering and mechanical engineering. *Problems of mechanical engineering and automation, 3,* 3-19.

Myamlin, S. V. (2002). *Simulation of rail carriage dynamics*. New Ideology Publ.

Rakov, D. L. & Sinev, A. V. (2011). Structural analysis and synthesis of new technical systems based on the morphological approach. *Problems of machine engineering and automation, 2*, 73-80.

Reznik, Y. N. (2001). *Basics of mineral training in the development of mineral deposits*. VLADMO Publ.

Salpin, S. N., Sinev, A. V., Lebedev, V. N., et al. (2002). *The device of angular stabilization of a suspended object on a transport vehicle.* RF No. 2181683 / 2002.

Saveliev, Yu. F. (2003). The method of effective vibration protection of rolling stock and carriage based on additional mechanical devices with alternating elasticity. Omsk: OmGUPS Publ. .

Shedlos', A. I. (2005). *The choice of rational parameters of low-frequency articulated stands of linear displacements.* [Dissertation, Tula University].

Sobolev, V. I. (2002). *Discrete-continual dynamic systems and vibration isolation of industrial screens.* ISU Publ.

Vargunin, V. N., Gusarov, V. N., Ivanov, B. G., Levchenko, A. S. (2006). Design and calculation of articulated devices and units (O.P. Mulyukin, Ed.). Samara: SamGAPS Publ.

ADDITIONAL READING

Eliseev, A. V. (2017). Dyads in mechanical systems: Features of dynamic properties. Part II. *Vestnik Irkutskogo Gosudarstvennogo Tehniceskogo Universiteta*, *21*(8), 22–37. doi:10.21285/1814-3520-2017-8-22-37

Eliseev, A. V., Artyunin, A. I., & Sitov, I. S. (2017). Method for determining conditions for maintaining contact under vibrational loads with account for non-holding connections. Modern technologies. System analysis. *Modeling*, *1*(53), 26–34.

Eliseev, A. V., Kuznetsov, N. K., & Mironov, A. S. (2022). Generalized approaches in problems of determining contact reactions in solids under static loads with account for non-holding connections: Multidimensional variant. Systems. Methods. *Technologies*, *1*(54), 7–19.

Eliseev, A. V., Mironov, A. S., & Eliseev, S. V. (2022). Formation of mathematical models of vibrational interactions of technical means of transport and technology. Bulletin of Voronezh State University. Series. *System Analysis and Information Technologies*, *1*, 32–42.

Eliseev, A. V., Mironov, A. S., & Eliseev, S. V. (2022). System approaches in assessing dynamic states of mechanical oscillatory systems: technology of structural transformations. Bulletin of Astrakhan State Technical University. Series: Management. *Computer Science and Informatics*, *1*, 7–15.

Eliseev, A. V., Sitov, I. S., & Kuznetsov, N. K. (2022). Systemic approaches to assessing dynamic states of technical objects under vibrational loads: Structural schemes, transfer functions, dynamic invariants. Systems. Methods. *Technologies*, *2*(54), 7–19.

Eliseev, A. V., Sitov, I. S., & Kuznetsov, N. K. (2022). Systemic approaches to assessing dynamic states of technical objects: Structural mathematical modeling methods. *Proceedings of the Institute of Systems Analysis of the Russian Academy of Sciences*, *72*(1), 93–104.

Eliseev, S. V., Bol'shakov, R. S., Eliseev, A. V., & Parshuta, E. A. (2014). Features of static and dynamic loading in mechanical oscillatory systems. Problems of vibration protection. *Bulletin of Ufa State Aviation Technical University*, *18*(1), 37–47.

Eliseev, S. V., Kargapol'tsev, S. K., Bol'shakov, R. S., & Eliseev, A. V. (2020). Mechanical chains in problems of correcting dynamic states of vibration machines. *Bulletin of Irkutsk State Technical University*, *24*(4), 718–727. doi:10.21285/1814-3520-2020-4-718-727

Eliseev, S. V., Kashuba, V. B., Kinash, N. Zh., & Eliseev, A. V. (2017). Dynamic damping of vibrations: Introduction of additional connections, lever interactions, and physical effects. *Vestnik Irkutskogo Gosudarstvennogo Tehniceskogo Universiteta*, *21*(1), 10–23. doi:10.21285/1814-3520-2017-1-10-23

Eliseev, S. V., Orlenko, A. I., & Eliseev, A. V. (2017). Structural formations in mechanical oscillatory systems: Dyads, their properties, opportunities for changing dynamic states. *Transport of the Urals*, *3*(54), 56–63. doi:10.20291/1815-9400-2017-3-56-63

Jakóbczak, D. (2009). Curve Interpolation Using Hurwitz-Radon Matrices. *Polish Journal of Environmental Studies*, *3B*(18), 126–130.

Jakóbczak, D. (2010). Application of Hurwitz-Radon Matrices in Shape Representation. In Z. Banaszak & A. Świć (Eds.), *Applied Computer Science: Modelling of Production Processes, 1(6), 63-74*. Lublin University of Technology Press.

Chapter 3
Theoretical Basics of Structural Mathematical Modeling in the Problems of Dynamics of Technological Machines and Transport Vehicles:
Structural Approaches

ABSTRACT

The chapter is devoted to the evaluation of dynamics and vibration protection of technical objects. The basic principles and methods of assessing the dynamic characteristics of objects, as well as various ways of protecting against vibrations, are considered. The main types of vibrations that can occur during the operation of technical facilities are described in detail. Recommendations are given on the choice of the most effective methods of vibration protection for each specific case. The chapter also provides examples of real objects where dynamics and vibration protection were evaluated, as well as the results of these works. The chapter will be useful both to specialists involved in the design and operation of technical facilities.

INTRODUCTION

The proposed research deals with the search and development of methods and means of assessment and the possibilities of changing the dynamic state of mechanical oscillatory systems. In solving the problems of the dynamics of machines in which the computational schemes are often mechanical oscillatory systems with one, two (or more) degrees of freedom, in recent years tendencies have evolved for more general approaches based on the use of structural interpretations of mathematical models of systems described by ordinary linear differential equations and differential equations with constant coefficients.

DOI: 10.4018/978-1-6684-7237-8.ch003

Structural analogs of differential equations obtained by known methods have the potential to search for new design solutions related to the expansion of a set of typical elementary units, the possibilities of building more complex structures that interact with each other as quasi-elements according to the rules of structural transformation management. The basis for the evaluation of dynamic properties in cases of periodic disturbances that are close to harmonic, are the transfer functions of the system. The subject of research with this approach is the peculiarities of establishing dependencies on input effects (vibrations of a different type) and coordinates that determine the dynamic state of the object of protection. In theory and practice of vibration protection systems, great attention is paid to the coordinates and velocities of the points of the protected objects in various variants of problems of estimating and controlling dynamic states. To a lesser extent, attention is paid to the force responses that occur in mechanical oscillatory systems between the elements of the system, as well as in contacts with supporting surfaces.

SOME PROVISIONS ON THE BASICS OF ASSESSING THE DYNAMIC STATES OF TECHNICAL OBJECTS

Many problems of the dynamics of machines, including the problems of vibration isolation and vibration protection, in their specific content are determined by the specificity of the structural and technical forms of the objects of protection and the conditions of interaction with sources of dynamic disturbances. At the same time, the separation of power and kinematic perturbations is rather conditional and can be reduced to generalized ideas about the effect of external perturbations on technical objects (Eliseev et al., 2008). The system of electromechanical analogies, developed in the theory of electrical circuits, also creates certain opportunities for expanding ideas about mechanical oscillatory systems within the structural theory of vibration protection systems (Eliseev et al., 2011). The latter implies, in particular, consideration of mechanical oscillatory systems as electrical circuits with all the features that are manifested when taking into account the properties of energy sources (or disturbances), the formation of input and output resistances, conditions for the passage of signals through the circuits, etc. (Khomenko et al., 2012).

In the general case, in the computational scheme of a vibration protection system, it is always possible to distinguish three main parts: the source of the disturbance, the object of protection and, in fact, the vibration protection device (or that part of the mechanical oscillation system that performs the functions of reducing the level of effects). In the simplest cases, when the computational scheme is described by one or several independent coordinates, the source and the object can be considered as solids. For further consideration, translational motions along a certain y axis (Figure 1) were chosen, although one may choose rotational motions with the possibility of obtaining the same results. At the same time, rotational motions have some characteristic features that require taking into account the features arising under certain conditions (gyroscopic effects, etc.).

Vibration protection device in Figure 1 is located between the source of the perturbation and the object of protection (Frolov, 1983). In most cases, the mass of one of the bodies of the system — a source or an object — substantially exceeds the mass of the other body — the object or source, respectively. Then the motion of the body of the "large" mass can be considered independent of the motion of the body of the "small" mass. If, in particular, an object has a "large" mass, then it is usually considered immobile, the motion of the system is caused in this case by external forces applied to the source, representing *force excitation* $Q = Q(t)$ (see Figure 1, *b*). If a source has a "large" mass, then the law of its motion $z = z(t)$ can be considered set; this motion plays the role of the *kinematic excitation* of the system (more precisely,

Figure 1. Schematic diagram of a vibration protection system with one degree of freedom: a – general case; b – power disturbance; c – kinematic disturbance (R – constraint reaction)

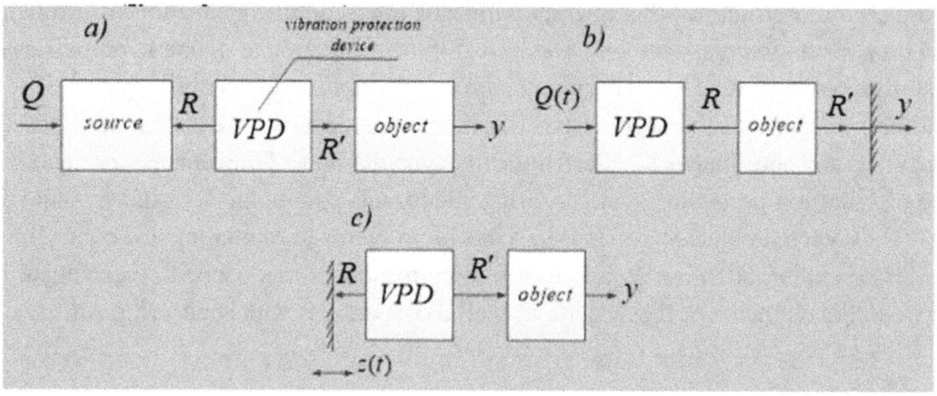

of the object) (see Figure 1, *c*). In both cases, the body of the “larger” mass is called the carrier, or base, and the body of the “small” mass is called carried. The scheme presented in Figure 1, *b*, is usually used when the protection of buildings, structures, ceilings or foundations from the dynamic effects induced by machines installed on them with unbalanced moving parts or other vibroactive equipment is considered. The scheme shown in Figure 1, *c* is used in problems of vibration protection of devices, apparatuses, precise mechanisms or machine tools, i.e. vibration-sensitive equipment installed on oscillating bases or moving objects. It can be assumed that the source of the perturbation is a form of interaction with a system that has incomparably large energy possibilities. In Figure 2 it is assumed that the partial system $M, k_{01} k_{02}$ reflects the properties of the source of power perturbation.

Figure 2. Formation of power perturbation in vibration protection system: a – computational scheme taking into account the existence of an energy source; b – structural diagram of a vibration protection system with a source of power disturbance

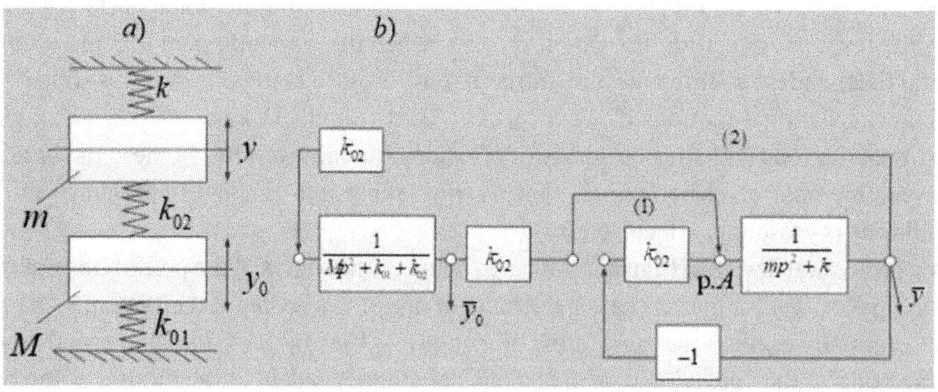

Since in the calculations of the action of vibrations the power disturbance is considered independent of the load, the source of the force action can be represented as contacts with an oscillatory system $(M,k_{01}k_{02})$ that performs continuous free harmonic oscillations. In this case, the condition $M>>m$ is satisfied that allows us to consider the action of the partial system (M,k_{01},k_{02}) only by the relation (1) (see Figure 2), which forms the force $Q=k_{02}y_0$ applied to the object of mass m. By virtue of the condition, the inverse effect on the relation (2) (see Figure 2) can be considered negligible.

The task of the study is to substantiate the approaches that allow developing a generalized methodology for determining the reaction of constraints, which is necessary for the dynamic calculations of vibration protection systems. The developed approach makes it possible to further take into account the limitation on the power of the source of power disturbance. Important for subsequent research is the fact that the force perturbation is implemented as a force applied to the object (p. *A*, see. Figure 2, *b*). In addition, the existence of such a type of perturbation is associated with the recognition of the asymmetry of the action of constraints (1) and (2) (Figure 3) between the partial systems (M,k_{01},k_{02}) and (m,k_{01},k). In the case of kinematic perturbation, the pattern of formation of actions takes place according to the same principle (see Figure 2, *b*), but the kinematic perturbation (see Figure 2, *a*) will be applied to p. *B*, as shown in Figure 3.

Figure 3. Schematic diagram of the formation of kinematic effects

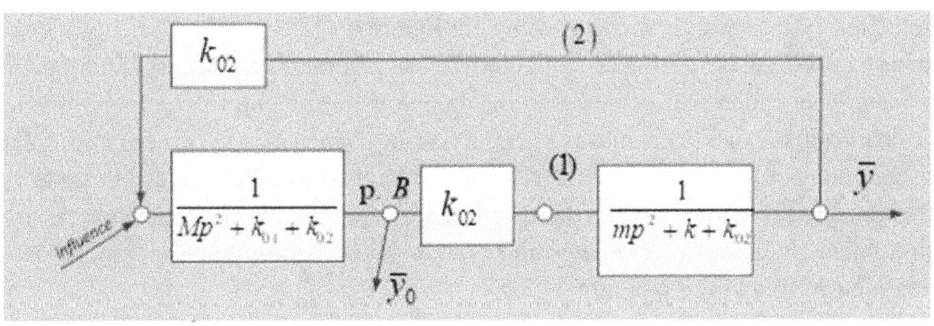

In contrast to the force perturbation (Figure 2, *b*), the displacement $\dot{y}_0$ acts through the element with the stiffness k_{02}. In this case, it is also assumed that the relation (2) does not affect the motion of the partial system (M,k_{01},k_{02}). Comparison of two forms of external influences in Figure 2, *b* and Figure 3 shows the relativity of ideas about the division of external disturbances into force and kinematic ones; in fact, they can be brought together. The latter, ultimately, served as the basis for generalized approaches (Eliseev et al., 2008).

Vibration protection device is the most important part of the vibration protection system; its purpose is to create such a mode of motion, initiated by given disturbances, in which the goal of protecting an object is implemented. In many cases, this is achievable with the use of an *inertialess* vibration isolation device, which, for the circuits shown in Figure 1, may represent a *uniaxial vibration isolator*. For such a vibration isolator, the reactions of the constraints *R* and *R*' coincide in magnitude ($R = R'$). Thus, in the simplest case, the response *R* can be considered proportional to the displacement and the velocity of displacement of the object:

Figure 4. Computational scheme of the vibration protection system with m, k, b parameters with one degree of freedom

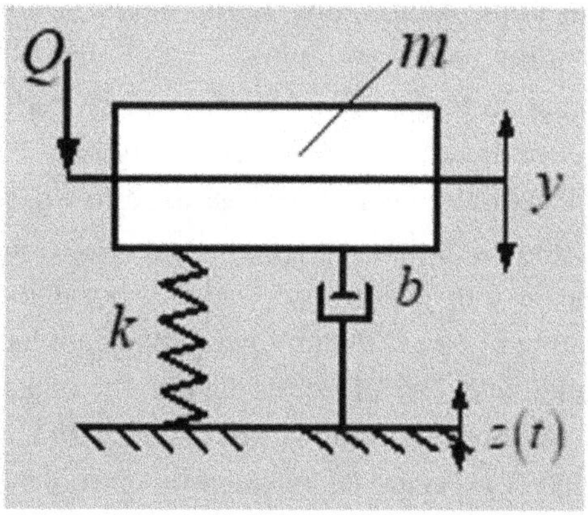

$$R = ky + b\dot{y}. \qquad (1)$$

In this case, it is assumed that the vibration isolator has a simple structure in which the motion of the protected object forms the dynamic state of the vibration isolator. The computational scheme of the vibration protection system has the form, as shown in Figure 4.

Dependence (1) describes the linear characteristic of a simple inertialess vibration isolator; the coefficients k and b are called the stiffness and the damping coefficient, respectively. When $b = 0$, (1) describes the characteristic of a linear elastic element (spring); when $k = 0$, it describes the characteristic of a linear viscous damper. Thus, the model of vibration isolator with characteristic (1) can be represented as a parallel connection of a spring and a damper (Figure 4). The stiffness coefficient k of a vibration isolator with a linear characteristic (1) determines the natural frequency of the system, as well as the static deformation (draft) y of the vibration isolator.

$$\omega_0 = \sqrt{\frac{k}{m}}. \qquad (2)$$

Damping properties of the system shown in Figure 4, characterized by damping factor

$$n = b/(2m) \qquad (3)$$

and relative damping

$$\nu = n/\omega_0 = b/2\sqrt{km}. \qquad (4)$$

When $\nu=1$, critical damping is implemented in the system (Frolov, 1983). The influence of damping, in general, has various forms and significantly changes the properties of vibration protection systems.

By the effectiveness of vibration protection we mean the degree of implementation by the vibration protection device of vibration protection aims. With force harmonic excitation

$$Q(t) = Q_0 \sin\omega t, \tag{5}$$

where Q_0 and ω are the amplitude and frequency of the driving force. The purpose of vibration protection may be to reduce the amplitude R_0 of the force transmitted to a fixed object,

$$R_0 = \frac{Q_0\sqrt{\omega_0^4 + 4n^2\omega^2}}{\sqrt{\left(\omega_0^2 - \omega^2\right)^2 + 4n^2\omega^2}}, \tag{6}$$

that follows from consideration of the equation of kinetostatics for the element m with $z(t)=0$:

$$m\ddot{y} + b\dot{y} + ky = Q. \tag{7}$$

Using the Laplace transforms (.1) with zero initial conditions, we find that

$$\dot{R}_0 = \dot{y}\left(bp + k\right). \tag{8}$$

In turn, the decrease in the amplitude of the steady-state forced oscillations of an object under the action of a force is determined by the formula

$$y_0 = \frac{Q_0}{m\sqrt{\left(\omega_0^2 - \omega^2\right)^2 + 4n^2\omega^2}}. \tag{9}$$

The transfer function of the system when the input signal $\dot{Q}$ and output signal $\dot{R}_0$ looks like:

$$W_R\left(p\right) = \frac{\dot{R}_0}{\dot{Q}} = \frac{bp + k}{mp^2 + bp + k}, \tag{10}$$

because

$$\dot{y} = \frac{\dot{Q}}{mp^2 + bp + k}. \tag{11}$$

Note that the reaction of the constraint is determined from the expression (8):

$$\dot{R}_0 = \dot{y}\left(bp + k\right) = \dot{Q} - mp^2\dot{y}. \tag{12}$$

Assuming that is determined from (11), we get:

$$\dot{R}_0 = \dot{Q} - \frac{mp^2\dot{Q}}{mp^2 + bp + k} = \frac{(bp+k)}{mp^2 + bp + k}\dot{Q}, \tag{13}$$

or

$$W_R(p) = \frac{\dot{R}_0}{\dot{Q}} = \frac{bp+k}{mp^2 + bp + k}, \tag{14}$$

which coincides with the expression (10). That is, in a vibration protection system with one degree of freedom the transfer function $W_R(p)$ determines the ratio of the magnitude of the constraint reaction $\dot{R}_0$ to the external harmonic force $\dot{Q}$ applied to the object of protection. The question is whether we can determine $W_R(p)$ directly from the structural diagram shown in Figure 5. This structural diagram corresponds to the calculation scheme in Figure 4.

Figure 5. Structural diagram of the vibration protection system with one degree of freedom

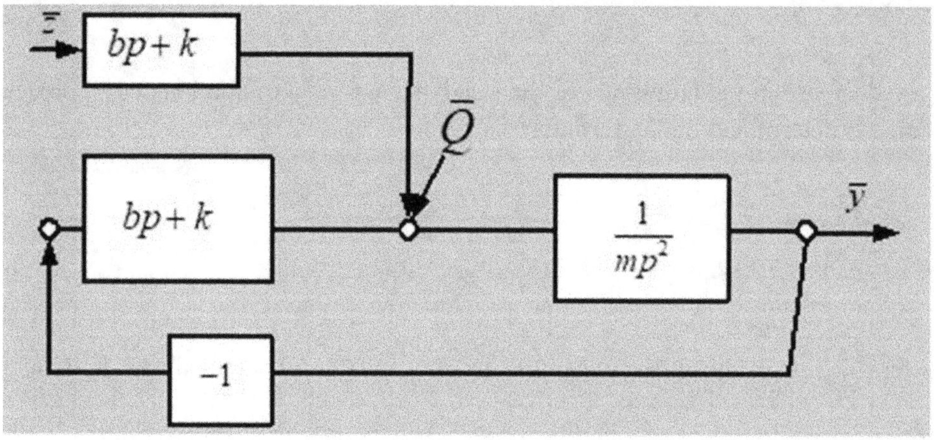

The structural diagram (Figure 5) is a graphical analog of the mathematical model of the vibration protection system in the form of a differential equation that can be obtained by known methods (Frolov, 1983). Kinematic perturbation $z(t)$ characterizes the motion of the base and is transmitted to the object of protection, which can be described using the transfer function

$$W(p)_z = \frac{\dot{y}}{\dot{z}} = \frac{bp+k}{mp^2 + bp + k}. \tag{15}$$

The physical meaning of (15) is that it corresponds to the ratio of the amplitude of the forced vibrations of an object to the amplitude of the vibrations of the base. Expressions (10) and (15) coincide, which indicates the possibility to find $W_R(p)$ through the definition of $W(p)_z$ using the structural diagram in Figure 5. In this case, we can bear in mind the possibility of equivalence of the action of two external perturbations, since $Q=kz$ is true. When $z(t)=z_0\sin\omega t$ and $Q(t)=0$ the task of vibration protection may be to reduce the amplitude of the absolute acceleration (overload) of the object

$$k_{ov} = \frac{z_0\omega^2\sqrt{\omega_0^2 + 4n^2\omega^2}}{\sqrt{\left(\omega_0^2 - \omega^2\right)^2 + 4n^2\omega^2}}, \tag{16}$$

and also in reducing the amplitude of oscillations of the object relative to the foundation, i.e in the coordinate system:

$$y' = \frac{z_0\omega^2}{\sqrt{\left(\omega_0^2 - \omega^2\right)^2 + 4n^2\omega^2}}. \tag{17}$$

Thus, for each task of vibration protection or a change in the dynamic state of an object one may choose dimensionless efficiency factors. Such coefficients are determined by the corresponding transfer functions of the initial (see Figure 5) structural model.

At power perturbation

$$k_R = \left|W_R(p)\right| = \left|W_z(p)\right|;\ k_y = \frac{ky_0}{Q};\ k_z = \left|W_z(p)\right| = \left|\frac{\dot{y}}{\dot{z}}\right|;\ k_R = k_z;$$

$$W_Q(p) = \frac{\dot{y}}{\dot{Q}} = \frac{1}{mp^2 + bp + k}. \tag{18}$$

In the case of kinematic effects, the following relationships are considered:

$$k_R = \frac{k_{ov}}{\omega^2 z_0};\ k_{y'} = \frac{y_0'}{z_0}. \tag{19}$$

In (De Silva, 2000; Karnovsky & Lebed, 2016) k_R is called the coefficient of vibration isolation, and k_y - the coefficient of dynamism. If to introduce a dimensionless parameter $\gamma = \frac{\omega}{\omega_0}$ and $\nu = \frac{n}{\omega_0}\left(2n = \frac{b}{m}\right)$, then

$$k_R = \sqrt{\frac{1 + 4\nu^2\gamma^2}{\left(1 - \gamma^2\right)^2 + 4\nu^2\gamma^2}};\ k_y = \frac{1}{\sqrt{\left(1 - \gamma^2\right)^2 + 4\nu^2\gamma^2}};\ k_{y'} = \frac{\gamma^2}{\sqrt{\left(1 - \gamma^2\right)^2 + 4\nu^2\gamma^2}}. \tag{20}$$

When calculating the simplest vibration protection systems, the condition of efficiency according to the criteria $k_R, k_y, k_{y'}$ are formed as inequalities:

$$k_R \leq 1;\ k_y \leq 1;\ k_{y'} \leq 1. \tag{21}$$

Since the indicated coefficients depend on the frequency, it is possible to consider the effectiveness of vibration protection at a given frequency γ or in a given frequency range $\gamma 1 \leq_{\gamma} \leq \gamma 2$. A$_n$alysis of relations (3.18) allows us to conclude that the effectiveness of vibration protection by the criterion kR≤1 $_i$s ensured at any level of damping in the frequency range

$$\gamma \geq \sqrt{2}. \tag{22}$$

For any of the range (22), the efficiency is the higher, the weaker the damping; the best performance has a perfectly resilient vibration isolator (v=0). The effectiveness of vibration protection by the criterion ky$_{\leq}$1 is also provided in the range (22) for any values of v. When $v = \frac{1}{\sqrt{2}}$, vibration protection is effective in the entire frequency range of $0<\gamma<\infty$; when $v < \frac{1}{\sqrt{2}}$, efficiency is provided in the range of:

$$\gamma > \sqrt{2\left(1-2v^2\right)}. \tag{23}$$

At a fixed value of γ, the efficiency increases with an increase in damping.

Vibration protection by criterion $k_{y'} \leq 1$ is effective in the entire frequency range, if $v > \frac{1}{\sqrt{2}}$, and when $v < \frac{1}{\sqrt{2}}$, in the range of

$$0 < \gamma < \frac{1}{\sqrt{2\left(1-2v^2\right)}}. \tag{24}$$

The degree of efficiency at a fixed frequency γ increases with increasing damping; in the worst case (with $\nu = 0$), the efficiency range corresponds to the band

$$0 < \gamma < \frac{1}{\sqrt{2}}. \tag{25}$$

The dependences $k_R = k_R(\gamma, v)$, $k_y = k_y(\gamma, v)$, $k_{y'} = k_{y'}(\gamma, v)$ for fixed v can be considered as the amplitude-frequency characteristics of the original system with respect to the corresponding input and output. Different frequency characteristics of the vibration protection system are respectively determined by:

$$\varepsilon = arctg\frac{2\nu\gamma}{1-\gamma^2}; \tag{26}$$

$$\eta = arctg\frac{2\nu\gamma^3}{1-\gamma^2\left(1-4\nu^2\right)}, \tag{27}$$

and represent the dependence of the phase shifts of the processes $y(t), y'(t)$ (26) and $R(t)$, $\ddot{y}(t)$ (27) with respect to perturbations $Q(t)$ and $z(t)$.

Vibration isolation device is often made in the form of a compound of several vibration isolators, forming a *complex vibration isolator*. Under certain conditions, the response R of such a compound can be approximated, and the complex vibration isolator is replaced by an equivalent one (Eliseev et al., 2011).

APPROACHES OF STRUCTURAL MATHEMATICAL MODELING IN PROBLEMS OF DYNAMICS AND VIBRATION PROTECTION OF TECHNICAL OBJECTS

A comparative review in the field of theoretical developments and experimental studies of vibration protection system shows that the elemental base of a set of typical elements of mechanical oscillatory systems is expanding. The development of the structural theory of the vibration protection system turns out to be a convenient approach, which could combine the ideas of digital control with ideas about typical active elements having transfer functions of an elementary form. The structural approaches use, as a graphic analogue, the system of differential equations of the motion of the vibration protection system in the steady state, dynamically equivalent systems of automatic control. The transfer functions of the vibration protection system in this case are determined on the basis of the usual transformation rules used in the theory of automatic control. However, the implementation of some applications of the concept of introducing additional constraints (Karnovsky & Lebed, 2016; Eliseev, 2018; Eliseev & Khomenko, 2014; Eliseev et al., 2019) requires consideration of a number of features.

First, the structural diagram of the vibration protection system is built on the basis of the previously obtained differential equations of motion, for the compilation of which the d'Alembert principle or the Lagrange formalism are used. Therefore, the constraints existing in the structural diagram have a specific physical content, and the points of application of force or kinematic perturbations in the computational scheme are uniquely correlated with the corresponding points on the basic elements of the vibration protection system structural diagram. The basic element has the form of a unit with the transfer function

$$W = \frac{1}{mp^2}. \tag{28}$$

The input of such a unit is force, and the output is displacement. All other units of the structural diagram of the vibration protection system (except for a single feedback) are elementary typical units, the input of which is the displacement, and the output is the force (Eliseev et al., 2008; Khomenko et al., 2012).

Secondly, typical elementary units, in accordance with the above, have different transfer functions, but all of them are essentially different types of springs, that is, all of the elementary units of the vibration protection system belong to the same class — the class of generalized springs. The usual elastic element is an elementary unit with the transfer function of the amplifying element. There are other types of elementary units, about which the author will further provide relevant information.

Thirdly, within the structural theory of the vibration protection system, additional constraints are introduced into the structural diagram of the vibration protection system, only as parallel chains, to the elastic element of the basic model of the vibration protection system. Thus, by definition, the additional chain is a "spring". Therefore, the reduced stiffness is determined, and the generalized spring is formed (as the physical essence of the additional constraint) based on the rules of parallel and consecutive connection of springs. Outside the aforementioned circumstances, the structural diagrams of vibration protection systems and automatic control systems follow the same rules of structural transformations.

Fourthly, typical elementary units of a vibration protection system consist of elementary units that cannot be simplified. All other units are the result of assembly (Eliseev, 2018). Typical units of automatic control systems for automatic control systems look completely different and are formed on the basis of ideas about the possibility of writing down a complex transfer function, a fractional rational type as a result of some multiplication of transfer functions of the elementary (typical) units of automatic control systems (Kim, 2003; Eliseev, 1978; Eliseev et al., 2015).

These differences are related to the fact that the structural diagrams of the vibration protection system are based on formalized methods for the compilation of differential motions. Whereas in the construction of automatic control systems structural diagrams more sophisticated methods (including heuristic ones) of compiling mathematical models are used. In this part of the equations reflects the physical features of building systems, the presence of various technical means of collecting, transmitting and processing information about the dynamic state of the object.

The Connection of the Elementary Units in the Chain of Additional Constraint

The basis of the vibration protection system is a mechanical oscillatory system. Its simplest form is an object having a mass (m) and an elastic element with a stiffness k. Such a system can be called the base model. It corresponds to a dynamically equivalent automatic control system (Figure 6).

Figure 6. The basic model of the vibration protection system (a) and its structural diagram (b)

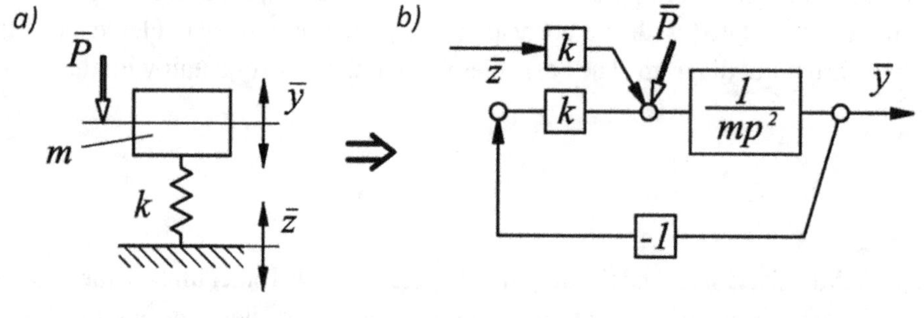

The structural diagram provides an idea on of the place and role of the elastic element, at the input of which a displacement (y) is applied. The output of the unit with the transfer function $W_1=k$ is force. This force is fed to the input of the second unit with the transfer function $W_2 = \frac{1}{mp^2}$. Such a unit in the structural diagram is an "adder," since all forces gather at the input. This unit performs a basic function and is not included in the number of elementary units and has another transfer function in dimension, since the output signal is a displacement (y). In such a system of minimal complexity, the unit $W_1=k$ creates the force of elastic interaction. From the standpoint of the theory of automatic control, this unit is called amplifying. The second basic unit is a second-order integrating unit, which "transforms" the elastic force into displacement. Both schemes in Figure 6, *a*, *b* correspond to the same differential equation:

$$m\ddot{y} + ky = kz + P(t)\,. \tag{29}$$

The structural diagram in Figure 6, b is a graphical form of displaying a differential equation (29). Getting the equation (29) can be implemented on the basis of the d'Alembert principle or using the Lagrange equation. For us, an important circumstance is to select the "key" element in the base model: this is an object with a mass m. More precisely, the "key" element is the mass point with a mass m, relative to which the kinetostatic equation is compiled. The elastic element with the stiffness coefficient k is thus an elementary unit.

The complication of the vibration protection system can go through a parallel introduction to the elastic element of the additional circuit, the transfer function of which can be represented as a fractional rational expression:

$$W_{add}(p) = \frac{a_0 + a_1 p + \ldots + a_n p^n}{b_0 + b_1 p + \ldots + b_m p^m}, \tag{30}$$

where m,n are integers ($n \le m$); a_i, b_j are the coefficients determined by the design features of the vibration protection system, $i = 1,\overline{n}$ ...

Figure 7. Types of additional constraints on the structural diagram (a) and computational scheme (b)

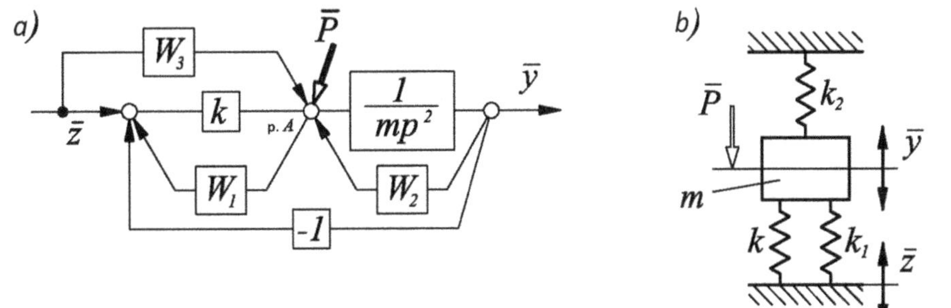

An additional circuit with a transfer function $W_{add}(p)$ is an additional feedback introduced into the structural diagram (see Figure 6, *b*). In this case, it is possible to use the principles of control introduced in theory of automatic control by the absolute and relative deviations, as well as by external disturbance (Figure 7, *a*, *b*).

If we accept that $W_1=k_1$, and $W_2=k_2$, then in Figure 7, *b* one can see the form of the physical implementation of such constraints. These are springs that are inserted between the object of protection and the base: mobile and stationary one. As regards the introduction of a constraint by perturbation (W_3), its implementation requires the construction of a special chain consisting of a number of units; this relationship can be shown in the structure of the transfer function of the vibration protection system

$$W = \frac{\bar{y}}{\bar{z}} = \frac{k + W_1 + W_3}{mp^2 + k + W_1 + W_2}. \tag{31}$$

If the system has several degrees of freedom, then the structural diagram of the vibration protection system will consist of partial systems, each having one degree of freedom, but these systems will be interconnected through elastic units. The latter implements the "action equals counteraction" principle in the system. The construction of structural diagrams can be carried out directly according to the computational scheme. However, preliminary preparation of the differential equations of motion can be useful. In particular, the coefficients matrices of the system of equations have certain symmetry properties, which makes it possible to control the correctness of the compilation of the equations of the vibration protection system.

The various types of additional constraints discussed above are essentially generalized springs, since the input for them is the coordinate of the object (or its displacement), the output is force. Point *A* (Figure 7, *a*) is an "adder" in which the sum of all applied forces is zero (in accordance with the d' Alembert principle) (Eliseev et al., 2008).

The number of elementary units can be expanded, and it is not limited to springs, as was shown in (Eliseev et al., 2015; Nasnikov, 2008). If by elementary unit we mean a unit, whose transfer function cannot be obtained by simple transformations, then the number of elementary units could include: units of single and double differentiation, single and double integration. We will assume that from such elementary units constituting the expanded elementary base of the vibration protection system, it will be possible to obtain other typical units, but they will already be units of the second level, since they will consist of the mentioned elementary units. In this sense, typical units in the theory of automatic control, with a few exceptions, are units of the second level from the standpoint of the structural theory of the vibration protection system. Thus, the set of elementary units of the vibration protection system in the first approximation consists of five elements:

- an elastic element (or unit) with a transfer function $W_1=k$;
- a differentiating unit of the first kind (or order) with a transfer function $W_2=Ap$;
- a differentiating unit of the second order $W_3=Ap^2$;
- an integrating unit of the first order $W_4 = \frac{A}{p}$;
- an integrating unit of the second order $W_5 = \frac{A}{p^2}$

- As a special case, you can consider a pure delay unit $W_6=Ae^{-ap}$.

All the transfer functions of the elementary units mentioned above can be obtained from the expression (3.26) with an appropriate choice of the values of the coefficients. The elastic element has a transfer function $W_1 = k = \frac{a_0}{b_0}$; first order differentiator $W = Ap = \frac{a_1 p}{b_0}$; second order differentiator $W_3 = Ap^2 = \frac{a_2 p^2}{b_0}$; first order integrator $W_4 = \frac{A}{p} = \frac{a_0}{b_1 p}$; second order integrator $W_5 = \frac{A}{p^2} = \frac{a_0}{b_2 p^2}$, etc. It would be possible to expand the element base and supplement it with other elements, the proposed approach formally allows it. The question is whether there are physically implementable forms of such units. Nowadays, in practice, springs, viscous friction dampers Ap, motion transformation devices Ap^2 are known (Eliseev et al., 2008). It should be noted that at the preliminary stage of consideration it seems appropriate to assess the possibilities of the so-called passive vibration protection means. Such means do not use external sources of energy. If the vibration protection system uses energy sources, this is connected with the construction of quite complex circuits, including measuring, amplifying, corrective units, etc. Such a system is called active vibration protection system and is essentially an automatic control system.

The theory of automatic control also uses typical units. However, the choice of typical units is based on other approaches. In particular, the transfer function of the entire system is considered as the result of some transformations that allow the transfer function of the system to be considered as the result of consecutive and parallel connection of typical units (Kim, 2003; Ivashchenko, 1993; Yurevich, 2007).

Taking into account this circumstance, the differences in the methods of dynamic synthesis of vibration protection system and automatic control systems are determined, although they have a common basis. Let us clarify one more circumstance, which is connected with the fact that the additional constraint introduced (additional chain) is essentially a generalized spring and consists of elementary units, each of which is also a peculiar spring: displacement is at the input and the force is at the output. In all other cases, automatic control systems and vibration protection system obey the general rules of structural transformations (Eliseev et al., 2008; Eliseev et al., 2011; Khomenko et al., 2012).

Elementary Units of Vibration Protection Systems

Elastic element, which is a spring with transfer function $W_1=k$, has the following frequency characteristics:

- real frequency response characteristic $U_1=k$;
- imaginary frequency response characteristic $V_1=0$;
- phase frequency response characteristic $\varphi 1\_0$, $arctg\varphi 1=_0$;
- amplitude-frequency response characteristic $A_1 = \sqrt{U_1^2 + V_1^2} = k$;
- the amplitude-phase frequency characteristic, which is a vector lying on the axis $0V$ and having a length k (it does not depend on the frequency), as shown in Figure 8.
- the amplitude-phase frequency characteristic, which is a vector lying on the axis $0V$ and having a length k (it does not depend on the frequency), as shown in Figure 8.

Figure 8. Amplitude-phase frequency characteristic of the elastic element

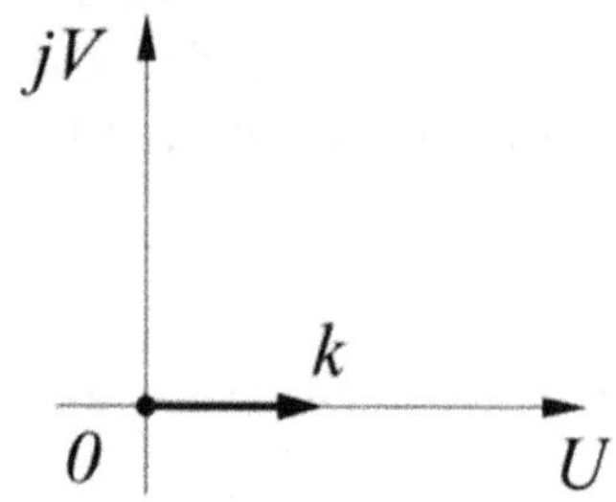

The elastic element with the transfer function $W_1=k$ in theory of automatic control corresponds to an amplifying or proportional unit (Kim, 2003; Ivashchenko, 1993). Elementary units can have various forms of physical implementation. The elastic element — a spring — is a widespread element of a vibration protection system and is usually represented by coil springs, leaf springs, etc.

Viscous friction damper consists of several parts (piston, cylinder), i.e. mechanism is used. The first-order differentiating unit has a transfer function $W_2=Ap$ and frequency characteristics:

$$W_2(j\omega)=jA\omega,\ V_2(\omega)=A\omega,\ U_2(\omega)=0,$$

$$arctg\varphi_2(\omega)=\infty,\ \varphi_2(\omega)=\frac{\pi}{2},\ A_2(\omega)=\sqrt{(A\omega)^2}=A\omega.$$

In Figure 9 it is shown that AFC has the form of a vector located on the axis $0jV$. The magnitude of the vector depends on the frequency ω, the phase shift is $\frac{\pi}{2}$.

Figure 9. Amplitude-phase frequency characteristic of the differentiating unit of the first order

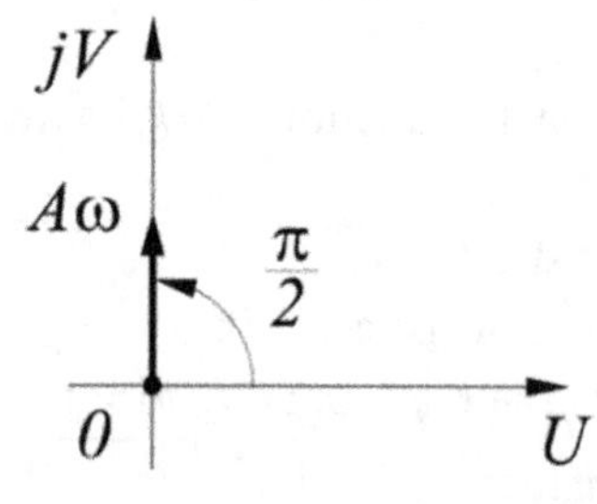

The differentiating unit of the first order is often called the differentiating unit.

The second-order differentiating unit has a transfer function $W_3=Ap^2$ and frequency characteristics

$$W_3(j\omega) = -A\omega^2,\ V_3(\omega) = 0,\ U_3(\omega) = -A\omega^2,\ \varphi_3(\omega) = arctg(0) = 0,\ A_3(\omega) = A\omega^2.$$

Figure 10 shows that the vector of the amplitude-phase frequency characteristic is located on the axis $0U$ and has a negative direction. The magnitude of the vector depends on the square of the frequency. The unit of double differentiation is usually not considered in theory of automatic control. Instead, second-order forcing units are considered, but the transfer function of such a unit has a different form: $W(p)= k(T^2p^2+2\xi Tp+1)$ $(0<\xi<1)$. In the framework of the proposed approach, such a forcing unit is a parallel connection of three units: a proportional unit W1($_{p}$)=k, a first-order differentiating unit W2($_{p}$)=Ap, and a second-order differentiating unit W3=$_{A}$p2.

Thus, the forcing unit (typical unit) is a second level unit in the mechanics of the vibration protection system, and can be considered assembled according to certain rules from elementary units. However, other approaches are possible, as reflected in (Kim, 2003; Ivashchenko, 1993).

Figure 10. Amplitude-phase frequency characteristic of the differentiating unit of the second order

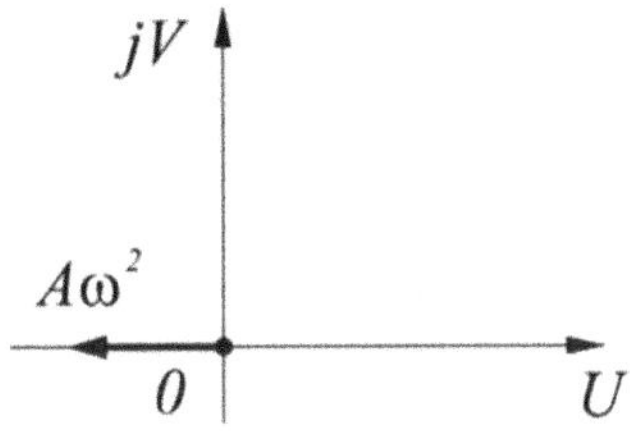

The integrating unit of the first order has a transfer function $W_4 = \dfrac{A}{p}$ and frequency characteristics:

$$W_4(j\omega) = \frac{A}{j\omega} = -j\frac{A}{\omega},\ V_4(\omega) = -\frac{A}{\omega},\ U_4(\omega) = 0,\ \varphi_4(\omega) = -\frac{\pi}{2},\ A_4(\omega) = \frac{A}{\omega}.$$

Figure 11 shows the amplitude-phase frequency characteristic of the first-order integrating unit. The amplitude-phase frequency characteristic vector has a negative direction, and its magnitude depends on ω in such a way that, for small ω, its magnitude has a large value, and with an increase of ω the modulus of the vector decreases (that is, there is a hyperbolic dependence).

The integrating unit of the second order has a transfer function of the form $W_5 = \dfrac{A}{p^2}$ and frequency characteristics:

$$W_5(j\omega) = -\frac{A}{\omega^2},\ V_5(\omega) = 0,\ U_5(\omega) = -\frac{A}{\omega^2},\ tg\varphi_5 = 0,\ \varphi_5(\omega) = 0,\ A_5(\omega) = \frac{A}{\omega^2}.$$

Figure 11. Amplitude-phase frequency characteristic of the integrating unit of the first order

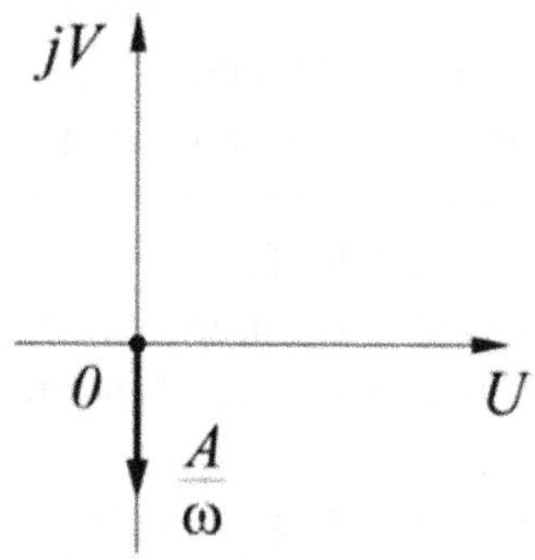

Figure 12 shows the amplitude-phase frequency characteristic of the second-order integrating unit. The vector is oriented in the negative direction along the axis $0U$, and its value is inversely proportional to the square of the frequency ω.

Figure 12. Amplitude-phase frequency characteristic of the integrating unit of the second order

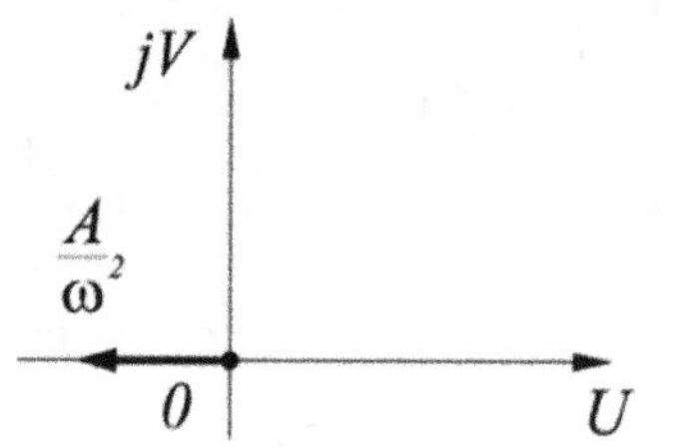

The unit of pure delay with the transfer function $W_6=Ae^{-ap}$ holds a special place among the elementary units. This is due to the fact that the transfer function of the elementary unit of pure delay cannot be obtained by simplifying the expression (30). The delay is manifested in the fact that when the input action (displacement) changes, the output quantity (force) starts changing not immediately, but after a certain period of time, characterized by a, called the time of pure or transport delay (Ivashchenko, 1993; Yurevich, 2007). The frequency characteristics of the pure delay unit are:

$$W_6(j\omega)=Ae^{-aj\omega}=A(\cos a\omega-j\sin a\omega); \tag{32}$$

$$V_6(\omega)=-A\sin a\omega;\ U_6(\omega)=A\cos a\omega. \tag{33}$$

This unit without distortion (with $A = 1$) reproduces the output amplitude as an ideal static unit, but with the difference that the output value is delayed relative to the input for a constant time.

The use of the transfer function of the elementary unit of pure delay in various situations, the formation of second-level units is not particularly difficult, however, the resulting frequency equations include exponential terms, which creates certain difficulties in the search for solutions. Therefore, a pure delay unit is often replaced by its simplification in the form of an aperiodic unit (Eliseev et al., 2008; Eliseev et al., 2011; Khomenko et al., 2012). In this case, it falls out of the general scheme of consideration. Note that the aperiodic unit is not an elementary unit in the vibration protection system mechatronics, but is a consecutive connection of the elastic element (proportional unit) and the first-order integrating unit

$$W_6 = \frac{k \cdot A/p}{k + A/p} = \frac{Ak}{kp + A}. \tag{34}$$

Thus, on the one hand, the pure delay unit can be considered as an elementary unit. On the other hand, its first approximation is a unit of the second level, consisting of the above considered elementary typical units. We would like to note that the emergence of specific problems of vibration protection can result in an increase in the number of elementary units. The question lies in the possibilities of their physical implementation. Currently, various forms of the physical implementation of the proportional unit, as well as elementary differentiating units of the first and second orders are known. These units can be attributed to the group of so-called passive means, since they can operate without the use of external energy sources. The units of the integrating type are of a more complex nature, which is apparently due to the use of external energy sources and the emergence of the more complex structure of the unit itself during the physical implementation. The latter in the mechatronics of the vibration protection system is associated with the use of active elements.

Typical Units in Automation Systems

Typical elementary units are also used in the theory of automatic control (Kim, 2003; Ivashchenko, 1993; Yurevich, 2007). The transfer function of the automatic control systems in the form of an arbitrary polynomial can also be decomposed into prime factors, that is, the transfer function of the system (unit) represented by expression (30) can always be written as a product of simple factors and fractions

$$k, p, \frac{1}{p}, Tp \pm 1, \frac{1}{Tp \pm 1}, T^2 p^2 \pm 2\xi Tp + 1, \frac{1}{T^2 p^2 \pm 2\xi Tp + 1}. \tag{35}$$

Recall that k in theory of automatic control is called a transfer coefficient, T is a time constant and $\xi (0<\xi<1)$ is a damping coefficient. From the above positions in the understanding of the elementary units in (35), one can see that $k, p, \frac{1}{p}$ is elementary ones, and the rest can be obtained by using the rules of consecutive and parallel connections, as is accepted in the theory of automatic control. The scheme of correlation of the transformation rules is shown in Figure 13. We would like to note a significant feature of the mechatronics of the vibration protection system. It is connected with the fact that an additional constraint or an additional chain has the dimension of a unit, at the input of which there is a displacement, and at the output there is force.

Figure 13. A diagram explaining the relationship between transformation rules in the additional feedback circuit and in the automatic control systems(ACS) structural diagram (the data are given for comparison with approaches in the theory of automatic control)

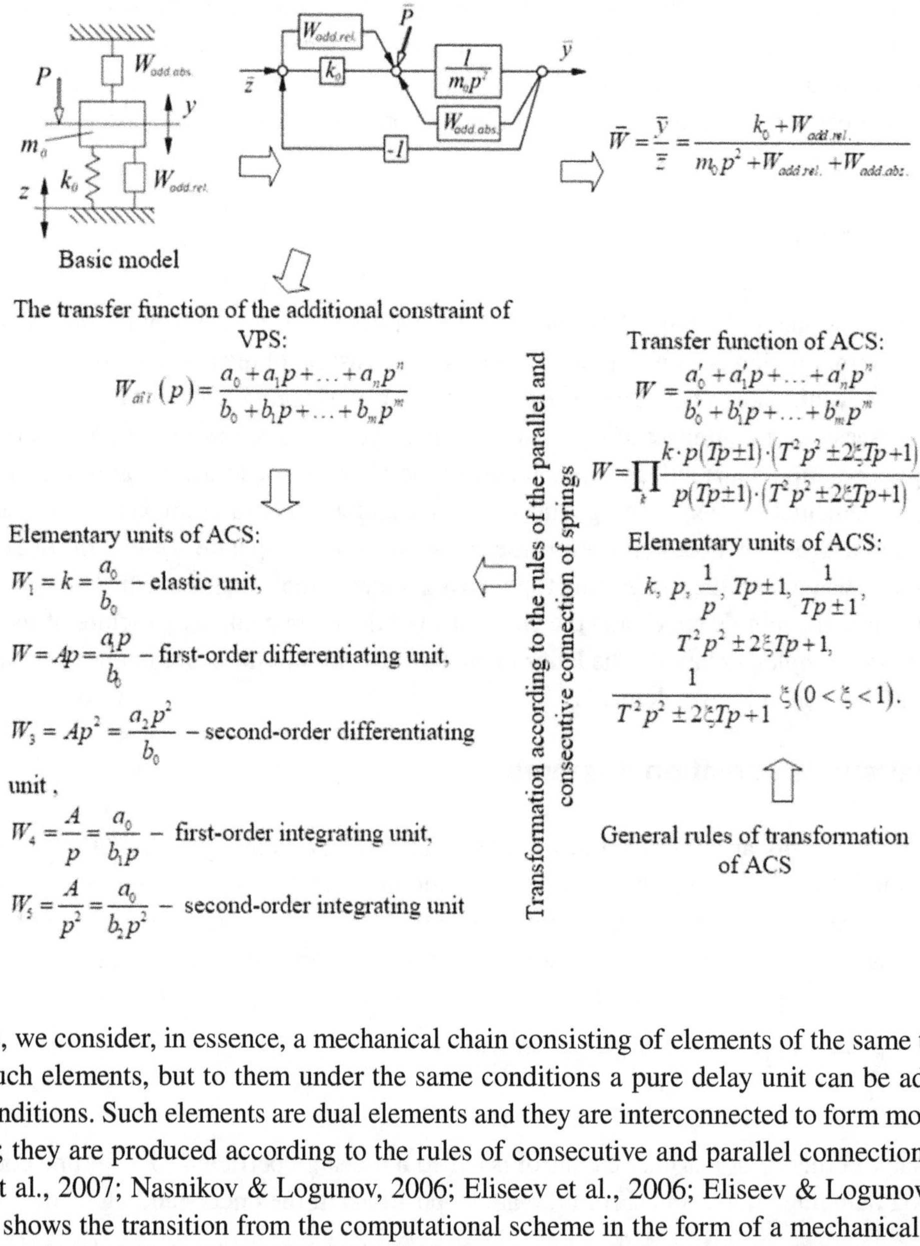

That is, we consider, in essence, a mechanical chain consisting of elements of the same type. There are five such elements, but to them under the same conditions a pure delay unit can be added, under certain conditions. Such elements are dual elements and they are interconnected to form more complex structures; they are produced according to the rules of consecutive and parallel connection of springs (Eliseev et al., 2007; Nasnikov & Logunov, 2006; Eliseev et al., 2006; Eliseev & Logunov, 2010). In Figure 13 shows the transition from the computational scheme in the form of a mechanical oscillatory system with one degree of freedom (such a system is called basic, since more complex systems can be reduced to such a form, with the exception of intermediate coordinates), which is interpreted a structural diagram (or structural mathematical model) of a dynamically equivalent automatic control systems.

In this case, the elastic elements (a spring with stiffness k_0 and more complex formations with transfer functions $W_{add.abs}$ and $W_{add.rel}$) are displayed on the structural diagram as a negative feedback circuit with respect to the object of protection.

The transfer function (30) in our case relates to an additional mechanical chain, introduced in a certain way into the structural diagram of a dynamically equivalent automatic control systems. The complexity of this chain determines the structure, the number of elementary units and the rules of their switching. When considering the automatic control systems, as such, the transfer function of the vibration protection system as a whole is determined, which leads to an expression having, in general, the same type as the expression (30). But in two cases, these expressions, for all external similarity, have a different physical meaning. Note that the elementary base of the vibration protection system is formed within the framework of an additional mechanical chain. In turn, the element base of the automatic control systems is formed for the entire system as a whole. The element base of a vibration protection system consists of dual elements of a well-defined dimension: all elements have an input – displacement and an output – force, i.e. each element is an elementary spring, but with its own dependencies on frequency, etc.

When considering the transfer function of the automatic control systems as a whole and the vibration protection system in particular, the typical elements proposed in theory of automatic control do not carry the limitations that are imposed in dual mechanical chains.

General Guidelines on Structural Approaches in Building Vibration Protection Systems

In the sections above, the directions of the development of the theory and practice of vibration protection systems are defined, including how to study in detail the dynamic properties of systems of a fairly simple type (with one, two or three degrees of freedom) with additional feedbacks, whose complexity increases as the measurement systems become more complex, technologies of information processing and servo drives reproducing control efforts. The latter implies the possibility of describing the dynamic properties of additional feedback using the transfer functions of the form corresponding to the expression (30).

The simplest basic model is a mechanical oscillatory system with one degree of freedom (Figure 6, *a*), which can be compared with a structural model (Figure 6, *b*). It is believed that all known methods of introducing the control is implemented by the simplest technical means. If the traditional set of elements forming vibration protection SYSTEM consists of a mass-and-inertia unit, a spring and a damper, then the extended set includes units that have transfer functions of an integrating unit of the first kind, integrating and differentiating units of the second kind, a unit of pure delay (although this list of elements apparently could be continued).

We will assume that the physical or constructive and engineering form of implementation of typical units is feasible on the basis of units of passive and active nature. The latter is connected with the use of mechanisms in the vibration protection system, which are set in motion from external sources.

Characteristic of the approach developed by the author of the dissertation is the perception of the mentioned typical elementary units, homogeneous in that all of them have a transfer function representing (in the field of Laplace transformations) the ratio of the output signal transform to the input signal. In this case, the input signal is a displacement, and the output signal is a force of the corresponding type. In this sense, typical elementary units are compatible, and their coupling is determined by the rules of parallel and consecutive connections as two elastic elements (or units). Figure 14 shows the ratio of traditional and extended sets.

Figure 14. The ratio of traditional and advanced approaches of typical elements of the vibration protection system

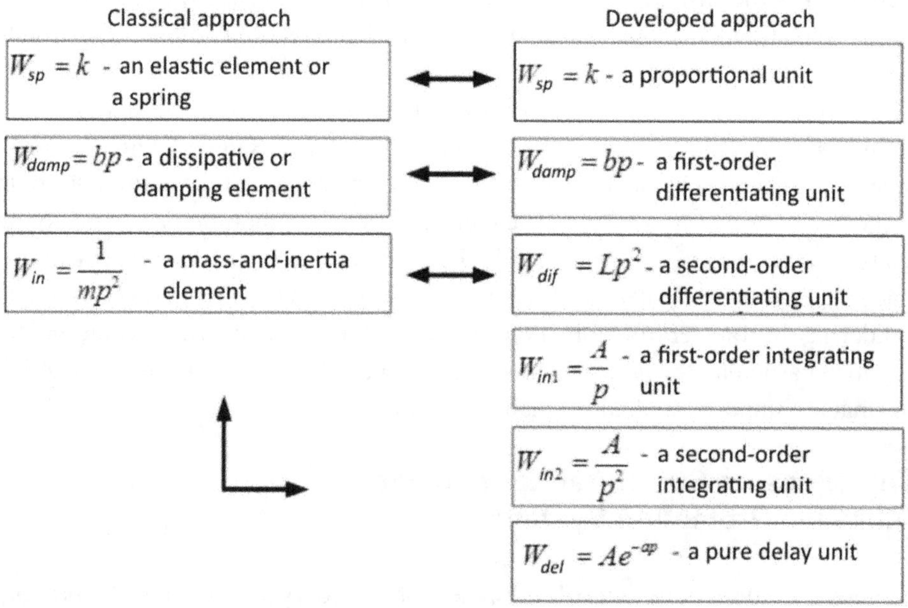

Since typical elements are homogeneous (in the above value), the transfer functions of typical second-level elements can be obtained from a combination or combination of two different elements, which consist of basic elements using the rules of "parallel and sequential" assembly (Eliseev et al., 2008; Eliseev et al., 2011; Khomenko et al., 2012). Although the number of combinations based on the combination of three typical elements is quite large, however, this set of combinations should be reduced due to the property of the combination of elements. It consists in the fact that parallel and sequential connections of the same type (spring with spring, dissipative unit with dissipative one, second-order differentiating unit with the same unit, etc.) provide units of the same type in combination form.

Consider a number of examples of compounds of not one type, but compounds of elements of different types. Let us represent the transfer function of the units of the second level, using the information about the possible combinations with a consecutive and parallel connection between the typical elements of the first level (Table 1).

Thus, by combining consecutive and parallel connections of typical elements from the first level set, you can get transfer functions of typical second level elements (see Table 1). In this case, the transfer functions of all types of elements, with the exception of the typical element of pure delay (and its compounds), can be obtained from the expression (30), as special cases, by "vanishing" the coefficients in the frequency equations of the denominator and numerator of the transfer function (30) of the additional constraint of general form. As for the transfer function of the pure delay unit, it can be simplified by decomposing the exponential function in a series, which makes it possible to introduce this typical element among the known ones.

Table 1. Possible types of standard units of the second level

S/p No.	Types of units	Parallel connection	Sequential connection	Notes
1.	$W_1=k,\ W_2=Ap$	$W'=k+Ap$	$W''=\frac{kAp}{Ap+k}$	W' – damper, W'' – first-order forcing unit
2.	$W_1=k,\ W_2=Lp^2$	$W'=k+Lp^2$	$W''=\frac{kLp^2}{Lp^2+k}$	$W'-W''$
3.	$W_1=k,\ W_2=\frac{A}{p}$	$W'=\frac{A+kp}{p}$	$W''=\frac{Ak}{A+kp}$	W',W'' – aperiodic unit
4.	$W_1=k,\ W_2=\frac{A}{p^2}$	$W'=\frac{p^2k+A}{p^2}$	$W''=\frac{Ak}{A+kp^2}$	W''
5.	$W_1=k,\ W_2=Ae^{-ap}$	$W'=kAe^{-ap}$	$W''=\frac{Ak}{ke^{-ap}+A}$	W''
6.	$W_1=bp,\ W_2=Lp^2$	$W'=bp+Lp^2=p(b+Lp)$	$W''=\frac{bLp^2}{Lp+b}$	-
7.	$W_1=bp,\ W_2=\frac{A}{p}$	$W'=\frac{A+bp^2}{p^2}$	$W''=\frac{Abp}{bp^2+A}$	-
8.	$W_1=Lp^2,\ W_2=\frac{A}{p}$	$W'=\frac{Lp^3+A}{p}$	$W''=\frac{ALp^2}{Lp^3+A}$	W''
9.	$W_1=bp,\ W_2=Ae^{-ap}$	$W'=bp+Ae^{-ap}$	$W''=\frac{bpAe^{-ap}}{bp+Ae^{-ap}}$	-
10.	$W_1=\frac{A}{p^2},\ W_2=Lp^2$	$W'=\frac{Lp^4+A}{p^2}$	$W''=\frac{ALp^2}{Lp^4+A}$	-
11.	$W_1=Lp^2,\ W_2=Ae^{-ap}$	$W'=Lp^2+Ae^{-ap}$	$W''=\frac{Lp^2Ae^{-ap}}{Lp^2+Ae^{-ap}}$	-
12.	$W_1=\frac{A_1}{p},\ W_2=\frac{A_2}{p^2}$	$W'=\frac{A_1p+A_2}{p^2}$	$W''=\frac{A_1A_2}{p(A_1p+A_2)}$	-
13.	$W_1=\frac{A_2}{p^2},\ W_2=Ae^{-ap}$	$W'=\frac{A_2+Ap^2e^{-ap}}{p^2}$	$W'=\frac{A_2A}{A_2e^{-ap}+p^2}$	-

A further increase in the complexity of combination construction leads to the concept of a mechanical chain, which is formed from connected typical elementary units on the basis of the mentioned rules of structural transformations. In the simplest form, such mechanical chains were considered in (Eliseev et al., 2007; Nasnikov & Logunov, 2006). Note that the additional feedback taken by itself is rather rationally transformed into compact structures based on two-terminal networks. As an example, we will consider how an oscillatory structure or mechanical chain (Figure 15) and its mathematical model (Figure 15, *b*), prepared for type-in on an analog computer, will look like.

Figure 15. Oscillatory structure and mechanical chain

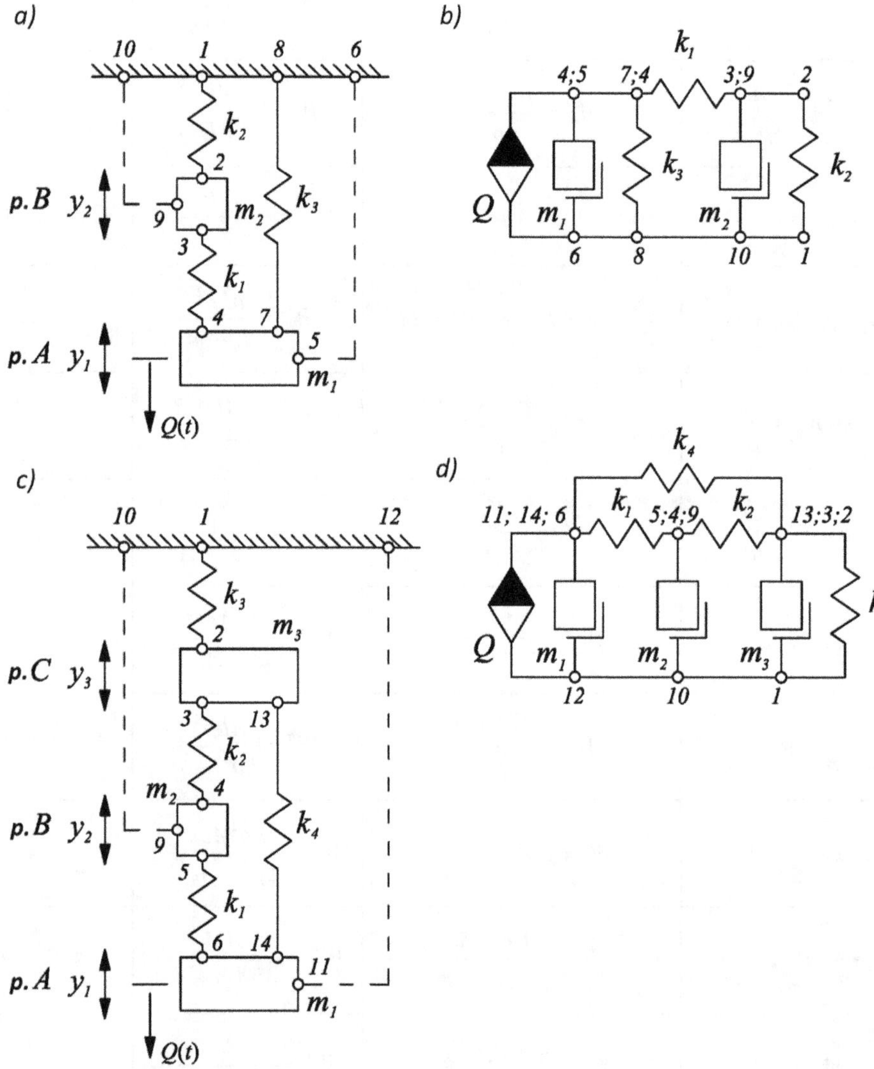

Figure 15 shows the dynamic constraints of two types:

1) constraints with a fixed surface (see Figure 15, *a*);
2) constraints between units (see Figure 15, *c*).

In the first case, there is a main system m_1–k_1–m_2–k_2; in addition, there is an element with stiffness k_3 that connects mass m_1 with a fixed surface. In a mechanical chain (see Figure 15, *b*) this element is shown parallel to the mass m_1 and next to it. In the second case, there is a main system m_1–k_1–m_2–k_2–m_3–k_3; moreover, there is an elastic element with stiffness k_4 connecting the masses m_1 and m_3. In a mechanical chain (see Figure 15, *d*), the elastic element is shown in a series circuit to the masses between terminals 13 and 14 and forms a so-called chain with intersecting constraints, which does not allow us to place linking lines between elements in one plane (hence there is a name "non-planar chain").

The use of circuits, as a methodological technique, reflects the positive aspects of structural approaches, which can be oriented towards the use of analog electromechanical modeling methods, almost without much additional cost, including in terms of taking into account the nonlinear properties that can have (and have small oscillations) typical elementary units.

The directions of the dynamic synthesis of additional feedback circuits can be, as shown above, built on combination constructions of a mechanical chain of typical elementary units, however, this methodical technique is not the only one. The results can be obtained, in particular, by reducing the expression (3.26). The latter is that by zeroing the corresponding coefficients of the expression (3.26), it is possible to obtain not only the transfer functions of the traditional and extended sets of typical elements, but also the transfer functions of the typical second-level structures, as well as more complex structures corresponding to the problems of active vibration protection and vibration isolation. The results of applying this approach for various cases are presented in (Eliseev et al., 2006; Eliseev & Logunov, 2010; Zasyadko & Nasnikov, 2008).

A characteristic circumstance that needs to be pointed out is that by choosing the transfer function of a particular type from the expression (30), you can at the same time get the opportunity of a preliminary assessment of the dynamic properties with the subsequent choice of methods for the physical implementation of such a chain. In this regard, at least three areas can be considered:

- combination of elementary units of the extended type set;
- combinations through typical units based on structural transformation rules;
- search for physical implementations of mechanical systems having a transfer function of a particular type from the expression (30) based on mechanisms of various types or servo drives.

Some of the results of research in this direction are reflected in the author's works (Belokobylsky et al., 2013; Dimov & Nasnikov, 2009).

THE POSSIBILITY OF SIMPLIFYING THE STRUCTURAL DIAGRAM OF THE INTRODUCTION OF A GENERALIZED SPRING

If we assume that the initial system has two generalized coordinates – the computational scheme (Figure 16, *a*) and the structural diagram (Figure 16, *b*) – then the differential equations of the system's motion take the form

$$m_1\ddot{y} + k_1 y_1 + k_2 y_1 - k_2 y_2 = k_1 z_1 + P_1, \tag{36}$$

$$m_2\ddot{y}_2 + k_2 y_2 + k_3 y_2 - k_2 y_1 = k_3 z_2 + P_2. \tag{37}$$

Figure 16. Computational scheme (a) and structural diagram (b) of mechanical oscillatory systems with two degrees of freedom

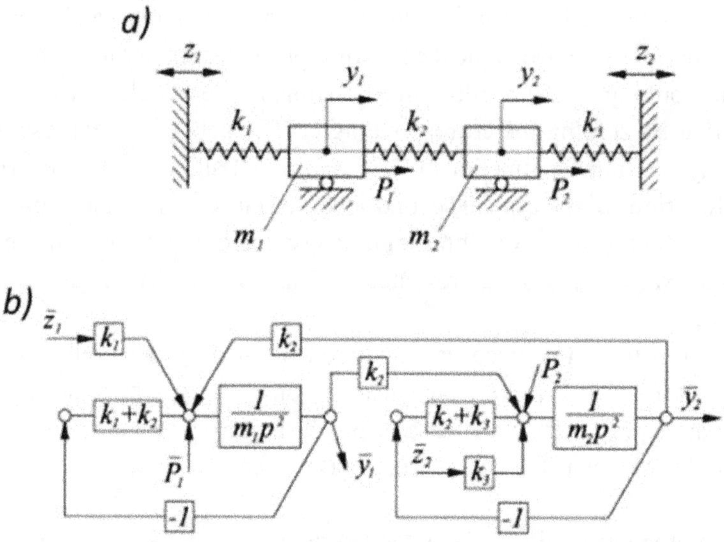

The transfer functions of the system with various types of external influences can be written:

$$W_1 = \frac{\bar{y}_1}{\bar{z}_1} = \frac{k_1\left(m_2 p^2 + k_2 + k_3\right)}{A}, \tag{38}$$

$$W_2 = \frac{\bar{y}_2}{\bar{z}_1} = \frac{k_1 k_2}{A}, \tag{39}$$

$$W_3 = \frac{\bar{y}_2}{\bar{z}_2} = \frac{k_3\left(m_2 p^2 + k_1 + k_2\right)}{A}, \tag{40}$$

$$W_4 = \frac{\bar{y}_1}{\bar{z}_2} = \frac{k_2 k_3}{A}, \tag{41}$$

$$W_5 = \frac{\bar{y}_1}{\bar{P}_1} = \frac{m_2 p^2 + k_2 + k_3}{A}, \tag{42}$$

$$W_6 = \frac{\bar{y}_2}{\bar{P}_1} = \frac{1}{A}, \tag{43}$$

$$W_7 = \frac{\bar{y}_2}{\bar{P}_2} = \frac{m_1 p^2 + k_1 + k_2}{A}, \tag{44}$$

$$W_8 = \frac{\bar{y}_1}{\bar{P}_2} = \frac{1}{A}, \tag{45}$$

where

$$A = \left(m_1 p^2 + k_1 + k_2\right)\left(m_2 p^2 + k_2 + k_3\right) - k_2^2 \tag{46}$$

is the characteristic equation for the system in Figure 16, *b*.

The considered system with two degrees of freedom can be simplified by using an approach based on the idea of sing a generalized spring, which forms (Eliseev, 2018) the reduced stiffness of a group or a block of elements, thereby simplifying the original system to a system with one degree of freedom. In this case, two options are possible, each of which is associated with its specific features of the coordinate systems and application of forces.

Using the results of (Dimov & Nasnikov, 2009), we write a series of expressions for the reduced stiffnesses

$$k_{red_1} = \frac{\left(k_3 + m_2 p^2\right) k_2}{k_2 + k_3 + m_2 p^2}, \tag{47}$$

$$k_{red_2} = \frac{\left(k_1 + m_1 p^2\right) k_2}{k_1 + k_2 + m_1 p^2}, \tag{48}$$

and find (with z_1=0, z_2=0, P_2=0)

$$W_1' = \frac{\overline{y}_1}{\overline{P}_1} = \frac{1}{m_1 p^2 + k_1 + k_{red_1}} = \frac{k_2 + k_3 + m_2 p^2}{\left(m_1 p^2 + k_1\right)\left(k_2 + k_3 + m_2 p^2\right) + \left(k_3 + m_2 p^2\right) k_2}. \tag{49}$$

Expression (49) coincides with expression (42), since in the transformed form the characteristic equation (46) takes the form

$$A = \left(m_1 p^2 + k_1\right)\left(m_2 p^2 + k_3 + k_2\right) + k_2\left(m_2 p^2 + k_3\right). \tag{50}$$

We write the transfer functions with kinematic perturbation $z_2 \neq 0$ (z_1=0, P_1=0, P_2=0):

$$W_2' = \frac{\overline{y}_1}{\overline{z}_2} = \frac{k_{red_1}}{m_1 p^2 + k_1 + k_{red_1}} = \frac{\left(k_3 + m_2 p^2\right) k_2}{\left(m_1 p^2 + k_1\right)\left(k_2 + k_3 + m_2 p^2\right) + \left(k_3 + m_2 p^2\right) k_2}. \tag{51}$$

Expression (51) differs from (41) in that an additional term $k_2 m_2 p^2$ appears in the numerator. This situation will be discussed below. Let $z_1 \neq 0$, z_2=0, P_1=0, P_2=0, then

$$W_3' = \frac{\overline{y}_1}{\overline{z}_1} = \frac{k_1}{m_1 p^2 + k_1 + k_{red_1}} = \frac{k_1\left(k_2 + k_3 + m_2 p^2\right)}{A}. \tag{52}$$

Comparing (3.48) with (3.34), we find a complete coincidence of the two expressions.

Considering the diagram in Figure 17, *b* and assuming that z_2=0, P_1=0, P_2=0, $z_1 \neq 0$, we get

$$W_4' = \frac{\overline{y}_2}{\overline{z}_1} = \frac{k_{red_2}}{m_2 p^2 + k_3 + k_{red_2}} = \frac{\left(k_1 + m_1 p^2\right) k_2}{A}. \tag{53}$$

Expression (3.49), as well as expression (3.47), contains in the numerator an extra term, which reflects the influence of the transference inertial forces arising from the vibration of the base. The latter requires some additional transformations in the structural diagram, which highlights the cases of direct kinematic perturbation on an element with reduced stiffness. Physically, this means that inside the element with reduced stiffness, due to external influences, additional forces appear to emerge, the influence of which compensates for the introduction of an additional constraint due to kinematic perturbation with a negative sign (this is how the action of the inertial forces of moving space manifests itself).

Find $\frac{\overline{y}_2}{\overline{z}_2}$, taking, and $z_2 \neq 0$, P_1=0, P_2=0, z_1=0, then

Figure 17. Options for simplifying the original system

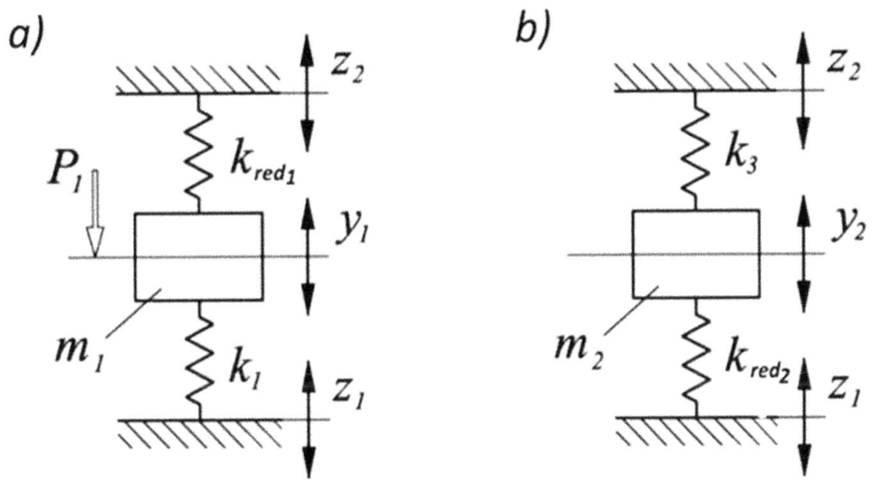

$$W_5' = \frac{\bar{y}_2}{\bar{z}_2} = \frac{k_3}{m_1 p^2 + k_3 + k_{red_2}} = \frac{k_3\left(k_1 + k_2 + m_1 p^2\right)}{A}. \tag{54}$$

The expression (54) completely coincides with (40).

Consider the cases of power perturbation, assuming that z_1=01, z_2=0, and the force and P_2 will be applied to the corresponding masses m_1 and m_2 in turn. We find that when P_1=0 –

$$W_6' = \frac{\bar{y}_2}{P_2} = \frac{k_1 + k_2 + m_1 p^2}{A}, \tag{55}$$

which coincides with the expression (44).

It can also be assumed that the intermediate coordinate in the generalized spring is perturbed by force P_1 or P_2, is "inside", then a discrepancy will also occur. As a way out of the situation, one can consider introducing an additional negative feedback $\frac{-k_2 m_1 p^2}{k_1 + k_2 + m_1 p^2}$ or $\frac{-k_2 m_2 p^2}{m_2 p^2 + k_2 + k_3}$, and expressions (51) and (3.49) are used for doing this, as shown in (Eliseev, 2018). The structural diagrams of the simplified system in the transmission of kinematic effects and are shown in Figure 18, *a* and Figure 18 *b*, respectively.

From the structural diagram in Figure 18, *a*, which corresponds to the calculation scheme in Figure 17, and, we will obtain transfer function

$$W_4'' = \frac{\bar{y}_1}{\bar{z}_2} = \frac{k_{red_1} - W_0}{m_1 p^2 + k_1 + k_{red_1}} = \frac{k_3 k_2}{A}, \tag{56}$$

Figure 18. Structural diagrams of a two-mass system with the introduction of generalized springs and kinematic perturbation of a generalized spring: a – perturbation z_2; b – perturbation z_1

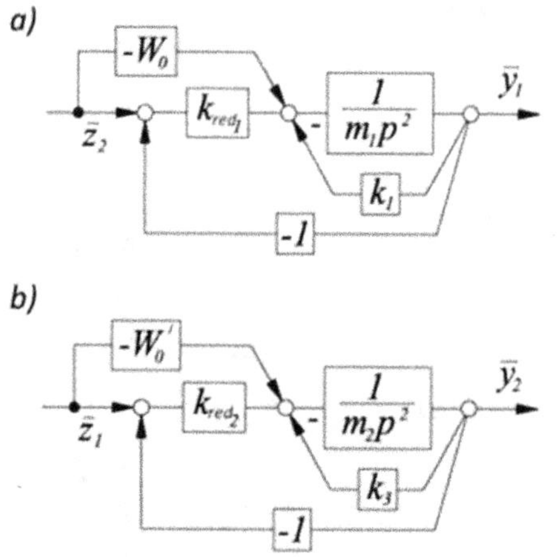

$$W_4''=\frac{\bar{y}_1}{\bar{z}_2}=\frac{k_{red_1}-W_0}{m_1p^2+k_1+k_{red_1}}.$$

$$W_2''=\frac{\bar{y}_2}{\bar{z}_1}=\frac{k_{red_2}-W_0'}{m_2p^2+k_2+k_{red_2}}.$$

where

$$W_0=\frac{k_2m_2p^2}{m_2p^2+k_2+k_3} \tag{57}$$

is determined by the transfer coefficient of the oscillation amplitude to m_2 from the side of the base z_2.

From the structural diagram in Figure 18, *b*, which corresponds to the calculation scheme in Figure 17, *b*.

$$W_2''=\frac{\bar{y}_2}{\bar{z}_1}=\frac{k_{red_2}-W_0'}{m_2p^2+k_2+k_{red_2}}=\frac{k_1k_2}{A}, \tag{58}$$

where

$$W_0'=\frac{k_2m_1p^2}{k_1+k_2+m_1p^2} \tag{59}$$

is determined by the transfer coefficient of the amplitude of oscillations of the base to the mass m_1 and in physical sense. Comparing expressions (57) and (41), we can note their coincidence.

Features of using the proposed approach requires a more detailed consideration of the influence of the kinematic effect on an object through a generalized spring. Let us assume for the computational scheme in Figure 16, *a* that the state of the system is described by two generalized coordinates: y_2– the mass coordinate in the fixed coordinate system, $y_1=y+z_1$, where z_1 is the base displacement, and y is the coordinate of the relative position of the element with the mass m_1, i.e. relative to the base. In this case, the expressions for the kinematic and potential energies take the form

$$T = \frac{1}{2} m_2 \dot{y}_2^2 + \frac{1}{2} m_1 (\dot{y} + \dot{z}_1)^2 \tag{60}$$

$$\Pi = \frac{1}{2} k_1 y^2 + \frac{1}{2} k_2 (y_2 - y - z_1)^2 + \frac{1}{2} k_3 y_2^2 . \tag{61}$$

Performing the appropriate transformations, we obtain the differential equations of motion

$$m_1 \ddot{y} + k_1 y + k_2 y - k_2 y_2 = -\ddot{z}_1 m_1 - k_2 z_1 , \tag{62}$$

$$m_2 \ddot{y}_2 + k_2 y_2 + k_3 y_2 - k_2 y = k_2 z_1 . \tag{63}$$

Figure 19 presents a structural interpretation corresponding to the new computational scheme in the form of a structural diagram of a dynamically equivalent automatic control system.

Figure 19. The structural diagram of the system in coordinates y, y_2

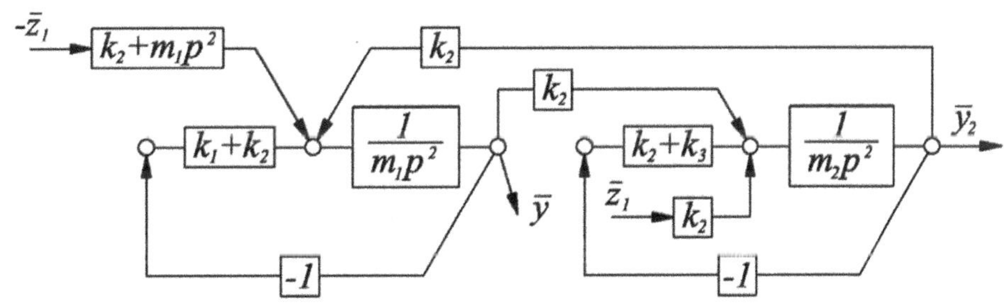

Comparison of the structural diagram in Figure 19 with the structural diagram in Figure 16, *b* (at z_2=0), shows that the effect of kinematic perturbation z_1 spreads through two channels. Firstly, z_1 acts on the mass m_1 not only by elastic force, but also has an inertial component from the portable motion. Secondly, the displacement of the base acts on the mass m_2 in the form of a corresponding elastic force. Find the transfer function of the system

$$W_2''' = \frac{\bar{y}_2}{\bar{z}_1} = \frac{-k_2\left(k_1 + m_1 p^2\right) + k_2\left(k_2 + k_1 + m_1 p^2\right)}{A} = \frac{k_2 k_1}{A} . \tag{64}$$

Expression (3.59) coincides with expression (39), however, in the structural diagram we already take an account of z_1 through two channels of influence, which was previously considered.

Let us show, successively in several steps, how the structural diagram in Figure 19 can be transformed to the structural diagram shown in Figure 18, *b*. Denote in the structural diagram (Figure 20) the nodal points by *a*, *b*, *c*, *d*, *e*, as shown in Figure 20, *a*. If we carry out the convolution by points (*a*), (*b*) and partially by (*d*) and (*e*), transfer the point of application z_1 to point (*e*), we obtain a scheme of a more general form (Figure 20, *b*). Given the fact that z_1 acts at point (*e*) through k_2, we can combine the points (*d*) and (*e*), then the single feedback disappears. If at the same time we transfer k_2 at points (*a*) (*b*) to the numerator of the transfer function $\dfrac{1}{m_2 p^2 + k_1 + k_2}$ and do the convolution of the function after z_1, then we obtain the scheme in Figure 20, *c*.

Figure 20. The successive stages of the transformation of the structural diagram of the source system to its simplified representation: a – convolution with notation of partial subsystems; b – transfer of the kinematic perturbation of z_1 on an element of mass m_2; c – introduction of the transfer function of the generalized spring

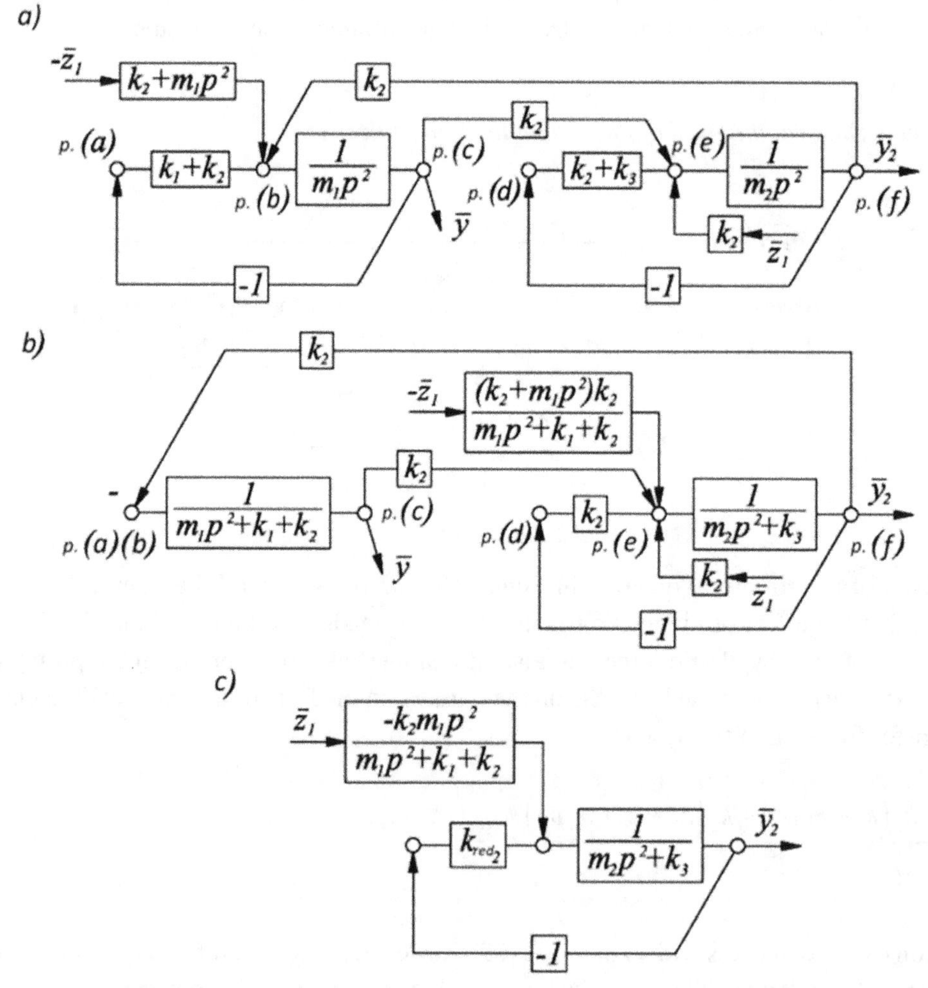

In Figure 20, c the structural diagram corresponds to a system with one degree of freedom. In this case, the elastic element k_{red_2}, represented in the form, enters a point in front of the mass m_2. Outputs are the disturbance y_2 and the single feedback that comes to the point of the unit k_{red_2}. The signal from z_1, which goes through k_{red_2} with a negative sign through the unit $\dfrac{m_1 p^2 k_2}{m_1 p^2 + k_1 + k_2}$, comes to the point before m_2. The physical meaning of this force is that it is the force of inertia, which arises in the mass m_1 from the action of the motion of the base, and is transmitted to the mass m_2.

When determining the transfer functions of the system as a whole, it is necessary to avoid situations in which the kinematic effect goes to the mass through a generalized spring. In this case inertial forces of moving space arise inside the generalized spring, taking into account the effect of which requires, as was shown above, the introduction of corrective impacts according to the principle of perturbation control (with a negative sign). If the observation coordinate or the force application coordinate is inside the structure of a generalized spring, then in this case, an appropriate structural rearrangement of the computational scheme is required. In the case of transmission of the action of the spring, based on a vibrating base, it is necessary to take into account the action of the inertial forces of moving space. If for the generalized coordinates we choose the coordinate of the relative motion and the coordinate of the absolute motion, then on the structural diagram (Figure 20) the effect of vibration of the base is reflected by two effects in the form of force $k_2 z$ applied at point (e) and $\ddot{z}_1 m_1 + k_2 z_1$ of force impact applied to the mass m_2 with negative sign at point (b).

CONCLUSION

The concept of constructing a vibration protection system on the basis of introducing additional constraints uses ideas about the basic models of vibration protection system with one, two, and three degrees of freedom. Basic models contain the object of protection (element by mass) and the elastic element. The expansion of the structure of the vibration protection system is achieved by the fact that an additional constraint (it can take the form of a mechanical chain) is introduced parallel to the elastic element of the base model. An additional chain in its simplest form can be in the form of an elementary unit. The complication of the additional constraint involves the introduction of a mechanical chain of elementary units. Inside an additional constraint (or additional circuit), the switching of elementary units is carried out according to the rules of parallel and consecutive connection of springs.

REFERENCES

Belokobylsky, S. V., Eliseev, S. V., & Kashuba, V. B. (2013). *Applied problems of the structural theory of vibration protection systems.*

De Silva, C. W. (2000). *Vibration. Fundamentals and Practice.* CRC Press.

Dimov, A. V., & Nasnikov, D. N. (2009). Mechatronics vibration protection systems. Features of structural transformations. Modern technologies. System analysis. *Modeling*, *24*, 75–85.

Eliseev, A. V., Kuznetsov, N. K. & Moskovskikh, A. O. (2019). *The Dynamics of Machines. System conception, structural schemes and connections of elements: monograph.* Moscow: Innovative mechanical engineering.

Eliseev, S. V. (1978). *Structural theory of vibration protection systems.* Nauka Publ.

Eliseev, S. V. (2018). *Applied system analysis and structural mathematical modeling (dynamics of transport and technological machines: connectivity of movements, vibration interactions, lever connections).* Irkutsk State Transport University.

Eliseev, S. V., & Khomenko, A. P. (2014). *Dynamic oscillation damping: feedback concept and structural methods of mathematical modeling.* Nauka Publ.

Eliseev, S. V., & Logunov, A. S. (2010). Mechatronics of pneumatic vibration protection systems. Systems. Methods. *Technologies, 7,* 9–21.

Eliseev, S. V., Lukyanov, A. V., Reznik, Yu. N., & Khomenko, A. P. (2006). *Dynamics of mechanical systems with additional ties.* Irkutsk State University Publ.

Eliseev, S. V., Reznik, Yu. N., & Khomenko, A. P. (2011). *Mechatronic approaches in the dynamics of mechanical oscillatory systems.* Nauka Publ.

Eliseev, S. V., Reznik, Yu. N., Khomenko, A. P., & Zasyadko, A. A. (2008). *Dynamic synthesis in generalized problems of vibration protection and vibration isolation of technical objects.* ISU Publ.

Eliseev, S. V., Trofimov, A. N., & Bolshakov, R. S. (2015). Structural formations as a part of mechanical oscillatory systems. In *Kulagin Readings: Technique and Technology of Production Processes: XV Intern. scientific and practical conf,* (pp. 273-277). Chita: Transbaikalian State University Publ.

Eliseev, S. V., Upir', R. Yu., Nasnikov, D. N., & Logunov, A. S. (2007). The system of new types in the structural representations of mechanical oscillatory systems. In Innovative technologies for transport and industry: *Proceed. of the 45th int. scientific conf.,* (*Vol. 3,* pp. 173-182). Irkutsk.

Frolov, K. V. (Ed.). (1983). Vibrations in technology: a reference book: in 6 vols.: Vol. 6. *Protection against vibrations and shocks.* Mashinostroenie Publ.

Ivashchenko, I. I. (1993). *Automatic adjustment. Theory and elements of systems.* Mashinostroenie Publ.

Karnovsky, I. A., & Lebed, E. (2016). *Theory of vibration protection.* Springer. doi:10.1007/978-3-319-28020-2

Khomenko, A. P., Eliseev, S. V. & Yermoshenko, Yu. V. (2012). *System analysis and mathematical modeling in the mechatronics of vibration protection systems.* Irkutsk: IrGUPS Publ.

Kim, P. D. (2003). The theory of automatic control: in 2 vols.: Vol. 1. *Linear systems.* Fizmatlit Publ.

Nasnikov, D. N. (2008). Active elements as typical units in controlled vibration protection systems. Modern technologies. *System analysis Modeling, 20,* 41-49.

Nasnikov, D. N., & Logunov, A. S. (2006). Typical units in structural interpretations of mechanical oscillatory systems. Modern technologies. System analysis. *Modeling, 12,* 78–93.

Yurevich, E. I. (2007). Theory of automatic control: educational ed. for universities (3rd ed.). St.Petersburg: BKhV – Petersburg Publ.

Zasyadko, A. A., & Nasnikov, D. N. (2008). Features of a hydraulic drive in active vibration protection systems. Modern technologies. System analysis. *Modeling*, *6*, 18–29.

ADDITIONAL READING

Eliseev, A. V. (2020). Development of system representations in dynamics of oscillatory structures: Frequency function and forms of consistency of element movements. Modern technologies. System analysis. *Modeling*, *4*(68), 40–49.

Eliseev, A. V. (2020). Frequency function and damping function in assessing dynamic processes in mechanical oscillatory systems with symmetry. *Advanced Engineering Research*, *20*(4), 360–369. doi:10.23947/2687-1653-2020-20-4-360-369

Eliseev, A. V., & Bolshakov, R. S. (2022). Systemic approaches in the generalized assessment of dynamic features of vibrational interactions of technical objects of transport and technology under connected power loads. Modern Technologies. System Analysis. *Modeling*, *1*(73), 39–52.

Eliseev, A. V., Kaimov, E. V., & Sitov, I. S. (2015). General approaches to determining contact reactions in solids under static loads with non-holding connections: Multidimensional variant. Systems. Methods. *Technologies*, *1*(25), 19–29.

Eliseev, A.V., Kuznetsov, N.K. & Eliseev, S.V. (2021). *Frequency energy function in assessing dynamic states of technical objects.* Moscow Aviation Institute, Moscow, Russia.

Eliseev, A.V., Kuznetsov, N.K. & Eliseev, S.V. (2021). *New approaches in assessing dynamic properties of oscillatory structures: frequency functions and consistency of element movements*. Proceedings of Moscow Aviation Institute, Moscow, Russia.

Eliseev, A. V., Kuznetsov, N. K., & Mironov, A. S. (2022). Systemic approach to assessing dynamic states of technical objects: Structural mathematical modeling methods. Mechatronics, Automation. *Control (Chicago, Ill.)*, *23*(5), 236–245.

Eliseev, A. V., Mironov, A. S., & Eliseev, S. V. (2022). Systemic approaches to assessing dynamic states of mechanical vibrational systems: technology of structural transformations. Vestnik Astrakhanskogo Gosudarstvennogo Tekhnicheskogo Universiteta. Seriya: Upravlenie. *Vychislitelnaya Tekhnika i Informatika*, *1*, 7–15.

Eliseev, A. V., Sitov, I. S., & Kuznetsov, N. K. (2022). System approach to assessing dynamic states of transport and technological objects: Structural schemes, transfer functions, dynamic invariants. Systems. Methods. *Technologies*, *2*(54), 7–19.

Eliseev, A. V., Sitov, I. S., & Nguyen, D. H. (2016). Some issues of forming the structure of the vibration field in a vibration stand: Features of the measurement system, tuning tools. Systems. Methods. *Technologies*, *2*(30), 17–26.

Jakóbczak, D. (2010). Implementation of Hurwitz-Radon Matrices in Shape Representation. In R. S. Choraś (Ed.), *Advances in Intelligent and Soft Computing 84, Image Processing and Communications: Challenges 2, 39-50.* Springer-Verlag Berlin Heidelberg. doi:10.1007/978-3-642-16295-4_4

Jakóbczak, D. (2010). Object Modeling Using Method of Hurwitz-Radon Matrices of Rank k. In W. Wolski & M. Borawski (Eds.), *Computer Graphics: Selected Issues* (pp. 79–90). University of Szczecin Press.

Sitov, I. S., Eliseev, A. V., & Kaimov, E. V. (2015). General approaches in problems of determining contact reactions in solids under static loads with account for non-holding connections: Multidimensional variant. Systems. Methods. *Technologies*, *1*(25), 19–29.

Chapter 4
Development of Generalized Approaches in the Problems of Evaluation and Control of the Dynamic State of Mechanical Oscillatory Systems:
Structural Forms of Mechanical Oscillatory Systems

ABSTRACT

The chapter is devoted to taking into account the design features of technical objects, the design schemes of which are mechanical oscillatory systems. General approaches to the construction of mathematical models of the rotating type and their dynamic properties are considered. The concepts of rotational-type systems with a massless rod with an intermediate fulcrum are detailed. A comparative analysis of the capabilities of vibration protection systems has been carried out. The analysis of systems with a protective object performing rotational movements is carried out, taking into account the discrepancy between the center of gravity and the center of rotation. The displacement of the center of gravity in a system with a lever of the first kind is considered.

INTRODUCTION

The questions of the constructive and engineering forms of interaction of elements of mechanical oscillatory systems are considered in the first chapter of the monograph. It is shown that many engineering objects (technical machines and transport vehicles) operating under conditions of intensive dynamic loading are displayed by mechanical oscillatory systems with several degrees of freedom. Usually,

DOI: 10.4018/978-1-6684-7237-8.ch004

plane interaction schemes are considered under the assumption of symmetry of the property of objects. It should be noted that the use of a fairly simple set of typical elements in the form of springs, dampers, and mass-and-inertia units, which can make motions as material points or solid bodies in a plane (rotation and translational displacements), is characteristic of the constructive and engineering forms of the engineering object. At the same time, the structural and technical features of the objects are manifested through the establishment of various kinds of connections. Such constraints are implemented through leverages, as well as using mechanisms that can be included in the structure of a mechanical oscillatory system. Often, in such situations, screw non-locking devices, toothed gearings, and lever mechanisms are used, which is also identified by the use of so-called motion transformation devices in structural mathematical modeling (Khomenko et al., 2013b).

In simplified schemes of preliminary calculations, motion transformation devices can be displayed as some typical elements with a transfer function of a second-order differentiating unit.

Such units in structural transformations have the same capabilities as linear springs. One of the advantages of structural mathematical modeling methods is the quality of the typical elements of the system, which is that the input to the standard elementary element is displacement, and the output is the corresponding force (elastic force, resistance force of viscous friction, inertial force).

ACCOUNTING FOR CONSTRUCTIVE AND ENGINEERING FEATURES OF OBJECTS

In many problems of the dynamics of machines and, in particular, the applied theory of vibration protection systems, the simplest basic models are usually represented by a mechanical oscillatory system with one degree of freedom; at the same time the object performs rectilinear small vertical oscillations. The object can be connected to the supporting surface with the help of connecting elements in the form of springs and dampers. The support surface for the object of protection can be represented by zones for contacts *I* and *II* (Figure 1), and the object of protection itself can have guides that provide vertical motion *III* (or motion along the same coordinate y).

Figure 1. Basic computational scheme of vibration protection system with one degree of freedom

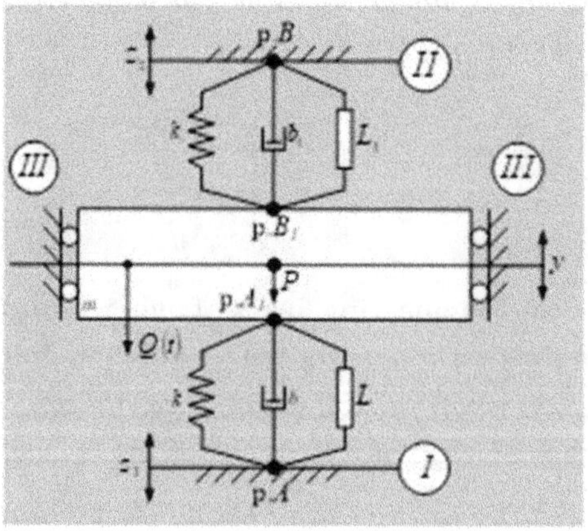

In the structure of the computational scheme (Figure 1), the elements have stiffnesses k and k_1; dissipative interactions are reduced to equivalent viscous friction dampers with coefficients b and b_1. In addition, in Figure 1 motion transformation devices are introduced into the system (Khomenko et al., 2013b).

Such devices can have various constructive and engineering forms, and their properties are reflected by the parameters L and L_1, determining efforts emerging at relative accelerations of the motion of elements.

Guides *III* in Figure 1 are shown in connection with the emphasis on the circumstances that, in the motion of an object, the motion along the coordinate y is chosen specifically and supported by certain means. In many cases, it is assumed that the motion along the coordinate y is dominant, while the other forms of motion are considered negligible in this regard. As external forces acting on the object of protection, the force of weight P, external force $Q(t)$ and motion of the supporting surfaces *I* and *II*, denoted respectively $z_1(t)$ and $z_2(t)$, are taken into account. The kinematic perturbations $z_1(t)$ and $z_2(t)$ can be reduced to equivalent force actions applied to the object of protection m. The points A, A_1, B and B_1 denote contact points, in which static and dynamic responses of constraints can be determined, which is reflected in a number of papers by the author (Eliseev et al., 2013a; Eliseev et al., 2012; Eliseev et al., 2013d).

Some Common Points: Building Mathematical Model

A mathematical model of a linear system (Figure 1) can be represented (for the general case) by the equations in the preliminary definition of expressions for the kinetic, potential, and also the energy dissipating function:

$$\mathrm{T} = \frac{1}{2}m(\dot{y})^2 + \frac{1}{2}L\cdot(\dot{y}-\dot{z})^2 + \frac{1}{2}L_1\cdot(\dot{y}-\dot{z}_1)^2; \tag{1}$$

$$\Pi = \frac{1}{2}k\cdot(y-z)^2 + \frac{1}{2}k_1\cdot(y-z_1)^2; \tag{2}$$

$$\Phi = \frac{1}{2}b\cdot(\dot{y}-\dot{z})^2 + \frac{1}{2}b_1\cdot(\dot{y}_1-\dot{z}_1)^2. \tag{3}$$

Using the usual approaches associated with the use of Laplace transforms, we find that:

$$\left[(m+L+L_1)p^2 + (b+b_1)p + k + k_1\right]\cdot\bar{y} = \left(mp^2 + bp + k\right)\cdot\bar{z} + \left(mp^2 + b_1p + k_1\right)\cdot\bar{z}_1 + \bar{P} + \bar{Q}, \tag{4}$$

where $p = j\omega$ ($j = \sqrt{-1}$) is a complex variable; $\bar{y}, \bar{P}, \bar{Q}$ is a symbolic Laplace transform of functions (Khomenko et al., 2012). If we assume that small motions of y occur relative to the position of static equilibrium, then with zero initial conditions and harmonic forms of external influences $\bar{Q}, \bar{z}_1$, and $\bar{z}_2$, the transfer functions of the system can be obtained:

$$\underset{\substack{Q=0\\ z_1=0}}{W(p)}=\frac{\bar{y}}{\bar{z}}=\frac{Lp^2+bp+k}{(m+L+L_1)p^2+(b+b_1)p+k_1+k}, \tag{5}$$

$$\underset{\substack{Q=0\\ z=0}}{W(p)}=\frac{\bar{y}}{\bar{z}_1}=\frac{L_1p^2+b_1p+k_1}{(m+L+L_1)p^2+(b+b_1)p+k_1+k}, \tag{6}$$

$$\underset{\substack{z_1=0\\ z=0}}{W(p)}=\frac{\bar{y}}{\bar{Q}}=\frac{1}{(m+L+L_1)p^2+(b+b_1)p+k_1+k}. \tag{7}$$

The denominator (5)–(7) is common to all transfer functions and is called the characteristic equation:

$$A_0 = (m+L+L_1)p^2+ (b+b_1)p+k_1+k. \tag{8}$$

The frequency of natural oscillations can be determined from (8), and using (5)–(7) – the corresponding frequency characteristics are constructed and the dynamic properties are estimated through responses (or reactions) to typical external influences. A more detailed methodology for constructing transfer functions is presented in (Eliseev et al., 2013d; Khomenko et al., 2012; Eliseev et al., 2013c).

The transfer functions of system (5)–(7) can also be determined using the structural diagram (Figure 2), which is equivalent to equation (4), in the sense that the structural diagram in the symbolism of the concepts of automatic control theory (Khomenko et al., 2013a) is graphic or structural analogue of the linear differential equation (4).

The structural diagram in Figure 2 reflects the dynamic interactions arising between the object of protection (mass-and-inertia element *m*) and support surfaces, the motion of which is given and forms kinematic perturbations. In particular, Figure 2, *c* shows that kinematic perturbations are transformed into force effects applied at p. *1* (Figure 2). At the same time, in contact of the object of protection with the fixed surface *III*, which provide vertical (in this case) motion, interactions may also occur, determined by the corresponding parameters of the units k_2, b_2 and L_2. Such units can conditionally be shown in Figure 3, *a*, and also on the structural diagram shown in Figure 3, *b*.

Comparison of structural diagrams in Figure 2 and Figure 3 shows that there is a certain difference in the reflection of the properties formed by the relative and absolute motion of the units. So, for example, the parameters k_2, b_2, L_2 are associated only with the parameters of the absolute motion of the object of protection. In the expressions defining the transfer functions (4.5)–(4.7), the numerators of the fractionally rational expressions do not change, but the characteristic equation (8) takes the form:

$$A_0' = (m+L+L_1+L_2)p^2+(b+b_1+b_2)p+k+k_1+k_2=0. \tag{9}$$

In this connection, features arise in the estimation of the dynamic properties of the system under the action of various external disturbances. It is necessary to take into account that the estimates made are

Figure 2. Variants of the construction of structural diagrams for the basic computational scheme in Figure 1: a – detailed structural diagram with the allocation of a single feedback; b – structural diagram with the allocation of feedbacks in the form of elastic elements with reduced stiffness and object of protection as the second-order integrating element; c – generalized scheme

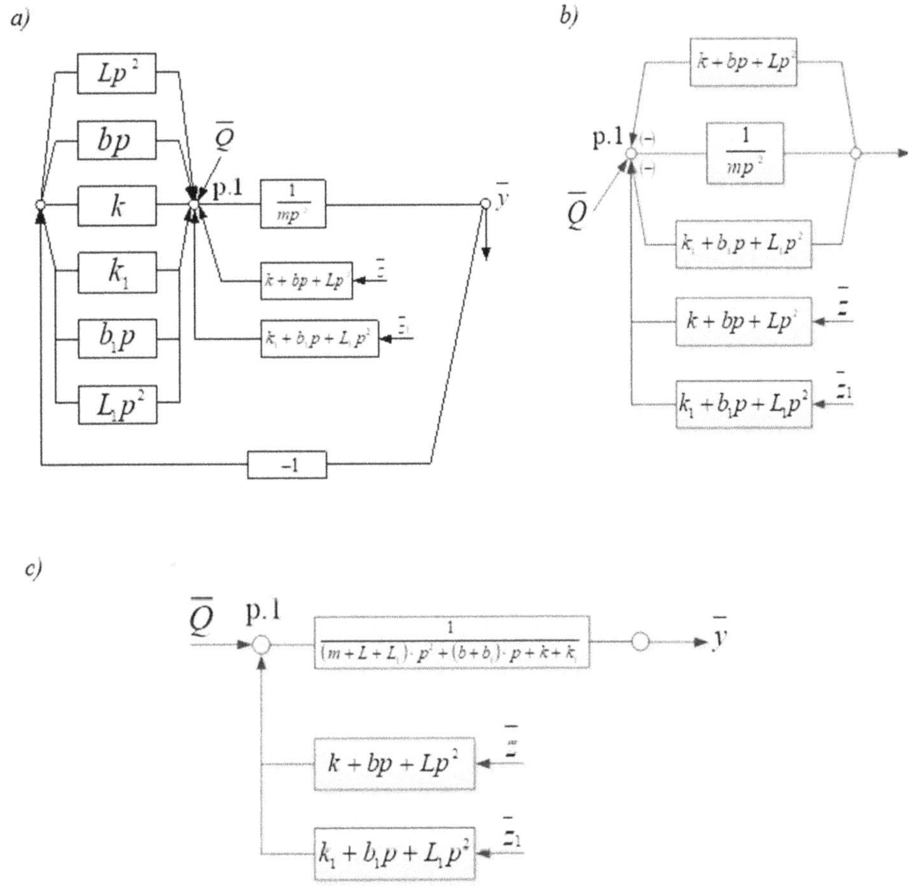

rather conditional in connection with the idealization of the real properties of the elementary units and the assumptions made about the features of motion.

Considered in Figure 2 and 3 structural diagrams are equivalent and can be obtained by conventional structural transformations. Point (1) in all variants of structural diagrams is symbolic, in the sense that it contains all the forces corresponding to the ideas of the d'Alembert principle. Essentially, the sum of all forces converging at p. 1 (Figure 2, *a–c*) corresponds to the writing of equation (4) in a symbolic form.

The above circumstance is associated with the object of protection as a solid body in the form of a material point. However, a solid can have finite dimensions, as shown in Figure 4.

Such situations arise when a solid body consists of several parts, and the task of vibration protection implies the possibility of contact failure with the emergence of gaps and subsequent collisions of elements.

To ensure contact and to avoid the emergence of gaps in the systems (see Figure 3), preloading conditions are created, for example, by pre-compression of the springs or using special additional impacts.

Figure 3. Diagrams presenting interaction of the object of protection with support surfaces III: a – computational scheme; b – structural diagram of the system

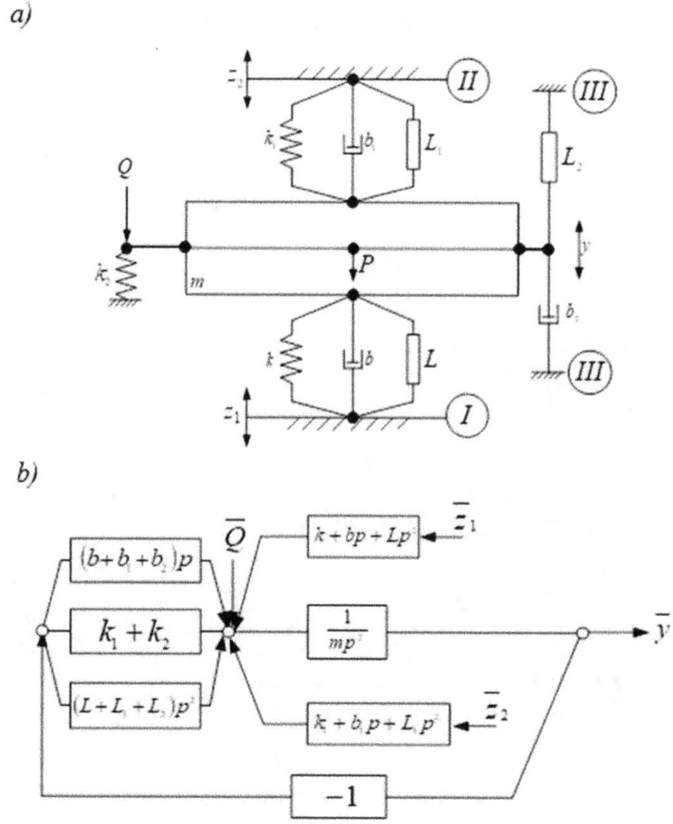

Figure 4. The computational scheme of the vibration protection system with a two-part protection object ($m = m_1 + m_2$)

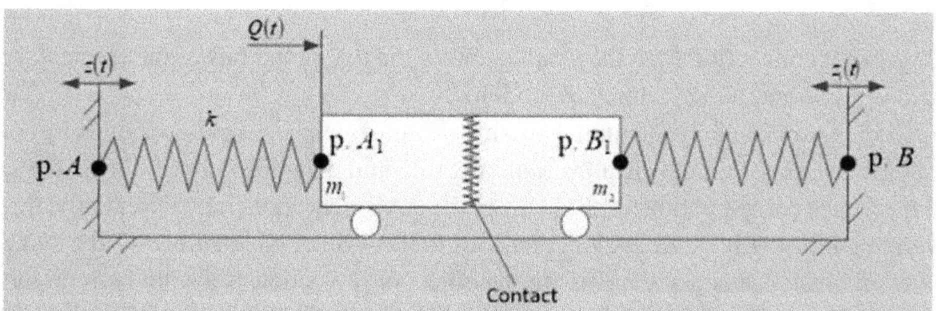

Mathematical Model of The Rotation Type System

The finite dimensions of the object of protection can manifest itself in computational schemes that reflect not transient (rectilinear) oscillatory motions, but rotating and oscillating motions, where the object in the form of a material point of mass m is connected to the center of rotation by an absolutely weightless absolutely rigid rod (Figure 5).

Figure 5. Vibration protection system with an object of mass m, making rotating and oscillating motions relative to a fixed point

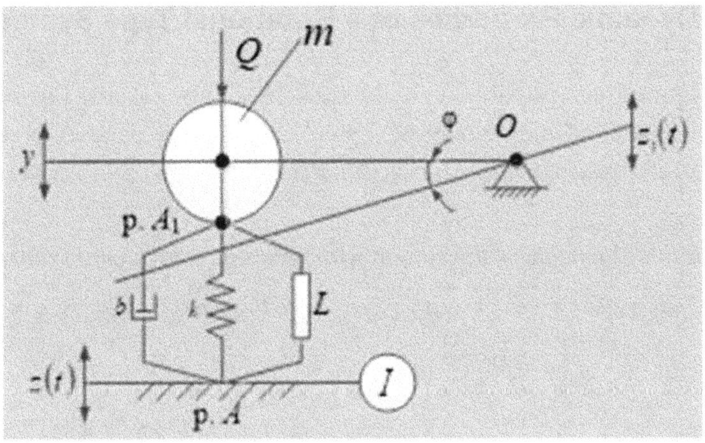

The mathematical model for the computational scheme in Figure 5 can be obtained on the basis of the usual approaches on the assumption that the angular motions φ and the corresponding vertical motions along the coordinate *y* (in a fixed coordinate system) are small. In the computational scheme (Figure 5), the interaction with the supporting surface *I*, moving according to the known law *z*(*t*), is considered. An external force is applied to the object of protection (mass *m*); The interaction of an element of mass *m* with a fixed basis (k_2, b_2, L_2) is taken into account similar to the above method. The equations of motion relative to the position of static equilibrium (with $z_1 = 0$) can be written in the coordinates *y* and φ, respectively:

$$\left[(m+L)p^2+bp+k\right]\cdot\bar{y}=\left(Lp^2+bp+k\right)\cdot\bar{z}+\bar{Q}, \tag{10}$$

$$\left[\left(J+Ll^2\right)p^2+bl^2p+kl^2\right]\cdot\bar{\phi}=l^2\cdot\left(Lp^2+bp+k\right)\cdot\bar{z}+\bar{M}. \tag{11}$$

where $J = ml^2$ is the moment of inertia of the object of protection relative to the fixed point *O*; *l* is the length of the rod; $y = \phi \bullet l$; $M = \bar{Q}\cdot l$.

From comparison (4.10) and (4.11), the relationship between the equation is visible through a constant coefficient l^2. If we take the equation in the form (10) as a basis, then the mathematical model of the system with the computational scheme in Figure 5 is not fundamentally different from equation (4). That is, in a certain sense, it is possible to computational schemes of two types (see. Figure 1, *a* and Figure 5) to be considered equivalent. However, in systems there are fundamental differences: the object of protection in Figure 1, and is a material point (and in this case does not have a spatial "metric"). As for the system of rotational type (see Figure 5), the object of protection is represented by a material point of mass m connected by a weightless rigid rod of length *l* with a fixed center of rotation at p. *O*. In this case the system has a spatial metric, which predetermines the emergence of a number of features, consisting in the fact that the attachment points of the elementary units may not coincide with the material point *m*, but be located along the length of the rod.

Features of The Dynamic Properties of a Rotational Type System

The rod length *l* in Figure 5 it is possible, in a certain sense, to be regarded as some kind of lever of the first kind, on which the attachment points of typical elements are parallel to each other (Figure 6). Mathematical model of the system at $Q = 0$, $z_1 = 0$, $z \neq 0$.

Figure 6. The computational scheme of the system with separate attachment points of elementary units

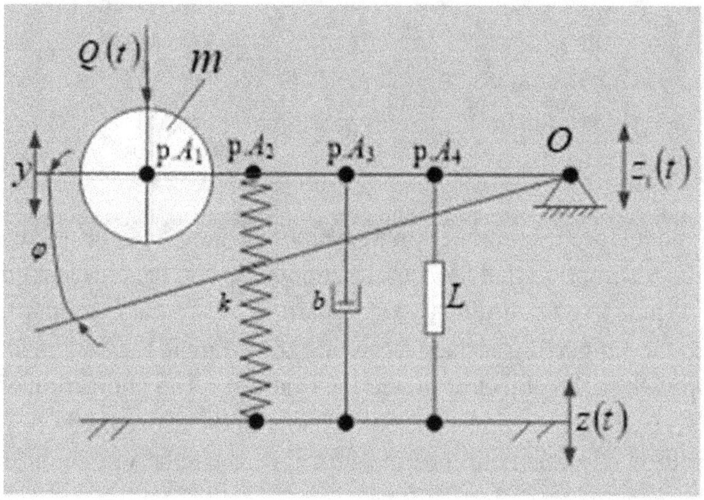

We denote for further calculations that $OA_1 = l$; $OA_2 = l_1$; $OA_3 = l_2$; $OA_4 = l_4$. We write the expressions for the kinetic and potentially energies, as well as for the energy dissipation function:

$$\mathrm{T} = \frac{1}{2} m (\dot{y})^2 + \frac{1}{2} L (\dot{\phi} l_3 - \dot{z})^2; \tag{12}$$

$$\Pi = \frac{1}{2} k \cdot (\phi l - z)^2; \tag{13}$$

$$\Phi = \frac{1}{2} b \cdot \left(\dot{\phi} l - \dot{z}\right)^2. \tag{14}$$

After a series of transformations, we obtain the equation of motion with the coordinate φ ($y=\phi \bullet l$):

$$\left[\left(ml^2 + L \cdot l_3^2\right) p^2 + \left(bl_2^2\right) p + kl_1^2\right] \cdot \bar{\phi} = \left(Ll_3^2 p^2 + bl_2^2 p + kl_1^2\right) \cdot \bar{z}. \tag{15}$$

The transfer function of the system in accordance with (15) takes the form:

$$\underset{\substack{Q=0 \\ z_1=0}}{W(p)} = \frac{\bar{\phi}}{\bar{z}} = \frac{Ll_3 p^2 + bl_2 p + kl_1}{\left(ml^2 + Ll_3^2\right) p^2 + bl_2^2 p + kl_1^2}. \tag{16}$$

From expression (16) you can go, using the ratio $y=\phi \bullet l$, to the transfer function

$$\underset{\substack{Q=0 \\ z_1=0}}{W'(p)} = \frac{\bar{y}}{\bar{z}} = \frac{Li_3 p^2 + bi_2 p + ki_1}{\left(m + Li_3^2\right) p^2 + bi_2^2 p + ki_1^2}, \tag{17}$$

where $i_3 = \frac{l_3}{l}$; $i_2 = \frac{l_2}{l}$; $i_1 = \frac{l_1}{l}$ are transfer ratios of lever linkages.

It should be noted that in this case, the transfer ratios have a positive sign, since in the levers of the first kind there is no change in direction during the transmission of motion.

If we assume that $z_1 \neq 0$, $z = 0$ and $Q = 0$, then the expressions (12) ÷ (14) will take the form, respectively:

$$\mathrm{T} = \frac{1}{2} m\left(\phi \cdot l + \dot{z}_1\right)^2 + \frac{1}{2} L\left(\phi \cdot l_3\right); \tag{18}$$

$$\Phi = \frac{1}{2} b\left(\phi \cdot l_2\right)^2; \tag{19}$$

$$\Pi = \frac{1}{2} k\left(\phi l_1\right)^2. \tag{20}$$

The equation of motion of the system is written:

$$\left[\left(ml^2 + Ll_3^2\right) p^2 + b \cdot l_2^2 + k \cdot l_1^2\right] \cdot \bar{\phi} = \left[ml \cdot p^2\right] \cdot \bar{z}_1. \tag{21}$$

The transfer function of the system at $z_1 \neq 0$ ($z = 0$; $Q = 0$) is determined from (21) and has the form:

$$W(p)=\frac{\bar{y}}{\bar{z}_1}=\frac{mp^2}{\left(m+Li_3^2\right)p^2+bi_2^2p+ki_1^2}. \tag{22}$$

From (22) it follows that the vibrations of the point of rotation O change the transfer function. We write (21) not in the coordinates of φ, but in the coordinates of y, assuming that for $y = \varphi l + z_1$. In this case (18)–(20) will take the form:

$$\mathrm{T}=\frac{1}{2}m\left(\dot{y}\right)^2+\frac{1}{2}L\left(\frac{\dot{y}-\dot{z}_1}{l}\right)^2\cdot l_0^2; \tag{23}$$

$$\Phi=\frac{1}{2}b\cdot\left(\frac{\dot{y}-\dot{z}_1}{l}\right)^2\cdot l_2^2; \tag{24}$$

$$\Pi=\frac{1}{2}k\cdot\left(\frac{\dot{y}-\dot{z}_1}{l}\right)^2\cdot l_1^2. \tag{25}$$

Then equation (4.21) is written:

$$\bar{y}\cdot\left[\left(m+L\frac{l_3}{l^2}\right)p^2+p\left(b\frac{l_2^2}{l^2}\right)+k\left(\frac{l_1^2}{l^2}\right)\right]=\bar{z}_1\cdot\left[\left(L\frac{l_3}{l^2}\right)p^2+p\left(b\frac{l_2^2}{l^2}\right)+k\left(\frac{l_1^2}{l^2}\right)\right]. \tag{26}$$

The transfer function of the system in the y coordinate will have a different form than that presented in (4.22):

$$W_y(p)=\frac{\bar{y}}{\bar{z}_1}=\frac{Li_3^2p^2+b_2i_2^2p+ki_1^2}{\left(m+Li_3^2\right)p^2+bi_2^2p+ki_1^2}. \tag{27}$$

Note that the denominators (22) and (27) are the same, but the numerators of the transfer functions will be different. A more general case is the simultaneous effect of external disturbances, when $z_1 \neq 0$, $z \neq 0$ ($Q = 0$). In this case, two external influences can lead to a single input signal. An assumption is made that $z = z_1$, although there may be situations when there are certain relations between z and z_1. The features of such interactions can be attributed to the known forms of control in automatic control systems by force. We write the expressions (18)–(20) assuming that $z \neq 0$; $z_1 \neq 0$ ($Q = 0$) and we get:

$$\mathrm{T}=\frac{1}{2}m\cdot\left(\dot{\phi}\cdot l+\dot{z}_1\right)^2+\frac{1}{2}L\cdot\left(\dot{\phi}l_3-\dot{z}\right)^2; \tag{28}$$

$$\Phi=\frac{1}{2}b\cdot\left(\dot{\phi}l_2-\dot{z}\right)^2; \tag{29}$$

$$\Pi=\frac{1}{2}k\cdot\left(\phi\cdot l-\dot{z}\right)^2. \tag{30}$$

The equation of motion will take the form:

$$\left[\left(ml^2+Ll_3^2\right)p^2+bl_2^2p+kl_1^2\right]\cdot\bar{\phi}=-mlp^2\cdot\bar{z}_1+\left[Ll_3p^2\bar{z}+bl_2p+kl_1\right]\cdot\bar{z}. \tag{31}$$

In accordance with equation (4.31), the system is under the influence of two external factors: $\bar{z}$ and $\bar{z}_1$. The principle of superposition can be used to build the system's responses to typical external inputs. If there is reason to believe that $\bar{z}=\alpha\cdot\bar{z}_1$, where α can be determined from the condition $-\infty<\alpha<\infty$ and is a real number, then (31) can be reduced to the form:

$$\underset{\bar{z}=\alpha\bar{z}_1}{W(p)}=\frac{\bar{\phi}}{\bar{z}_1}=\frac{\left(\alpha Ll_3-ml\right)p^2+\alpha bl_2p+\alpha kl_1}{\left(ml^2+Ll_3^2\right)p^2+bl_2^2p+kl_1^2}. \tag{32}$$

Note that the introduction of a constraint between $\bar{z}$ and $\bar{z}_1$ affects only the numerator of the transfer function (31), but does not change the denominator. Such features of control of the dynamic state in the theory of automatic control are associated with the principle of perturbation control, which, in particular, was considered in (Eliseev et al., 2013b; Eliseev et al., 1988).

Similarly, (31) can be transformed into an equation using the coordinate y. In this case, the transfer function will also have a difference from (32).

Taking into account the action of a power disturbance, when $Q\neq0$, and $z_1=0$, $z=0$, we get:

$$\underset{\substack{\bar{z}=0\\ \bar{z}_1=0}}{W_y(p)}=\frac{\bar{y}}{\bar{Q}}=\frac{1}{\left(m+Li_3^2\right)p^2+bi_2^2p+ki_1^2}. \tag{33}$$

In expressions (17), (27) and (33), the denominators of the transfer functions are the same, but the numerators depend on the type and nature of the action of external forces. Note that in systems with rotational motion external influences, such as external force factors, can have points of application not only on the object of protection, but also on a weightless rod. At the same time, the transfer functions will change accordingly, which creates the basis for the corresponding adjustment of the vibration protection equipment. Note also that the rotational system shown in Figure 5, may have a weightless rod, which has a pivot point that creates lever linkages characteristic of levers of the second kind.

Rotary Type System With a Weightless Rod With an Intermediate Point of Support

Let us consider the computational scheme of the vibration protection system (Figure 7) of the rotational type with a pivot point providing the effects of different signs.

Figure 7. The computational scheme of the vibration protection system with the separation of the points of attachment of typical elements on a weightless rod with an intermediate center of rotation

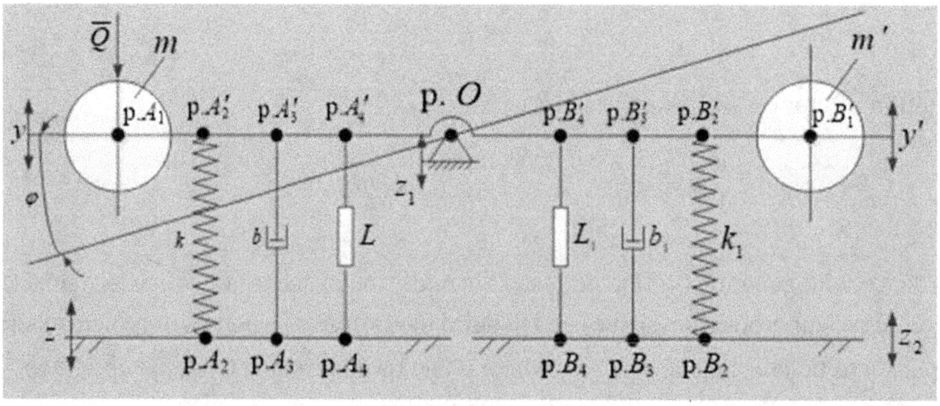

We assume that $Q = 0$, $z_1 = 0$, and $z \neq 0$ and $z_2 \neq 0$. Let the motion of the protected object be defined by the coordinate φ. Then the expressions for the kinetic and potential energies, as well as for the dissipation energy, have the form:

$$T = \frac{1}{2} m \cdot \left(\dot{\phi} l\right)^2 + \frac{1}{2} L\left(\dot{\phi} l_3 - \dot{z}\right)^2 + \frac{1}{2} m' \cdot \left(\dot{\phi} l'\right)^2 + \frac{1}{2} L_1 \cdot \left[-\dot{\phi}\left(l_3'\right) - \dot{z}_2\right]^2; \tag{34}$$

$$\Phi = \frac{1}{2} b_1 \left(-\dot{\phi} l_2' - \dot{z}_2\right)^2 + \frac{1}{2} b\left(\dot{\phi} l_2 - \dot{z}_2\right)^2; \tag{35}$$

$$\Pi = \frac{1}{2} k\left(\dot{\phi} l_1 - \dot{z}\right)^2 + \frac{1}{2} k_1 \left(-\dot{\phi} l_1' - \dot{z}_2\right)^2. \tag{36}$$

where $OB_l = l'$, $OB_2' = l_1'$, $OB_3' = l_2'$, $OB_4' = l_3'$, are the parameters of the attachment points (Figure 7) of the elements m', L_1, b_1 and k_1, respectively.

The equation of motion of the system in Figure 7 for the case $Q = 0$, $z_1 = 0$, $z \neq 0$, $z_2 \neq 0$ has the form:

$$\bar{\phi}\cdot\left[p^2\left(ml^2+m'(l')^2+Ll_3^2+L_1\cdot(l_3')^2\right)+p\left(bl_2^2+b_1(l_2')^2\right)+k\left(l_1^2\right)+k_1(l_1')^2\right]= \\ =\bar{z}\cdot\left[Ll_3p^2+pbl_2+kl_1\right]+\bar{z}_2\cdot\left(-L_1l_3'p^2-pb_1l_2'-k_1l_1'\right) . \qquad (37)$$

If we assume that $\bar{z}=\bar{z}_2$, then (4.37) is converted to the form:

$$\bar{\phi}\cdot\left[p^2\left(ml^2+m'(l')^2+Ll_3^2+L_1\cdot(l_3')^2\right)+p\left(bl_2^2+b_1(l_2')^2\right)+k\left(l_1^2\right)+k_1(l_1')^2\right]= \\ =\bar{z}\cdot\left\{p^2\cdot\left[Ll_3p^2-L_1l_3'\right]+p\cdot\left(bl_2-b_1l_2'\right)+kl_1-k_1l_1'\right\} . \qquad (38)$$

The transfer function of the system in coordinates of φ and external perturbation $\bar{z}_1=\bar{z}_2=z$ takes the form:

$$W_\phi'(p)=\frac{\bar{\phi}}{\bar{z}}=\frac{p^2\cdot\left[Ll_3-L(l_3')\right]+p\cdot\left(bl_2-b_1l_2'\right)+kl_1-k_1l_1'}{p^2\cdot\left[ml^2+m'(l')^2+Ll_3^2+L_1(l_3')^2\right]+p\left[bl_2^2+b_1(l_2')^2\right]+kl_1^2+k_1(l_1')^2} . \qquad (39)$$

The expression (4.39) can be reduced to the coordinate $\bar{y}$, then:

$$W_y'(p)=\frac{\bar{y}}{\bar{z}}=\frac{p^2\cdot\left[Li_3-L(i_3')\right]+p\cdot\left(bi_2-b_1i_2'\right)+ki_1-k_1i_1'}{p^2\cdot\left[m+m'(i')^2+Li_3^2+L_1(i_3')^2\right]+p\left[bi_2^2-b_1(i_2')^2\right]+ki_1^2+k_1(i_1')^2} , \qquad (40)$$

where $i'=\frac{l'}{l};\ i_1'=\frac{l_1'}{l};\ i_2'=\frac{l_2'}{l};\ i_3'=\frac{l_3'}{l}$.

The expressions (4.39) and (4.40) for transfer functions differ from each other by the value of the constant factor l, since $\varphi = y / l$.

However, this situation changes if $z_1 \neq 0$. In this case:

$$\phi=\frac{y-z_1}{l} . \qquad (41)$$

Expressions (4.34)–(4.36) respectively will take the form:

$$\mathrm{T}=\frac{1}{2}m\left(\dot{\phi}l+\dot{z}_1\right)^2+\frac{1}{2}m'\cdot\left(-\dot{\phi}l'+\dot{z}_1\right)^2+\frac{1}{2}\left(\dot{\phi}l_3+\dot{z}_1-\dot{z}\right)^2+\frac{1}{2}L_1\cdot\left(-\dot{\phi}l_3'+\dot{z}_1-\dot{z}_2\right)^2; \qquad (42)$$

$$\Phi=\frac{1}{2}b\cdot\left(\dot{\phi}l_2+\dot{z}_1-\dot{z}\right)^2+\frac{1}{2}b_1\cdot\left(-\dot{\phi}l_2'+\dot{z}_1-\dot{z}_2\right)^2; \qquad (43)$$

$$\Pi = \frac{1}{2}k\left(\phi l + z_1 - z\right)^2 + \frac{1}{2}k_1\left(-\phi l' + z_1 - z_2\right)^2. \tag{44}$$

The equation of motion of the system in coordinates of $\bar{\phi}$ and taking into account the compatible input action ($z = 0$, $z_1 \neq 0$, $z_2 \neq 0$) has the form:

$$\begin{aligned}&\bar{\phi}\cdot\{p^2\cdot[ml^2 + m'(l')^2 + Ll_3^2 + L_1(l_3')^2] + p\cdot[bl_2^2 + b_1(l_2')^2] + kl_1^2 + k_1(l_1')^2\} =\\ &= \bar{z}_1\cdot[p^2\cdot(-ml + m'l' - Ll_3 + L_1 l_3') + p\cdot(-bl_2 + b_1 l_2') - kl_1 + kl_1'] +\\ &+ \bar{z}\cdot[p^2 Ll_3 + pbl_2 + kl_1] + \bar{z}_2\cdot[-L_1(l_3')^2 p^2 - p\cdot(b_1 l_2') - k_1 l_1'].\end{aligned} \tag{45}$$

Obtaining a response to the total effect of kinematic excitations is possible based on the application of the principle of superposition.

Suppose that $z = z_1 = z_2$, then, based on (45), the transfer function of the system takes the form:

$$W_\phi(p)\Big|_{\substack{z=z_1=z_2\\ Q=0}} = \frac{\bar{\phi}}{\bar{z}} = \frac{p^2\cdot(m'l' - ml)}{p^2\cdot[ml^2 + m'(l')^2 + Ll_3^2 + L_1(l_3')^2] + p\cdot[bl_2^2 + b_1(l_2')^2] - kl_1^2 + k_1(l_1')^2}. \tag{46}$$

In simplified form (46) it will be written:

$$W(p)\Big|_{\substack{z=z_1=z_2\\ Q=0}} = \frac{\bar{\phi}}{\bar{z}} = \frac{p^2\cdot(m'l_1' - ml)}{p^2\cdot[ml^2 + m'(l')^2 + Ll_3^2 + L(l_3')^2] + p\cdot[bl_2^2 + b_1(l_2')^2 + kl_1^2 + k_1(l_1')^2]}. \tag{47}$$

In the coordinate system of the expression (42) –(44) will be:

$$\mathrm{T} = \frac{1}{2}m\cdot(\dot{y})^2 + m'\cdot\left[\left(-\frac{\dot{y}+\dot{z}_1}{l}\right)\cdot l' + z_1\right]^2 + \frac{1}{2}L\cdot\left[\frac{(\dot{y}-\dot{z}_1)}{l}l_3 + \dot{z}_1 + \dot{z}\right]^2 + \frac{1}{2}L_1\cdot\left[-\frac{(y-z_1)}{l}l_3' + z_1 - z_2\right]^2. \tag{48}$$

$$\Phi = \frac{1}{2}b\cdot\left|-\frac{(\dot{y}-\dot{z}_1)}{l}l_2 + \dot{z}_1 - z\right|^2 + \frac{1}{2}b_1\cdot\left|-\frac{(y-z_1)}{l}\cdot l_2' + \dot{z}_1 - \dot{z}_2\right|^2. \tag{49}$$

$$\Pi = \frac{1}{2}k\cdot\left|\frac{(y-z_1)}{l}\cdot l_1 + z_1 - z\right|^2 + \frac{1}{2}k_1\left|-\frac{(y-z)}{l}\cdot l_1' + z_1 - z_2\right|^2. \tag{50}$$

We write the equations of the system using the coordinate y:

$$\begin{gathered}\bar{y}\cdot\{p^2\cdot[m+m'(i')^2+Li_3^2+L_1(i_3')^2]+p[bi_2^2+b_1(i_2')^2]+ki_1^2+k_1(i_1')^2\}=\\=\bar{z}_1\cdot\left\{\begin{matrix}[-m'\cdot i'\cdot(1-i')-L_3i_3\cdot(1-i_3)+L_1i_3'\cdot(1+i_3')]\cdot p^2+\\+p\cdot[-bi_2\cdot(1-i_2)+b_1\cdot i_2'\cdot(1+i_2')]+k_1i_1'\cdot(1+i_1')-ki_1(1-i_1)\end{matrix}\right\}+\\+\bar{z}\cdot[Li_3p^2+bi_2+ki_1]+\bar{z}_2\cdot[-L_1i_3p^2-b_1\cdot i_2'p-k_1\cdot i_1].\end{gathered}\tag{51}$$

Using (4.51), we obtain the expression for the transfer function, assuming that $z = z_1 = z_2$ ($Q = 0$):

$$\underset{\substack{z=z_1=z_2\\Q=0}}{W(p)}=\frac{\bar{y}}{\bar{z}}=\frac{\begin{matrix}-m'i'\cdot(1-i')\cdot p^2+p^2\cdot[L_1i_3'\cdot(1+i_3')-Li_3\cdot(1-i_3)+Li_3-L_1i_3]\\+p\cdot[-bi_2\cdot(1-i_2)+b_1i_2'\cdot(1+i_2')+bi_2-b_1i_2']+k_1i_1'\cdot(1+i_1')-ki_1\cdot(1-i_1)+ki_1-k_1'i_1'\end{matrix}}{[m+m'(i')^2+Li_3^2+L_1(i_3')^2]p^2+p\cdot[bi_2^2+b_1(i_2')^2]+ki_1^2+k(i_1')^2}.\tag{52}$$

Let's make some simplifications and get:

$$\underset{\substack{z=z_1=z_2\\Q=0}}{W(p)}=\frac{\bar{y}}{\bar{z}}=\frac{[-m'i'\cdot(1-i')\cdot p^2+L_1i_3'+Li_3]\cdot p^2+p\cdot[b_1i_2^2+b_1(i_2')^2]+ki_1^2+k_1(i_1')^2}{[m+m'(i')^2+Li_3^2+L_1(i_3')^2]p^2+p\cdot[bi_2^2+b_1(i_2')^2]+ki_1^2+k(i_1')^2}.\tag{53}$$

Comparative Analysis of the Possibilities of Vibration Protection Systems of Two Types

In relation to the basic models of vibration protection systems, presented in the form of mechanical oscillatory structures with one degree of freedom in Figure 1 and 7, it can be noted that they are among the most common ones. The following characteristics are common for models of translational and rotational types:

1. The minimum complexity of the configuration consists of mass-and-inertia and elastic elements.
2. The object of protection in the form of a solid body during the forward motion in an idealized representation is considered as a material point.
3. The elastic element connects the material point (or the object of protection) with the spring, which, in turn, is connected to the supporting surfaces.
4. There may be several supporting surfaces. When considering rectilinear motions of stiffness of the springs are grouped according to the rules of parallel addition.
5. When an object is exposed to several elastic elements at different angles, the total stiffness of the springs can be considered as a reduced characteristic. This takes into account the number of elements and the angles of inclination of the lines of action of elastic forces.
6. Joints of elastic elements with the object of protection and supporting surfaces are considered to be two-way or two-way bonds.
7. A mass-and-inertia element (object of protection) can be viewed in the mode of interaction with a fixed base (surface *III* in Figure 1).

Such an interaction reflects the idea of the object of protection as a unit that performs translational motion in contact with a fixed rack of the so-called initial mechanism. In this case, the kinematic pair of class V (translational) is considered. This circumstance is due to the fact that the object of protection, being, in general, a solid body (with an appropriate level of abstract presentation) realizes only motion along one coordinate.

8. External forces may be in the form of force factors. In this case, they are applied directly to the protected object. With kinematic perturbation, the motion of the support surface is known. Kinematic perturbation can be considered as a separate action or "reduced" to an equivalent force (Eliseev et al., 1988).
9. To describe the motion of a vibration protection system, the following can be used: a coordinate system associated with a fixed base; coordinate system in relative motion. The supporting surface, as already noted, may have a known law of motion. Then the relative coordinate is the sum of the motion of the base and the object of protection.
10. A mathematical model of the basic vibration protection system may include dissipative elements. In this case, we do not go around taking into account the features of the location of dissipative elements. In the general case, dissipative elements (meaning elements of viscous friction or a damper) can be introduced as a parallel addition to elastic elements when using an extended set of standard elements, which is reflected in the structural theory of vibration protection systems; introduction and accounting of additional constraints is carried out in a similar way.

Some Generalizations About the Basic Models

In the framework of the traditional approach to mathematical models of vibration protection systems of translational motion with one degree of freedom, the structure of the system of interaction of elements does not imply the existence of mechanisms that transform motion. Although the object of protection, as mentioned, can be considered as the initial unit of the mechanism. It should be noted that ideas about mechanical systems change if the structure of the units of the mechanism, as a mechanical chain, includes not only solid, but also elastic bodies. Note also that the considered elements of models of vibration protection systems are abstract representations of real objects with integral properties, therefore the division into elementary typical units is conditional (Nasnikov et al., 2010; Parshuta & Kaimov, 2013).

Figure 8 presents the options for complicating the basic computational scheme of a vibration-type translational system with one degree of freedom. Analysis of structural diagrams (Figure 8, *b* and *c*, as well as Figure 8, *d* and *e*) shows that taking into account the peculiarities of dynamic interactions in the oscillatory system is associated with the emergence of units with the transfer function of a second-order differential unit among traditional elastic and dissipative elements. Such a unit, as well as a spring (k) and a damper (bp), has an input signal in the form of an displacement, and an output signal in the form of a force. Thus, the structural model of the basic computational scheme has three typical elements (k, bp and mp^2), which have the displacement as an input signal, and the force as the output signal. Essentially, these elements are a kind of "spring"; such forms of generality of dynamic properties can be correlated with properties of generalized springs. Such substantiations are connected with the methodological basis of the structural theory of vibration protection systems (Parshuta, 2013a; Parshuta, 2013b).

The object of protection, as follows from the structural models, is represented by an integrating element of the second order. The input of such a unit is force, and the output signal is the displacement. Such a structural interpretation and functional differences between the typical elements of the structural diagram of the original mechanical oscillatory system are explained by the peculiarities of the analytical tools of the theory of automatic control. As for the properties of mass-and-inertia elements, they can manifest themselves in two forms: the mass-and-inertia element is the object of protection (or control) or the mass-and-inertia element works as a generalized spring (Wolfson, 2008).

The search for ways to expand the set of typical elements results in the introduction of additional constraints implemented by the mechanisms, as shown in Figure 9.

Computational schemes shown in Figure 9 were considered in (Harris & Creed, 2002; Eliseev & Artyunin, 2016), in which it was shown that, in a mechanized form, the properties of additional constraints (or mechanisms) are interpreted by a typical unit that implements the functions of a second-order differentiating unit. Further complication of additional constraints introduced into the basic scheme leads to structures that have transfer functions in the form of fractional rational expressions, which allows even in simple models displaying complex dynamic interactions of real technical objects with a spatial metric. The development of these provisions is reflected in the works (Eliseev, 2018; Parshuta & Gordeeva, 2010; Parshuta et al., 2012).

Figure 8. Variants of complicating the basic model of a vibration protection system with one degree of freedom: a – basic computational scheme; b – structural diagram of the basic system (fixed coordinate system); c – structural diagram of the system in relative coordinates ($y_1 = y - z$); d – computational scheme of the basic system with additional elastic and dissipative units; e – structural diagram of the system in absolute coordinates (y); f – structural diagram of the system in relative coordinates ($y_1 = y - z$)

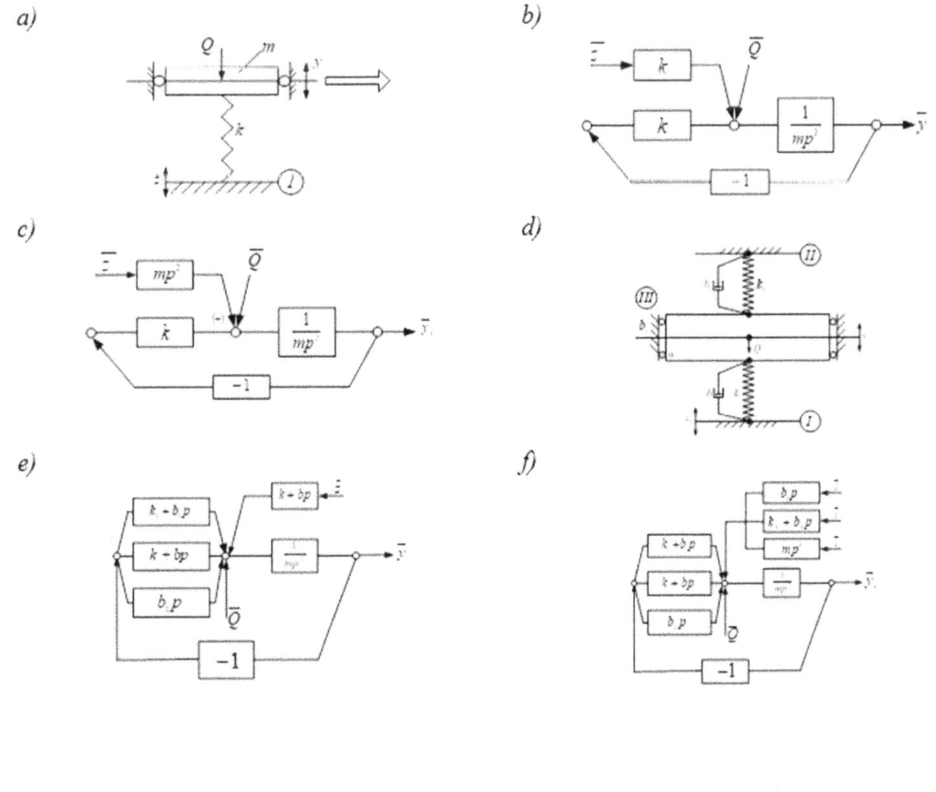

Figure 9. Computational schemes of the basic vibration protection system with additional constraints (or typical elements): a – articulated mechanism; b – lever of the first kind; c – toothed gearing

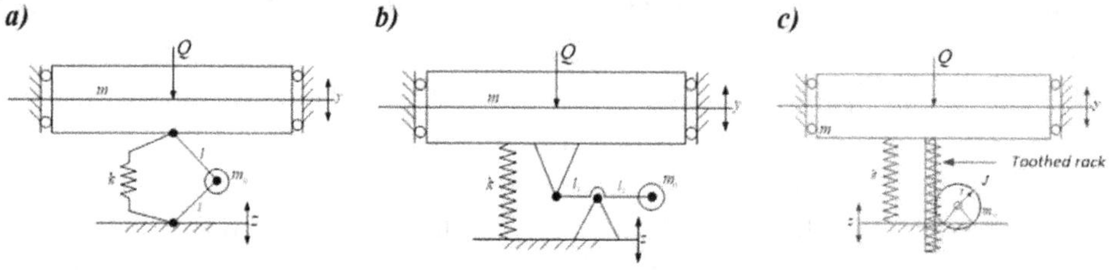

If in a system of translational motion of an object of protection, a mass-and-inertia element can be represented as a material point, then in a system of rotational motion, the object of protection has an inertia moment, that is, it has a "spatial" metric.

For a solid body moving in a plane, one can distinguish (as opposed to a system of translational type) two characteristic points. First, there is a fixed point (Figure 5), around which the body oscillates. Secondly, the solid body also has a center of gravity. These points do not always coincide, which forms certain features of the motion. The idealized motion of a solid body with a fixed point of rotation can be represented as a material point attached to one end of a weightless rigid rod. The second end of this rod (Figure 5) is mounted at a fixed point. This compound forms a kinematic rotational pair of class V, whereas during translational motion of this kind, the connection of an object with a rack has the form of a progressive kinematic pair of class V (Eliseev et al., 2019).

The motion of the system can be described using the angular coordinate φ, as well as using the coordinates of the rectilinear vertical motion y and y_1, as was discussed earlier.

In Figure 7 it is possible to distinguish support surfaces with which elastic or other typical elements come into contact. The peculiarity of the rotational type of motion is the possibility of forming an additional kinematic perturbation through the motion of a point of rotation of a solid body.

Since the characteristic feature of the system is the presence of a weightless rigid rod or lever having spatial dimensions, in such a system, the coordinates of the fixing points of typical elements become essential. Such features form dynamic interactions depending on geometrical parameters. A significant role is acquired by the shape of the lever, which predetermines the location of the elements not only on one side, but also on both sides of the pivot point, and this provides the possibility of mutual compensation when summing up the power factors. The transfer functions of the systems in angular and linear coordinates will also be different.

The most important circumstance is that the vibration protection system of the rotational type (or mechanical oscillatory system) as a characteristic feature suggests the presence of a lever or a lever mechanism. The consideration of such factors does not negate the analogy between the two types of vibration protection systems, but focuses attention on the existence of features. In any case, a complete analogy should suggest the possibility of manifestation in systems of the translational type of similar effects. In addition, the difference of rotational type systems is in consideration of the relative position of the center of gravity and the center of rotation. This circumstance results in the fact that a solid body in a plane motion relative to a fixed point can be represented as two material points located on a weightless rigid rod with a center of rotation between the selected material points. Some questions of the evaluation of dynamic systems of this kind are reflected in (Eliseev et al., 1988).

Thus, issues of reducibility of systems of one type to another (translational motion to rotational and vice versa) require attention to the features of systems whose manifestation is associated with manifestations of lever linkages inherent in this type of motion.

DETERMINATION OF REDUCED STIFFNESSES IN SYSTEMS WITH A ROTATING OBJECT OF PROTECTION

In section 4.1 when considering the basic computational scheme of the vibration protection system, it was assumed that the center of gravity and the center of rotation coincided, which made it possible to obtain the necessary transfer functions, both for force and for kinematic perturbations. At the same time, the center of gravity and the center of rotation of a solid body may not coincide, which, in general, results in the need to take into account a number of features in the dynamic interactions of the elements of the system.

The Case of Mismatch of the Center of Gravity and the Center of Rotation, a Lever of the Second Kind

Let us consider a system consisting of a lever mechanism of the second kind, as shown in Figure 10.

Figure 10. Schematic diagram of the system rotational type with one degree of freedom

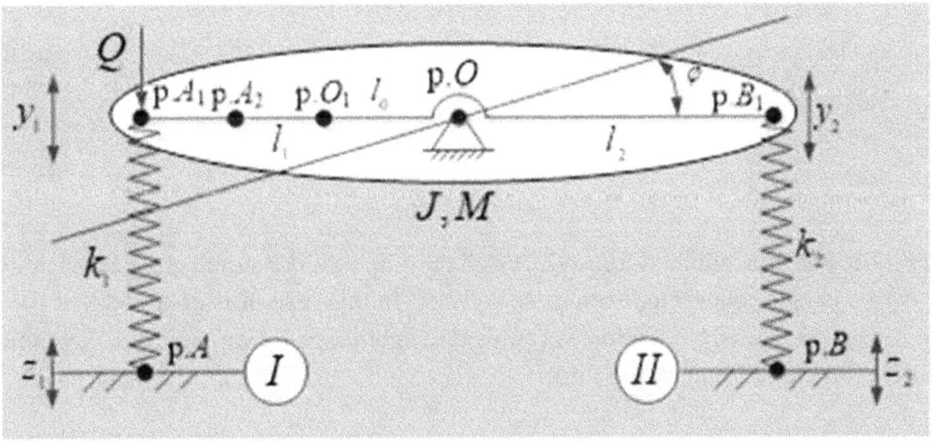

We write the expressions for the kinetic and potential energies when moving relative to the position of static equilibrium, assuming that the body performs rotating and oscillating motions relative to p. *O*: $l_0 = OO_1$); $Q = Q_0 \sin\omega t$:

$$T = \frac{1}{2}\left(J + Ml_0^2\right)\cdot\dot{\phi}^2 ; \tag{54}$$

$$\Pi = \frac{1}{2}k_1 \cdot (\phi l_1)^2 + \frac{1}{2}k_2 \cdot (\phi l_2)^2. \tag{55}$$

The equation of motion of the system in the coordinate φ has the form:

$$\left(J + Ml_0^2\right)\cdot\ddot{\phi} + \left(k_1 l_1^2 + k_2 l_2^2\right)\cdot\phi = l_1 Q_0 \sin\omega t. \tag{56}$$

Using the Laplace transform and at zero initial conditions, we find the transfer function of the system with the "input signal" $M_{dv}=l_0Q_0\sin\omega t$ and "output signal" – φ:

$$W(p) = \frac{\bar{\phi}}{\bar{M}_{dv}} = \frac{l_1^2}{[(J + Ml_0^2)p^2 + k_1 l_1^2 + k_2 l_2^2]}. \tag{57}$$

With fixed bases *I* and *II* ($z_1 = 0$, $z_2 = 0$), we can write down that $y_1 = \varphi \cdot l_1$, $y_2 = \varphi \cdot l_2$, then from (57) we can find:

$$W_1(p) = \frac{\bar{y}_1}{\bar{Q}} = \frac{l_1^2}{[(J + Ml_0^2)p^2 + k_1 l_1^2 + k_2 l_2^2]}. \tag{58}$$

Assuming that the lever is a weightless rigid rod, we get:

$$W_1'(p) = \frac{\bar{y}_1}{\bar{Q}} = \frac{1}{k_1 + k_2 i^2}, \tag{59}$$

where $i = \frac{l_2}{l_1}$ are transfer ratios of the lever of the second kind.

With respect to the concentrated force *Q*, under the imposed constraints, the lever system can be considered as two parallel-connected springs k_1 and $k_2 i^2$. In this case, it is assumed that the force *Q* is applied at p. A_1 (see. Figure 10). If the force is applied in another p. A_2 (see. Figure 10), where $l_{00} = OA_2$, then the transfer function (58) takes the form:

$$W_1''(p) = \frac{\bar{y}_1}{\bar{Q}_{pA_2}} = \frac{l_1 \cdot l_{00}}{\left(J + Ml_0^2\right)p^2 + k_1 l_1^2 + k_2 l_2^2}. \tag{60}$$

With the same restrictions as in the original case, we get:

$$W_1''(p) = \frac{\bar{y}_1}{\bar{Q}_{pA_2}} = \frac{i_0}{k_1^2 + k_2 i^2}, \tag{61}$$

where $i_0 = \frac{l_{00}}{l_1}$ characterizes the place of the application, the place of application of the concentrated force Q.

Note that the reduced stiffness of the system, determined by the relation $\bar{Q}$ and displacement $\bar{y}_1$, can be found inverse of expression (61), then:

$$k_{\text{redA}_2} = \frac{k_1}{i_0} + \frac{k_2 i^2}{i_0}. \tag{62}$$

From (62) it follows that the smaller i_0, the greater will be the reduced stiffness, and for $i_0 = 0$, $k_{redA2} \to \infty$. In a physical sense, this situation is quite explicable. Note also that the reduced stiffness from expressions (58) and (60) can be obtained under the condition $p = 0$ ($p = j\omega$ is a complex variable). This corresponds to the transformation of the periodic force Q into a constant. Thus, the reduced stiffness, determined from (58) and (60) under the condition $p = 0$, allows us to estimate the elastic properties of the system under static force action. It should be noted another characteristic circumstance: the expression (58) is used to determine the reduced stiffness in the case of the application of force Q at p. A_1. If the force Q is applied at p. A_2. In the first case, $i_0 = 1$. In the second case, $0 \le i_0 \le 1$. If i_0 becomes such that $i_0 > 1$, then the reduced stiffness k_{redA2}, defined by the expression (62) will decrease. This is understandable, since the moment of force will increase and at the same time the displacement of p. A_1 increases. Note that in this case an attempt is made to separate the point of application of the concentrated force and the point to which the elastic element joins.

That is, when $i_0 = 1$, the stiffness determines the deformation of the elastic element with stiffness k_1. If $i_0 < 1$, then the force Q is applied to p. A_2, and the deformation of the spring is determined by the motion of p. A_1 (or the coordinate y). When $i_0 > 1$, p. A_2 of the application of force moves beyond the length of the lever arm l_1 (Figure 10), although the deformation of the elastic element at p. A_1 is taken into account. Similar features were considered in Section 4.1 when assessing the possibility of distribution of fixing points for typical elements (springs, dampers, etc.) over the arms of the first and second kind levers.

Accounting for the Location of Elastic Elements

Static responses in a system with objects of rotation are considered for a solid body (see Figure 10), which has a fixed point of rotation and rests on the elastic constraints k_1 and k_2. If we assume that the force of gravity falls on the point of rotation O, then in order to maintain a state of equilibrium in the absence of external disturbance Q, it is necessary to consider two options for a constructive design (Figure 11).

We assume that for $l_0 = 0$, the springs k_1 and k_2 are undeformed, and the lever occupies a position parallel to surfaces *I* and *II*. In the initial position, the static responses R_A, R_{A1} and R_B, R_{B1} will be zero: the reaction at p. $O - R_0 = Mg$. When the lever is deflected (see Figure 11, *a*) at an angle φ at pp. A and A_1, reactions $|R_A| = |R_{A_1}| = \phi \cdot l_1 \cdot k_1$ will arise. The reactions will be equal in magnitude, but directed in opposite directions, as shown in Figure 12, *a*. In turn, at the same time, reactions $|R_B| = |R_{B_1}| = \phi \cdot l_2 \cdot k_2$ will also occur at pp. B and B_1. Reactions R_A and R_B, as well as R_{A1} and R_{B1} reactions will have opposite directions.

In the absence of pre-deformation of the springs k_1 and k_2 in order to create a "tension", the elastic element k_1 will be compressed and provide the appropriate reactions at pp. A and A_1. At the same time, the elastic element k_2 will stretch and reactions will occur at p. B and B1. In such a situation, it is necessary that at p. B and B_1 there should be bilateral constraints, otherwise there would be a contact failure at p. B and B_1. A similar situation occurs (Figure 12, *a*) if the deviation of the angle φ (swinging of the object) will be implemented in the other direction, i.e. clockwise. Characteristic of the variant of the load of supports (see Figure 12, *a*) is that the supports will always work in different modes: one is compressed and the other is stretched, that is, the springs are working asynchronously.

Figure 11. Variants of computational schemes of rotational vibration protection systems: a – asynchronous interaction of elements; b – synchronous impact

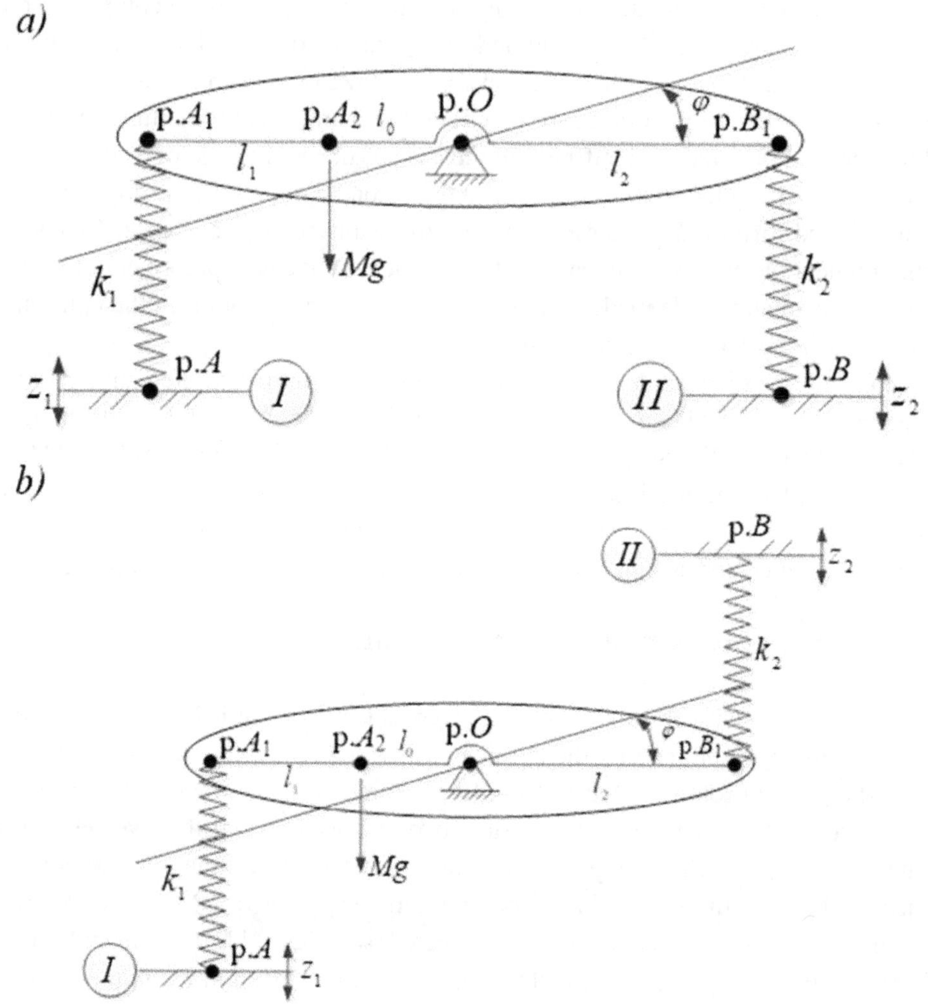

In turn, according to the variant of the arrangement of elastic elements (see Figure 12, *b*), when the body is rotated clockwise, the elastic elements k_1 and k_2 will simultaneously be compressed; when rotated clockwise, they will be simultaneously stretched, which can be called a synchronous effect. In any cases of support loading (see Figure 12) in the absence of pre-deformation of the springs, the reactions at the points of contact of the elastic element, both with the supporting surface (*I* and *II*) and with the solid body, will be alternating, and with the lever in parallel will be zero.

We will assume that the elastic element k_1 (see Figure 12, *a*) has a sufficient length for pre-compression by Δ_1. If one puts in such a compressed spring into the opening of AA_1 and tries leaving the lever positions (see Figure 12, *a*, *b*) parallel to the supporting surfaces *I* and *II*, then the spring k_2 must be deformed by compression by the quantity Δ_2, while to maintain static equilibrium the following condition will be fulfilled:

$$\Delta_1 \bullet k_1 \bullet l_1 = \Delta_2 \bullet k_2 \bullet l_2, \tag{63}$$

or

$$\Delta_1 = \frac{l_2}{l_1} \cdot \frac{k_2}{k_1} \cdot \Delta_2 = \frac{k_2}{k_1} \cdot i \cdot \Delta_2, \tag{64}$$

where $i - \frac{l_2}{l_1}$ is the transfer ratio of the lever of the second kind.

Pre-contraction results in static compression reactions; these reactions will be of the same sign:

$$R_A = R_I = k_1 \bullet \Delta_1; \tag{65}$$

$$R_B = R_{II} = k_2 \cdot \Delta_2 \cdot \frac{k_2}{k_1} \cdot l. \tag{66}$$

If the center of gravity of the solid body (Figure 12, *a*) is shifted by the value l_0, then the force of the weight $P = Mg$ will lead to a change in the reactions at pp. A_1 and *B*.

Given the static nature of the loading, it can be found that the application of the force *P* at p. A_2 will cause the lever to rotate from the initial (parallel) position by the angle φ:

$$\phi = \frac{Mg \cdot l_0}{k_1 l_1^2 + k_2 l_2^2}. \tag{67}$$

The reaction at p. *A*, i.e. on the support surface *I*, will be determined:

$$R_A = \phi \cdot l1_{\bullet}k1_{+}\Delta 1_{\bullet}k1_{=}k1_{\bullet}(\phi l1 +_{\Delta} 1)_{.} \tag{68}$$

In turn:

Figure 12. Variants of the location of the supporting surfaces and the corresponding reactions between elastic elements: a–k_1 and the support surface I–R_{AI} and R_{IA}, as well as between k_1 and a solid body– R_{AIr} and R_{rAI} ; b – k_2 and the support surface II – R_{BII} and R_{IIB} ; between k_2 and a solid body – R_{BIr} and R_{rBI}

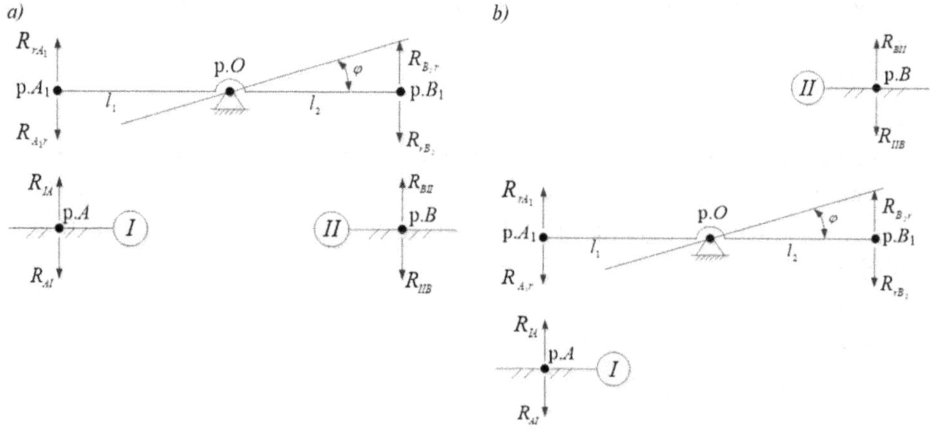

$$R_B = \Delta_2 k_2 - \phi l2 k2 = k2(\Delta 2 - \phi l2) \tag{69}$$

From (69) it follows that at RB = 0 a critical case occurs, determined by the relation:

$$\Delta_2 = \frac{Mg \cdot l_0 \cdot l_2}{k_1 l_1^2 + k_2 l_2^2} = \frac{Mg \cdot i_0}{k_1 + k_2 i^2}. \tag{70}$$

Knowing Δ_2, one can find the critical value l_0, assuming that:

$$i_0 = \frac{l_0}{l_1} = \frac{\Delta_2 \cdot \left(k_1 + k_2 i^2\right)}{Mg}, \tag{71}$$

$$l_0 = \frac{\Delta_2 \cdot l_1 \cdot \left(k_1 + k_2 i^2\right)}{Mg}. \tag{72}$$

Note that the fulfillment of condition (72) or the equality of R_B to zero in the presence of two-way constraint results in the situation when the contact between the elastic element and the supporting surface *II* is maintained, and with a further change in the parameters, the reactions at pp. B and B_1 will change the signs and the spring k_2 will be stretched. If the connection between k_2 and supporting surface *II* is one-sided, then the system loses its static stability.

For the variant of the location of the elastic elements k_1 and k_2 in accordance with the scheme in Figure 12, *b*, taking into account the forces of weight (Mg) through the definition of the angle φ results in the following reaction values:

$$R_A = \Delta_1 \cdot k_1 + \phi \cdot l_1 \cdot k_1 = k_1 \cdot (\Delta_1 + \phi l_1) =$$
$$= k_1 \cdot \left(\Delta_1 + \frac{Mgl_0 \cdot l_1}{k_1 l_1^2 + k_2 l_2^2} \right) = \frac{k_1 \cdot \left[\Delta_1 \left(k_1 + k_2 i^2 + Mgl_0 \right) \right]}{k_1 + k_2 i^2}; \tag{73}$$

$$R_A = \frac{Mgl_0 \cdot l_2}{k_1 l_1^2 + k_2 l_2^2} - \Delta_2 k_2 . \tag{74}$$

Assuming that $R_B = 0$, we find that the conditions defined by (74) coincide with conditions (72) with all the conclusions that follow from these relationships about the need to use bilateral constraints.

Displacement of the Center of Gravity in the System With a Lever of the First Kind

Regarding the use of levers of the first kind, as shown in Figure 13, *a*, *b*, the method of calculating static responses when taking into account the forces of the weight will remain the same.

Figure 13. Calculated scheme (a) and structural diagram (b) of a vibration protection system with lever linkages (levers of the first kind)

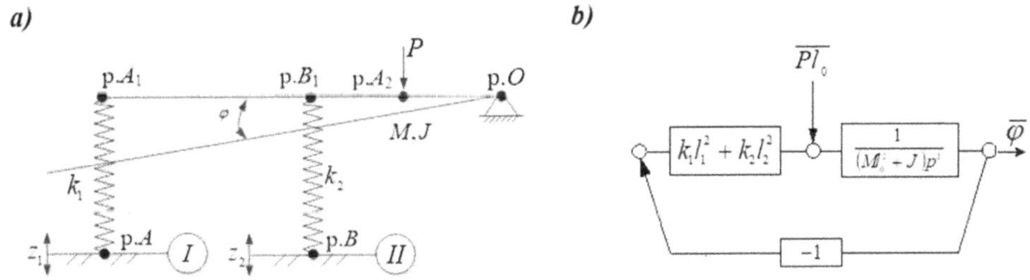

In Figure 13, *a* the following notation is taken: $l_0 = A_2O$, $l_1 = A_1O$, $l_2 = A_2O$. The computational scheme (Figure 13, *a*) can also be considered in two versions: synchronous deformation and asynchronous. In the latter case, surface *II* in Figure 13, and should be rotated 180° along the axis of the elastic element k_2.

Definition of Static Responses Under the Action of Constant Force

In addition to the weight forces in the computational schemes shown in Figure 10–4.13, can take into account the constant force, which may coincide with the force of weight or have the opposite direction. In this case, the application of a constant force with respect to a fixed point of rotation is essential. With an appropriate arrangement of a constant force and the choice of its direction, the discrepancy between the point of rotation and the center of gravity can be compensated, that is, the following condition must be met:

$$Mgl_0 + Q' \cdot l' = 0$$

where Q' is a constant force, l' is the distance to the center of rotation.

Such an approach to adjusting the properties of a mechanical oscillatory system with one degree of freedom coincides with the methods of static balancing known in machine science (in the theory of mechanisms and machines in particular) (Eliseev & Eliseev, 2020).

Figure 4.14 shows the location of the additional constant force Q', which can be used to compensate for the discrepancy between the center of gravity and the center of rotation of the object in the form of a solid body.

In Figure 14, *a* the constant force Q' is directed in the direction opposite to the force of the weight $P = Mg$, and the sum of the moments of the forces Q' and P with respect to p. O must be zero. On the structural diagram, the force factors will be applied to one point, but have different signs. In Figure 14 force Q_1 is on the other arm of the lever and has the same direction as the force of the weight P. Equilibration can be made in this case with additional weight, but this will change the moment of inertia of the solid body. When moving to the other shoulder, the force Q' may have the opposite direction with respect to P.

After determining the static responses at pp. A_1, B_1 (see Figures 10 and 13), you can find the reaction at p. O, using the equation of statics:

$$\bar{R}_{A_1} + \bar{R}_{B_1} + \bar{P} + \bar{R}_0 = 0,$$

that is, the reaction at p. O (R_O) is defined as the result of a geometric summation of the reactions of the constraints and external forces (forces, weight, and constant forces). In this case, it is assumed that all power factors are parallel vectors, which makes it possible to proceed to algebraic sums, assuming possible displacements of the protected object to be small (Eliseev et al., 2013e).

Figure 14. Principle diagrams of the location of a constant force Q' for static balancing: a – for the lever of the first kind; b – for the lever of the second kind

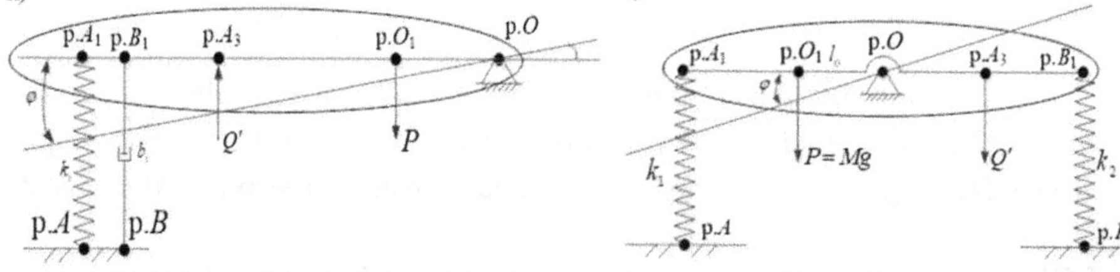

When a time-dependent force perturbation acts on the object of protection, the object starts moving relative to the static equilibrium position and in this case dynamic responses are formed at the points of contact of connecting elements with supporting surfaces, as well as at the points of connection of typical elements with the object of protection, and also connections of the latter with a rack. The definition of dynamic responses will be discussed below. The sum of dynamic and static responses gives an idea of the total reaction. Since the total reaction may take zero values. Such situations, as was shown above, with unilateral constraints can lead to contact disturbance and the emergence of gaps with subsequent collisions of contact surfaces. In any case, knowledge of the total reaction in the contacts of the elements of mechanical systems makes it possible to perform the necessary power calculations for determining the parameters of springs, dampers and other elements.

Possible Forms of Introducing Additional Constraints

The basic model of the vibration protection system can be extended by the addition of elements that implement the more complex motion transformation functions and force interactions.

Figure 15 shows how the original model (see Figure 14) can be transformed if the technical object is quite complex and, in its simplifications, one has to take into account a number of structural and technical features.

In the most general form, the problem of vibration protection can be represented as an installation between an object and a supporting surface of a vibration protection device (Figure 15, *a*). The external forces are the disturbing force applied to the object by the mass m (force Q), as well as the disturbance from the sides of the supporting surface or the base (kinematic disturbance $z(t)$). At the points of contact the vibration protection device with the object of protection (p. A_1) and the supporting surface, dynamic responses occur. In addition to dynamic responses at pp. A, A_1, static responses also occur, caused by constant forces, including the gravity of the object (and elements of the vibration protection device). In the general case, the reactions at p. A and p. A_1 are not equal to each other. However, in the simplest situations, when the vibration protection device consist only of elastic and dissipative elements and do not contain mass-and-inertia units, reactions at pp. A and A_1 are the same.

External disturbances may also be applied to the elements of the vibration protection systems.

In the simplest form, the vibration protection device can be represented (by set or one by one) by typical elementary units in the form of springs, dampers and motion transformation devices, having simple transfer functions of the amplifying unit, as well as differentiating units of the first and second orders (see. Figure 15, *b*). In Figure 15 *c–f* it is shown that the vibration protection device can be a mechanical circuit or a more complex mechanical oscillatory structure. In this case, the vibration protection device may have internal sources of disturbance. Note that linear systems that perform small oscillations are considered. As for the vibration protection device in Figure 15, *c–f*, they can be folded into generalized springs with transfer functions of a fractionally rational form.

A special class of vibration protection device are mechanisms, for example, non-locking screw lever mechanisms (Figure 15, *l*), lever mechanisms with fixed (Figure 15, *h*) and movable points of support (Figure 15, *i*), as well as different-type articulated mechanisms (Figure 15, *j–l*).

Consideration of mechanisms in the structure of a vibration protection system is quite justified for many reasons, including because in a real design, typical elements are mechanisms (shock absorbers, vibration dampers, springs, dampers, torsions, etc.).

These examples reflect only a small part of the possible solutions. Note that the same large variety of different vibration protection device in rotary type systems, in which toothed, cam and lever mechanisms are widely used. For the device shown in Figure 15, *h*, a patent for a utility model was obtained (Eliseev et al., 2013e).

The representation of the object of protection in the form of a solid body with one degree of freedom reflects, as a rule, preliminary statements of problems. Many real objects have two or more degrees of freedom. Note that the vibration protection of objects with several degrees of freedom, which is typi-

Figure 15. Variants of the constructive and engineering design of vibration protection devices in the system of translational rectilinear motion of an object are protected with one degree of freedom; variants of vibration protection device(VPD) are described in the text

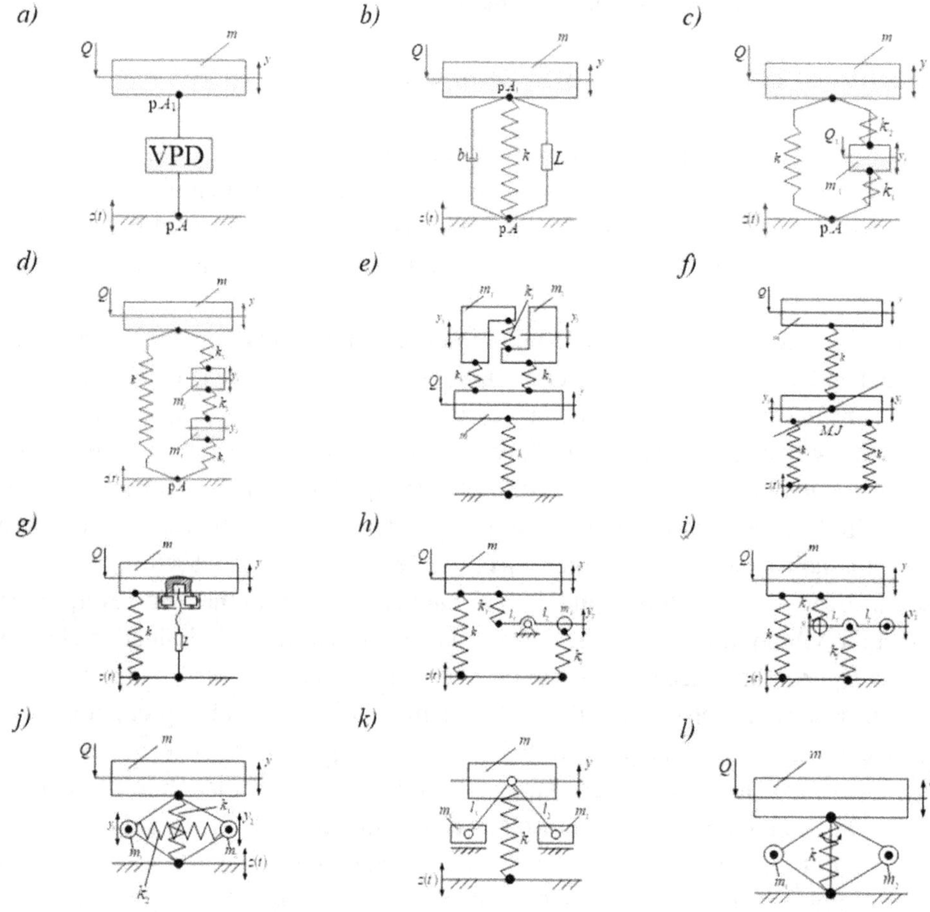

cal, in particular, for transport dynamics, is a complex task that requires the search and development of reasonable approaches and methods for assessing the dynamic state.

In the context of the development of structural approaches in the theory of vibration protection and vibration isolation of machines, equipment, instruments, and equipment, it was proposed to take into account the peculiarities of motion of the object of protection. In the simplest cases, the object of protection can be considered as a material point or a system of material points that make a straight forward motion. If the object of protection performs rotating and oscillating motions around a certain point, then it is necessary to take into account the dimensions of the object of protection. In this case, a solid body is considered, on which the points of attachment of elements of the vibration protection system are fixed.

A method for constructing mathematical models of mechanical oscillatory systems is proposed, which allows one to take into account not only the features of the spatial arrangement in the vibration protection system of typical elementary units, but also the dynamic effects arising from the operation of elastic, dissipative and mass-and-inertia elements within a certain structure. The geometry of the placement of attachment points of elements is of particular importance for the protection of objects that perform rotating and oscillating motion. It is shown that mathematical models of systems that perform rotating and oscillating motions should be built taking into account the emerging lever linkages. Such linkages can be implemented with the help of levers of the first and second kinds.

There are analogies between the basic computational schemes of vibration protection systems, but questions about the adequacy of models require consideration of a number of specific features. In a rotating and oscillating type system, static and dynamic balancing modes can occur. There are also significant differences in the estimates of the effect of external disturbances of the force and kinematic types. In systems with several degrees of freedom, basic vibration protection systems can be divided into several groups: systems with elements that perform rectilinear translational motion, systems of rotational type, and combined systems. Each of the distinguished classes of systems has its own characteristics that must be considered when determining static and dynamic responses in the contacts of the elements of the systems between themselves and with the supporting surfaces. Based on the above, it is recommended to determine the static and dynamic responses at the considered points of vibration protection systems based on the use of structural diagrams and the determination of the corresponding transfer functions.

A methodological basis for the dynamic synthesis of vibration protection systems using the ideas about the possibilities of new types of additional constraints created by mechanical chains and mechanisms, in particular, has been proposed. The latter is due to the fact that in many cases, elastic and dissipative and mass-and-inertia constraints in real vibration protection systems take the form of special devices and mechanisms.

REFERENCES

Eliseev, A. V., Kuznetsov, N. K., & Moskovskikh, A. O. (2019). *The Dynamics of Machines: System Conception, Structural Schemes and Connections of Elements*. Innovative Mechanical Engineering.

Eliseev, S. V. (2018). *Applied System Analysis and Structural Mathematical Modeling (Dynamics of Transport and Technological Machines: Connectivity of Movements, Vibration Interactions, Lever Connections)*. Irkutsk State Transport University.

Eliseev, S. V., & Artyunin, A. I. (2016). *Applied Theory of Oscillations in Problems of Dynamics of Linear Mechanical Systems*. Science.

Eliseev, S.V. & Eliseev, A.V. (2020). Theory of Oscillations: Structural Mathematical Modeling in Problems of Dynamics of Technical Objects. *Studies in Systems, Decision and Control*, 252. Springer International Publishing.

Eliseev, S. V., Kovyrshin, S. V., & Parshuta, E. A. (2013a). Some questions of the theory of vibration isolation: Justification of structural approaches. *News of ISEA*, (3), 121–127.

Eliseev, S.V., Parshuta, E.A. & Bolshakov, R.S. (2012). On the possibilities of mechatronic approaches to the tasks of vibration protection of technical objects. *Vibration machines: measurement, reduction, protection, 4*(31), 46-49.

Eliseev, S. V., Parshuta, E. A., & Bolshakov, R. S. (2013b). Features of vibration protection system with the object of protection in the form of a solid: dynamic responses. In *Proceedings of the International Scientific Conference "Science in the Modern Information Society"* (Vol. 1, pp. 101-116). Moscow: Academic Publishing.

Eliseev, S. V., Parshuta, E. A., & Bolshakov, R. S. (2013c). Transfer functions of mechanical oscillatory systems: Possibilities for assessing reduced stiffness. *International Journal of Applied and Fundamental Research, 1*, 11–18.

Eliseev, S. V., Parshuta, E. A., & Kaimov, E. V. (2013d). Vibration protection systems: dynamic responses and methods for their determination. In *Proceedings of the International Scientific Conference "Engines 2013"* (pp. 279-284). Khabarovsk: Pacific National University Press.

Eliseev, S. V., Savchenko, A. A., Trofimov, A. N., Parshuta, E. A., & Artyunin, A. I. (2013). A Device for Damping. Patent RU 133232, IPC F16F7/10 F16F15/04. No. 2013105925/11, 12.02.2013; publ. 10.10.2013, Bull. No. 28.

Eliseev, S. V., Zasyadko, A. A., Karpukhin, E. L., Kukharenko, V. P., & Metzger, P. K. (1988). Software and algorithmic complex optimization design of complex vibration protection systems. In *Proceedings of the Sixth All-Union Conference on Control in Mechanical Systems* (pp. 61-63). Lviv: Institute of Applied Mechanics of the Academy of Sciences of the Ukrainian SSR.

Harris, C. M. (2002). *Shock and Vibration Handbook*. McGraw-Hill Book Co.

Khomenko, A. P., Eliseev, S. V., Artyunin, A. I., Eliseev, A. V., Bolshakov, R. S., & Kaimov, E. V. (2013a). Elastic elements in mechanical systems: structural interpretations. Irkutsk: Irkutsk State Technical University Press.

Khomenko, A. P., Eliseev, S. V., Artyunin, A. I., Parshuta, E. A., & Kaimov, E. V. (2013b). Mechanisms in elastic oscillatory systems: features of taking into account dynamic properties, problems of vibration protection of machines, devices and equipment. Irkutsk: Irkutsk State Technical University Press. .

Khomenko, A. P., Eliseev, S. V., & Yermoshenko, Yu. V. (2012). *System analysis and mathematical modeling in the mechatronics of vibration protection systems*. Irkutsk State Technical University Press.

Nasnikov, D. N., Parshuta, E. A., Trofimov, A. N., & Sorin, V. V. (2010). Features of dynamic interactions in systems with an extended set of typical elements. *Bulletin of the Irkutsk Regional Branch of the Academy of Sciences of the Higher School, 2*(17), 170–186.

Parshuta, E. A. (2013a). Determination of dynamic responses in mechanical oscillatory systems: structural approaches. In Information and Mathematical Technologies in Science and Management: *Proceedings of XVIII of the Baikal All-Russian Conference* (Vol. 1, pp. 216-221). Irkutsk: MESI SB RAS.

Parshuta, E. A. (2013b). Mechanisms in mechanical oscillatory systems as a form of introducing additional bonds. In Intellectual and Material Resources of Siberia: Materials of the Region Scientific-Practical Conference (pp. 42-48). Irkutsk.

Parshuta, E. A., & Gordeeva, A. A. (2010). Mathematical Modeling in Problems of Dynamic Oscillation Damping. Modern Technologies. System Analysis. *Modeling, 1*(25), 149–153.

Parshuta, E. A., & Kaimov, E. V. (2013). Possible forms of changing the dynamic state of vibration protection systems. *Metallurgical Processes and Equipment, 1*, 55–59.

Parshuta, E. A., Kaimov, E. V., & Mironov, A. S. (2012). Some Approaches to Dynamic Oscillation Damping. In *Technique and Technology: New Development Prospects: Proceedings of VII Int. Scientific-Practical Conference* (pp. 36-40). Moscow: Sputnik + Publishing.

Wolfson, I.I. (2008). *Oscillations in Machines: A Textbook.* St. Petersburg: SPbGUTID Publishing.

ADDITIONAL READING

Eliseev, A. V., & Eliseev, S. V. (2017). Dynamic damping of vibrations: Introduction of additional connections, lever interactions, and physical effects. *Vestnik Irkutskogo Gosudarstvennogo Tehniceskogo Universiteta, 21*(1), 10–23. doi:10.21285/1814-3520-2017-1-10-23

Eliseev, A. V., & Eliseev, S. V. (2017). Dynamic properties of dyads with heterogeneous partial systems. *Vestnik Irkutskogo Gosudarstvennogo Tehniceskogo Universiteta, 21*(5), 32–53. doi:10.21285/1814-3520-2017-5-32-53

Eliseev, A. V., Kashuba, V. B., & Kaimov, E. V. (2014). Measurement devices for fixing conditions of vibrational interaction with non-holding connections. Systems. Methods. *Technologies, 4*(24), 13–20.

Eliseev, A.V., Kuznetsov, N.K. & Eliseev, S.V. (2021). Dynamic properties of oscillatory structures: frequency functions and consistency of element movements. *Proceedings of Moscow Aviation Institute*, 120. Moscow Aviation Institute.

Eliseev, A. V., Kuznetsov, N. K., & Mironov, A. S. (2022). Generalized approaches to determining contact reactions in solids under static loads with non-holding connections: Multidimensional variant. Systems. Methods. *Technologies, 1*(54), 7–19.

Eliseev, A. V., Orlenko, A. I., & Sitov, I. S. (2017). Mathematical models of contact interactions during movement with detachment stages. Systems. Methods. *Technologies, 4*(36), 18–24.

Eliseev, A. V., Sitov, I. S., & Kuznetsov, N. K. (2022). System approach to assessing dynamic states of transport and technological objects: Structural schemes, transfer functions, dynamic invariants. Systems. Methods. *Technologies, 2*(54), 7–19.

Eliseev, A. V., Sitov, I. S., & Nguyen, D. H. (2016). Some issues in forming the structure of the vibration field in a vibration stand: Features of the measurement system, tuning tools. Systems. Methods. *Technologies, 2*(30), 17–26.

Jakóbczak, D. (2011). Curve Parameterization and Curvature via Method of Hurwitz-Radon Matrices. Image Processing & Communications-. *International Journal (Toronto, Ont.), 1-2*(16), 49–56.

Jakóbczak, D. (2011). Object Recognition via Contour Points Reconstruction Using Hurwitz-Radon Matrices. In J. Józefczyk & D. Orski (Eds.), *Knowledge-Based Intelligent System Advancements: Systemic and Cybernetic Approaches* (pp. 87–107). IGI Global. doi:10.4018/978-1-61692-811-7.ch005

Chapter 5
Lever Linkages in Mechanical Oscillatory Systems:
Forms of Lever Interactions

ABSTRACT

The chapter displays various forms and possibilities of lever interactions in the system of connections of typical elementary links. Elastic links in lever connections formed by devices for converting movements are considered. Systems with two and three lever connections of the second kind are considered. A new approach has been developed to assess the possibilities of sequential connection of typical elements in structural interpretations of vibration protection systems. A method for determining the reduced stiffness is presented, taking into account the introduction of devices with motion transformation. A generalized approach has been developed that takes into account the support of the lever on the link.

INTRODUCTION

As studies have shown, lever linkages in mechanical oscillatory systems are quite diverse and are essential. Levers can be considered as special units of oscillatory structures – they provide "scaling" of any signal, that is, they change the modulus of the input signal. On the other hand, not only the signal modulus can change at the same time, but also the direction of motion of the force application point the changes, that is, the direction changes, the sign of the transfer ratio changes. In specific tasks, this circumstance of transmission of interaction should be taken into account. And, finally, in planar levers, not only the signal module, but also its direction can change, that is, the lever converts the vector if the signal has such a shape. A distinctive feature of the lever is that in this unit there is no change in the physical nature of the signal, only its "geometric" transformation occurs. In this regard, determining the place of the lever in relation to the typical set of elementary units, it can be assumed that the lever linkage is a unit that implements the functions of the connection, thereby ensuring the structural commonality of the branch structure of the mechanical oscillatory system, including connection of partial blocks that implement their own kind of oscillatory motions. Consideration of lever interactions is associated with

DOI: 10.4018/978-1-6684-7237-8.ch005

another new circumstance, the essence of which is the emergence of the so-called connected motions. The latter is observed in the oscillations of solids having an elastic support on the base, but having a connection between them in the form of a kinematic pair of one or another class. Such a connection may be accompanied by other couplings in accordance with the capabilities of the extended typical set of elementary units. The use of lever linkages is often encountered in the tasks of ensuring a dynamic state in vibratory technological machines and in transport.

Consideration of the elemental base of modern vibration protection systems is associated not only with the implementation of the ideas of controlling vibrations with the help of active elements, but also with the introduction of elementary units in the structure of systems and detailing ideas about and means of interconnecting the units on the basis of lever linkages and mechanisms.

POSSIBILITIES AND FORMS OF LEVER INTERACTIONS IN THE SYSTEM OF CONNECTIONS OF TYPICAL ELEMENTARY UNITS

If we take the usual mechanical oscillatory system, then the expressions for the kinetic and potential energy usually have a complete quadratic form for the potential energy and an incomplete quadratic form for the expression determining the kinetic energy of the system. Choosing the appropriate coordinate systems, it is possible to carry out an inversion and obtain the complete quadratic form for expressing the kinetic energy, and incomplete form for the potential energy. The solution of the problem of simultaneous reduction of expressions to incomplete form means a transition to the main coordinates, which is possible only for some special cases. Expressions for the kinetic and potential energies simultaneously in a complete quadratic form are possible, however, this requires the introduction of additional constraints, as shown, for example, in Eliseev et al. (2006). Essentially, the possibility of reducing the expressions for kinetic and potential energies to full quadratic forms predetermines the fundamental possibilities of the "legitimacy" of introducing additional feedbacks and expanding the typical element base of the vibration protection systems.

Virtual Lever Interactions in the Structure and Dynamics of Mechanical Oscillatory Systems

If we assume that the basic models of vibration protection systems exist in the form of two variants, as shown in Figure 1, it makes sense to consider how expressions for kinetic and potential energies are formed in this case, taking into account the use of ideas about quadratic forms.

Let the kinetic and potential energy of a vibration protection systems have the form in a system of generalized coordinates y_1, y_2 (see Figure 1a)

$$\mathrm{T} = \frac{1}{2}(a_{11}\dot{y}_1^2 + 2a_{12}\dot{y}_1\dot{y}_2 + a_{22}\dot{y}_2^2) \tag{1}$$

$$\Pi = \frac{1}{2}(c_{11}y_1^2 + 2c_{12}y_1y_2 + a_{22}y_2^2) \tag{2}$$

Figure 1. Basic computational schemes of the vibration protection systems with two degrees of freedom: a) a chain model, b) a solid body on elastic supports

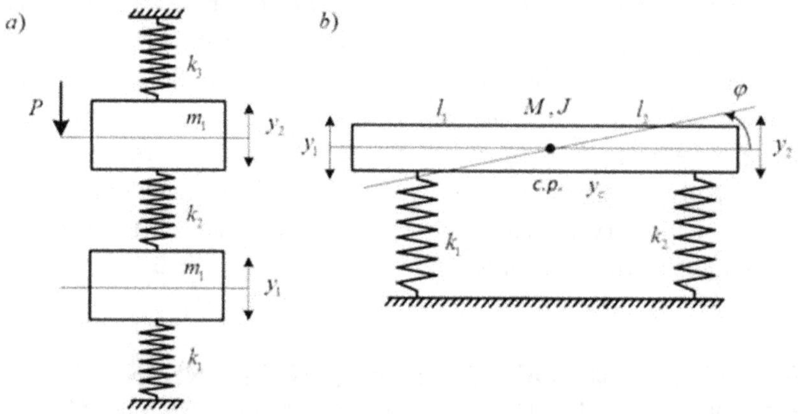

Consider the expression (1), assuming (Figure 1a) that:

$$T = \frac{1}{2}(m_1\dot{y}_1^2 + m_2\dot{y}_2^2);\ a_{11}=m_1,\ a_{22}=m_2,\ a_{12}=0 \tag{3}$$

Let us introduce into the computational scheme in Figure 1a an additional constraint of a form Lp^2 (Eliseev et al., 2008) between masses m_1 and m_2, and determine the values of the coefficients in expression (1). The expression (3) will change its form:

$$T = \frac{1}{2}m_1\dot{y}_1^2 + \frac{1}{2}L(\dot{y}_2 - \dot{y}_1)^2 + \frac{1}{2}m_2\dot{y}_2^2 \tag{4}$$

where L is the reduced moment of inertia of the motion transformation device.

Having made a number of transformations, we can write that:

$$T = \frac{1}{2}(m_1 + L)\dot{y}_1^2 + \frac{1}{2}(m_2 + L)\dot{y}_2^2 - L\dot{y}_1\dot{y}_2 \tag{5}$$

$$\Pi = \frac{1}{2}(k_1y_1^2 + k_2(y_2 - y_1) + k_2y_2^2) \tag{5$'$}$$

$$a_{11}=m_1+L,\ a_{22}=m_2+L,\ a_{12}=-L; \tag{6}$$

$$c_{11}=k_1+k_2,\ c_{12}=-k_2,\ c_{22}=k_2+k_3. \tag{6$'$}$$

Figure 2. The computational scheme of the basic model (see Figure 1a) with the introduction of an additional constraint in the form of a motion transformation device (a) and the dynamically equivalent structural scheme of ACS (b)

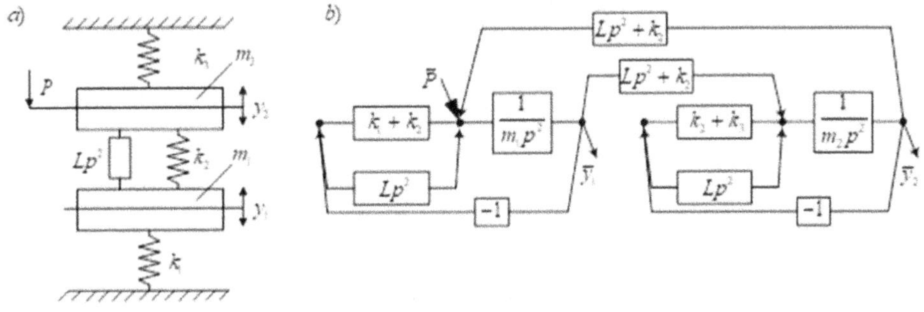

The computational scheme of the basic model in the form of a chain system (see Figure 1a) is converted in this case into a mechanical oscillatory system, as shown in Figure 2a and b.

Note that the expression for the kinetic energy can be reduced to the form (1), in which a_{12} ¹ 0, can be carried out, but in this case the corresponding additional constraint should be introduced. At the same time, the transformation of cross-couplings (see Figure 1b), which takes an elastic-inertial character (and not a purely elastic one), occurs.

Consider, in connection with the opening possibilities, ways of changing the structure of expressions for kinetic energy due to the introduction of an additional constraint, and other approaches that are associated with certain algebraic transformations of the original expressions (1), (2) (Genkin & Ryaboy, 1988).

We introduce the entry for potential energy in the form of:

$$2\Pi = (c_{11} + c_{12})y_1^2 - c_{12}(y_2 - y_1) + (c_{22} + c_{12})y_2^2 \quad (7)$$

that corresponds to the dynamic scheme in Figure 3.

The stiffnesses of the elastic elements (from the expression (7)) have the form:

Figure 3. The computational scheme of the vibration protection system taking into account the transformations of T and Π

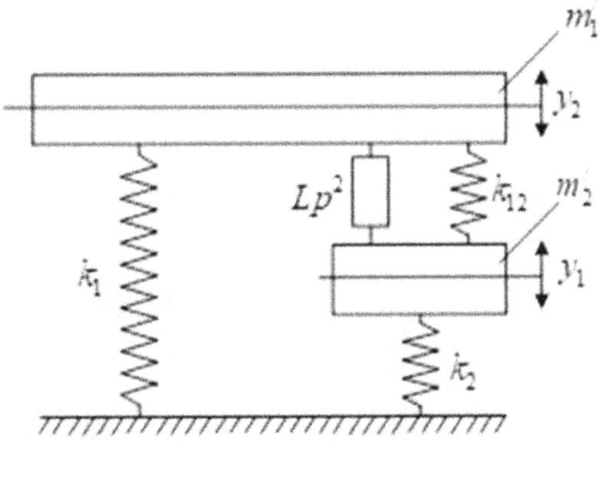

$$k_1=c_{11}+c_{12};\ k_2=c_{22}+c_{12};\ k_{12}=-c_{12}. \tag{8}$$

The construction of such a scheme (Figure 3) is not possible for any positive definite stiffnesses matrix k, since the inequalities must be satisfied

$$c_{11}+c_{12}\geq 0;\ c_{22}+c_{12}\geq 0;\ c_{12}\leq 0. \tag{8'}$$

In particular, for the mechanical system in Figure 1, where the coefficients of the quadratic form of the potential energy are determined by the expression (6′), which corresponds to the conditions (8′). In turn, for the kinetic energy of the system with the introduction of an additional constraint Lp^2, it can be noted that the expression for the kinetic energy is not associated in this case with the expression for the potential energy. Thus, for a system with additional constraint, we have:

$$m_1' = a_{11} + a_{12}; m_2' = a_{12} + a_{22}; a_{12} = -L. \tag{9}$$

At the same time, conditions (8′) are satisfied:

$$m_1' = m_1 + L - L = m_1 \geq 0;$$
$$m_2' = m_2 + L - L = m_2 \geq 0;$$
$$a_{12} = -L \leq 0.$$

By appropriate transformations, the potential energy can be decomposed into a sum of squares of the form:

$$2\Pi = \left(\frac{c_{12}y_1 + c_2 y_2}{c_{22}}\right)^2 + cy_1^{\ 2}, \tag{10}$$

where

$$c = \frac{c_{11}c_{22} - c_{12}^{\ 2}}{c_{22}}, \tag{11}$$

which leads to the scheme shown in Figure 4. Note that T has the original value (i.e. $a_{12}=0$), and the expression for the kinetic energy does not change.

In this case:

$$k_1 = c; k_2 = \frac{(c_{12} + c_{22})^2}{c_{22}}; \frac{l}{r} = \frac{c_{22}}{c_{12}}. \tag{12}$$

Figure 4. Changing the type of the computational scheme of vibration protection systems in the transition to another form of Π (expression 10)

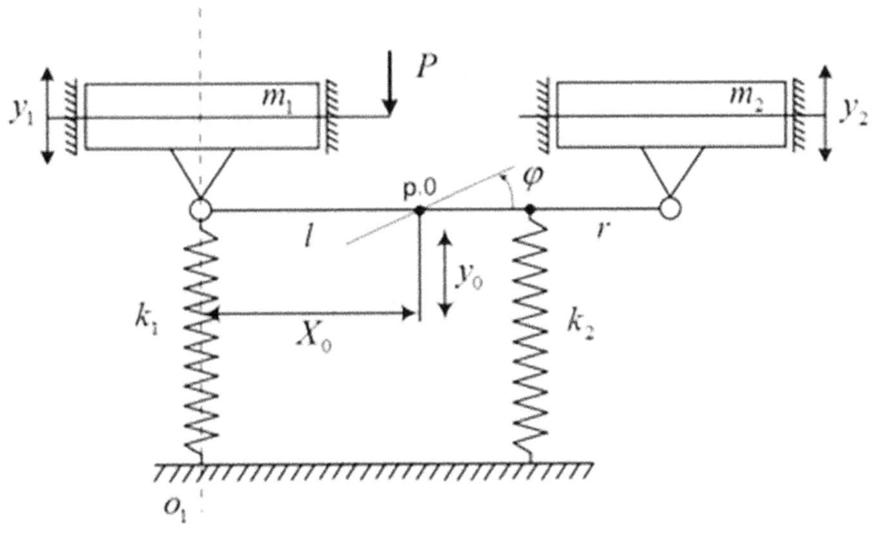

The physical meaning of the expression k_1=c is that the potential energy of the system in Figure 4, can be defined in two stages. At the first stage, the reduced stiffness in the system is determined with m_2=0, that can be written in the form:

$$k_{red} = k_1 + \frac{k_2 k_3}{k_2 + k_3} = \frac{k_1 k_2 + k_1 k_3 + k_2 k_3}{k_2 + k_3} \tag{13}$$

which corresponds to the expression (11).

Expression (5.11) reflects the potential energy of elastic elements when introducing a “virtual lever” of the first kind with arms c_{22}=l and c_{12}=r (Genkin & Ryaboy, 1988).

The introduction of a “virtual lever” requires compliance with certain conventions in the consideration of the motion of the system. This is due to the fact that the “virtual lever” does not introduce spatial representations into the system, which is connected, in turn, with the need to introduce the concept of the center of masses, etc. However, this approach (Belokobylsky et al., 2013) is very symbolic, because

Figure 5. The computational scheme of the vibration protection systems with a virtual lever having a fixed point of support

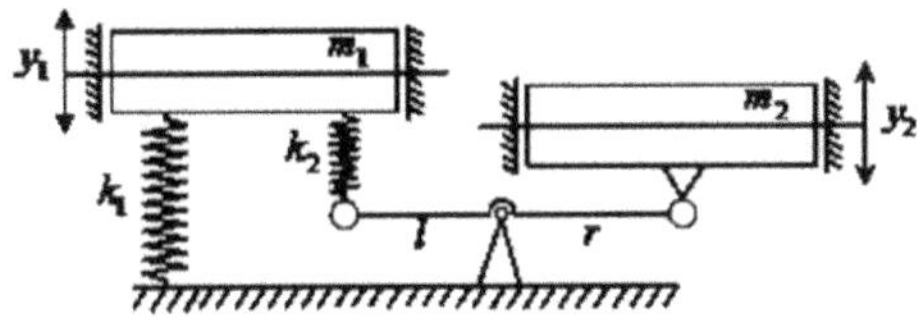

Figure 6. The computational scheme of the vibration protection systems with the introduction of the elastic element k_2 *between the lever and mass* m_2

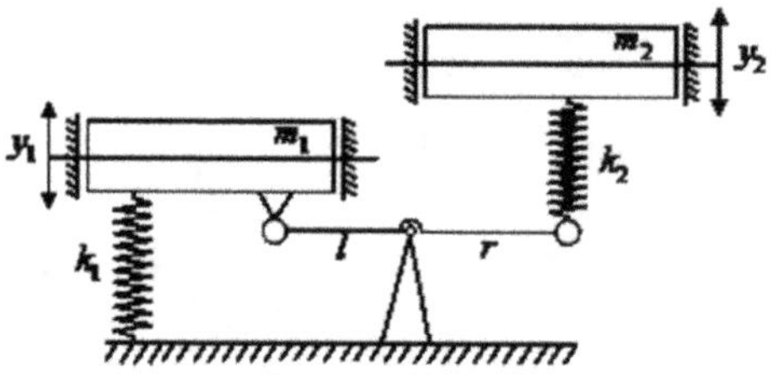

it raises the question of the existence of lever linkages in mechanical oscillatory systems, which has been developed in recent years (Eliseev et al., 2009; Eliseev & Upir, 2008).

In the diagram in Figure 5, unlike expression (12), $k_2 = c_{12}^2 / c_{22}$.

In the computational scheme, shown in Figure 6, $k_2 = c_{22}$. Computational schemes in Figures 4-6 have a physical meaning for any positive-definite matrix k, except in the case $c_{22} = -c_{12}$.

For all cases, $l/r<0$ means a different arrangement of attachment points m_1 and m_2 both in relation to the lever and points of support: the points of attachment of masses are on one side of the point of support (lever of the second kind).

Taking into account the real properties of the lever mechanism leads to completely different results, which can be shown using traditional approaches. For the computational scheme in Figure 4, we find the position of the center of masses, assuming that the lever has linear dimensions:

$$X_0 = \frac{m_2 \cdot (l+r)}{m_1 + m_3} \tag{14}$$

$$Y_0 = \frac{y_2 \cdot x_0 + y_1 \left(l + r - x_0\right)}{l + r} \tag{15}$$

Accepting that $a = X_0/(l+r)$, $b = 1-a$, we write:

$$y_0 = y_2 a + y_1 b \tag{16}$$

In addition, we introduce some relations:

$$y_1 = y_0 - X_0 \varphi, y_2 = y_0 + \left(l + r - X_0\right), \varphi = \frac{y_1 - y}{l + r} = c\left(y_1 - y_2\right) \tag{17}$$

where $c = 1/(l+r)$.

The expressions for the kinetic and potential energy in this case take the form of:

$$T=\frac{1}{2}(m_1+m_2)\dot{y}_0^2+\frac{1}{2}m_1X_0^2\dot{\varphi}^2+\frac{1}{2}m_2r^2\varphi^2, \tag{18}$$

$$\Pi=\frac{1}{2}k_1y_1^2+\frac{1}{2}k_2\left(y_2'\right)^2, \tag{19}$$

where

$$y_2'=y_0+(l-X_0)\varphi \tag{20}$$

We transform (18), (19) to the form:

$$T=\frac{1}{2}(M)\left(\dot{y}_2^2a^2+2\dot{y}_2\dot{y}_1ab+\dot{y}_1^2b^2\right)+\frac{1}{2}m_1X_0^2c^2\left(\dot{y}_1-\dot{y}_2\right)+\frac{1}{2}m_2\left(l+r-X_0\right)^2c^2\left(\dot{y}_1-\dot{y}_2\right), \tag{21}$$

$$\Pi=\frac{1}{2}k_1y_1^2+\frac{1}{2}k_2\left(y_1^2a^2+2y_1y_2ab+y_2^2b^2\right). \tag{22}$$

Introducing $a=l-X_0$, $M=m_1+m_2$ and making the usual transformations, we find the differential equations of motion.

$$\begin{gathered}\ddot{y}_1\left[Mb^2+m_1X_0c^2+m_2(l+r-X_0)^2c^2\right]+k_1y_1+k_2y_1(b+a_1c)^2+\\+y_2'\left[Mab-m_1X_0c^2-m_2(l+r-X_0)^2c^2\right]-y_2k_2(a-a_1c)(b+a_1c)=P,\end{gathered} \tag{23}$$

$$\begin{gathered}\ddot{y}_2\left[Ma^2+m_1X_0c^2+m_2(l+r-X_0)^2c^2\right]+k_2y_2(a-a_1c)^2+\\+\ddot{y}_1\left[Mab-m_1X_0c^2-m_2(l+r-X_0)^2c^2\right]-y_1k_2(b+a_1c)(a-a_1c)=0.\end{gathered} \tag{24}$$

The structural diagram of the system corresponding to the computational scheme in Figure 4, it is resulted in Figure 7.

$$X_0=\frac{m_2(l+r)}{m_1+m_2};M=m_1+m_2;c=\frac{1}{l+r};a=\frac{X_0}{l+r};a=1-a;a_1=l-x_0$$

From consideration of the structural diagram of the system in Figure 7, it can be concluded that elastic-inertial constraint emerge in the system, which can "vanish" at certain frequencies, as for the expressions for kinetic (21) and potential (22) energies, they have a complete quadratic form.

Figure 7. The structural diagram of the system in Figure 4 with a lever with arms l and r

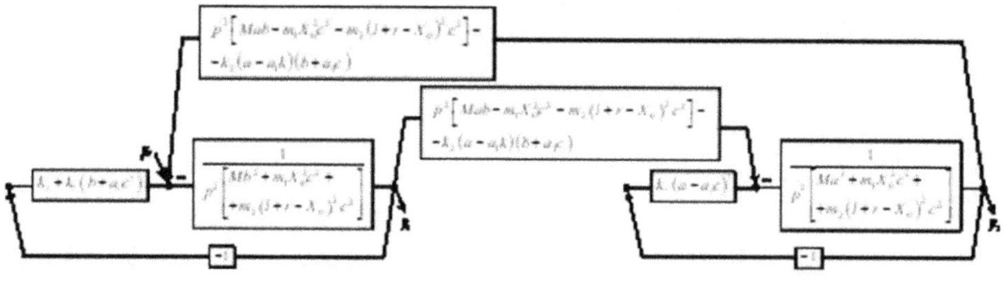

Thus, a detailed consideration of the structure of mechanical oscillatory systems is associated with the use of ideas about the properties manifested in lever interactions. When considering oscillations in mechanical systems, it is quite possible to introduce the concept of "virtual levers or virtual lever mechanisms", however, the concept of virtuality itself suggests the possibility of the emergence of "lever linkages" in dynamic interactions of elements of mechanical systems. Such linkages may have physical form, but they may not have (Genkin & Ryaboy, 1988). In the case when lever mechanisms or levers of the first or second kind are used in a mechanical oscillatory system, the dynamic state of a mechanical oscillatory system is described by a completely different system of differential equations, which requires different physical representations and is associated with the study of other features of the dynamic system states (Eliseev & Khomenko, 2009; Ermoshenko & Fomina, 2009).

Lever interactions are implemented through the inclusion in the oscillatory structures of the levers of the first and second kind. These levers differ from each other in that in the levers of the first kind the input and output signals have the same direction, although they may be different in modulus. In the lever of the second kind, the direction of the signals under the same conditions will be different (Belokobylsky et al., 2013). When considering it is assumed that the lever is made in the form of a rectilinear rod, although in practice there are levers, the body of which can be a flat figure. Examples of the use of lever linkages in the structures and suspensions of rolling stock are given in the computational schemes in (Upir, 2009; Ivanov, 2007).

The Elastic Link in Leverages With the Motion Transformation Device

The capabilities of leverages have been considered in a number of papers, for example (Eliseev et al., 2009; Eliseev & Upir, 2008). However, these studies did not include an assessment of the change in dynamic properties with the introduction of new types of typical constraints, which in some cases may be of independent interest.

1. Consider a mechanical oscillatory system with a lever of the second kind and a kinematic perturbation at p. *A* and p. *B*, as shown in Figure 8a, where L_1 and L_2 are the mass-and-inertia characteristics of devices for converting relative motion (assuming small values of the masses of these devices). Figure 8b shows a system of the same type, but with a lever of the first kind.

Figure 8. The computational scheme of the vibration protection systems with lever linkages of the second kind (a) and the first kind (b)

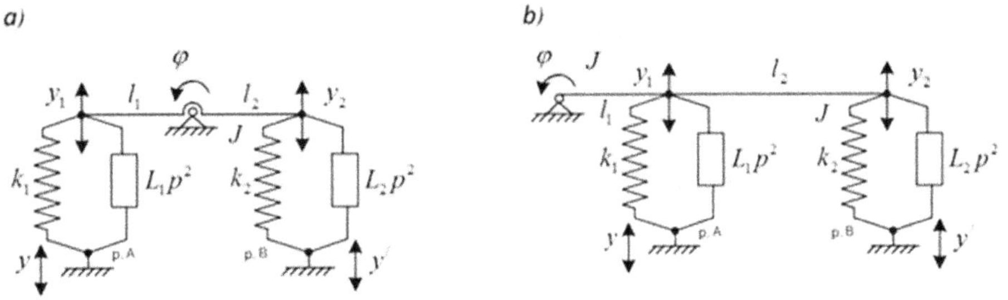

The kinetic and potential energy of the system can be written as:

$$\mathrm{T}=\frac{1}{2}J\dot{\varphi}^2+\frac{1}{2}L_1\left(\dot{y}_1-\dot{y}\right)^2+\frac{1}{2}L_2\left(\dot{y}_2-\dot{y}'\right)^2, \tag{25}$$

$$\Pi=\frac{1}{2}k_1\left(y_1-y\right)^2+\frac{1}{2}k_2\left(y_2-y'\right)^2. \tag{26}$$

Let us take a number of obvious correlations $y_1=\phi\bullet l1_;\ y2_=-\phi\bullet l2$, that is characteristic for the lever of the second kind. Then (25) and (26) can be represented as:

$$\mathrm{T}=\frac{1}{2}J\dot{\varphi}^2+\frac{1}{2}L_1\left(\dot{\varphi}l_1-\dot{y}\right)^2+\frac{1}{2}L_2\left(\dot{\varphi}l_2+\dot{y}'\right)^2; \tag{27}$$

$$\Pi=\frac{1}{2}k_1\left(\varphi l_1-y\right)^2+\frac{1}{2}k_2\left(\varphi l_2+y'\right)^2. \tag{28}$$

Making the usual transformations, where y and y' are the corresponding ki-nematic perturbations at the points A and B, we get:

$$\frac{\partial T}{\partial\dot{\varphi}}=J\dot{\varphi}+L_1l_1^2\dot{\varphi}-L_1l_1\dot{y}+L_2l_2^2\dot{\varphi}+L_2l_2\dot{y}';\frac{\partial\Pi}{\partial\varphi}=k_1l_1^2\varphi-k_1l_1y+k_2l_2^2\varphi+k_2l_2y'.$$

from which we find that:

$$J\ddot{\varphi}+\varphi\left(k_1l_1^2+k_2l_2^2\right)+L_1l_1^2\ddot{\varphi}+L_2l_2^2\ddot{\varphi}-L_1l_1\ddot{y}+L_2l_2\ddot{y}'=k_1l_1y-k_2l_2y' \quad (29)$$

We will accept that $y = y'$ and obtain the transfer function "angle of rotation φ by offset y":

$$W(p)=\frac{\bar{\varphi}}{\bar{y}}=\frac{l_1\left(k_1+L_1p^2\right)-l_2\left(k_2+L_2p^2\right)}{\left(J+L_1l_1^2+L_2l_2^2\right)p^2+k_1l_1^2+k_2l_2^2}. \quad (30)$$

From the expression (30) it is possible to determine the frequency of the system's natural oscillations:

$$\omega_{\text{nat}}^2=\frac{k_1l_1^2+k_2l_2^2}{J+L_1l_1^2+L_2l_2^2}=\frac{k_1+k_2i^2}{\frac{J}{l_1^2}+L_1+L_2i^2}, \quad (31)$$

as well as the frequency of dynamic absorbing:

$$\omega_{dyn}^2=\frac{k_1l_1-k_2l_2}{L_1l_1-L_1l_1}=\frac{k_1-k_2i}{L_1-L_1i} \quad (32)$$

which makes it possible to compare the system shown in Figure 8b with the usual system. Note that in a conventional system, the mode of dynamic oscillation suppression always occurs at a frequency less than the resonance frequency.

Ratio of frequencies *R* is determined by the expression:

$$R=\frac{\left(k_1+k_2i^2\right)\left(L_1-L_2i\right)}{\left(\frac{J}{l_1^2}+L_1+L_2i^2\right)\left(k_1-k_2i\right)}, \quad (33)$$

where $i=\frac{i_2}{i_1}$ (the sign is taken into account when forming expressions (27), (28)).

If *i*=1, then:

$$R=\frac{\left(k_1+k_2\right)\left(L_1-L_2\right)}{\left(\frac{J}{l_1^2}+L_1+L_2\right)\left(k_1-k_2\right)}. \quad (34)$$

The study of (33) shows that the mode of dynamic oscillation damping is observed in the resonance region, and the structural diagram of the *vibration protection system* has the form, as shown in Figure 9a.

Figure 9. The structural diagram: a) of the vibration protection systems corresponding to the computational scheme in Figure 8a; b) structural diagram for the computational scheme in Figure 8b

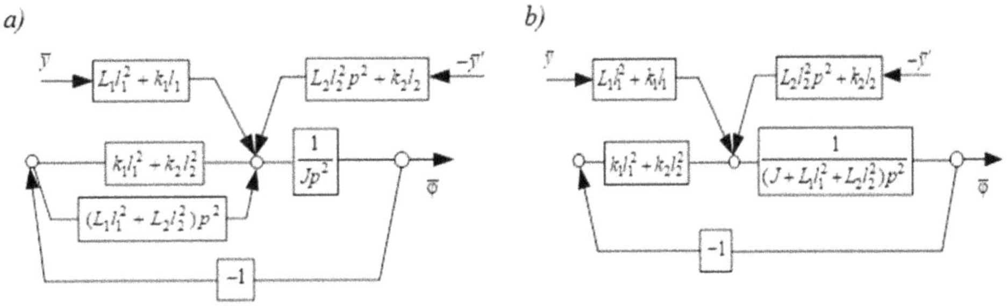

The transfer function of the system when has the form:

$$W(p) = \frac{\bar{\varphi}}{\bar{y}} = \frac{p^2\left(L_1l_1 - L_2l_2\right) + k_1l_1 - k_2l_2}{\left(J + L_1l_1^2 + L_2l_2^2\right)p^2 + k_1l_1^2 + k_2l_2^2}, \tag{35}$$

which makes it possible to assess the nature of the influence of the second kind of leverages on the dynamic properties of a *vibration protection system* of this type, as compared with conventional systems that execute rectilinear rather than swinging motions.

Consider a system with similar parameters and notation, shown in Figure 8a, but with a lever of the first kind, as shown in Figure 8b. The kinetic and potential energies of the *vibration protection systems* are also determined by expressions (25), (26), but in them y_1=ϕ•l1 y2_ϕ•l2. After appropriate transformations, we obtain the system of differential equations of motion in the form of:

$$\ddot{\varphi}\left(J + L_1l_1^2 + L_2l_2^2\right) + \varphi\left(k_1l_1^2 + k_2l_2^2\right) = L_1l_1\ddot{y} + L_2l_2\ddot{y}' + k_1l_1y + k_2l_2y'. \tag{35'}$$

The structural diagram of the equivalent automatic control system is shown in Figure 8b. The difference with respect to the diagram in Figure 8a is that the external influence has a sign (–).

The transfer function of the system in this case for $y = y'$ will take the form:

$$W(p) = \frac{\bar{\varphi}}{\bar{y}} = \frac{p^2\left(L_1l_1 + L_2l_2\right) + k_1l_1 + k_2l_2}{\left(J + L_1l_1^2 + L_2l_2^2\right)p^2 + k_1l_1^2 + k_2l_2^2}. \tag{35''}$$

If you introduce transfer ratio $i = \frac{i_2}{i_1}$, you can get an expression to determine the natural frequency of the system, similar to (31):

$$\omega^2_{nat} = \frac{k_1 + k_2 i^2}{\frac{J}{l_1^2} + L_1 + L_2 i^2}$$

and also for dynamic absorbing frequency –

$$\omega^2_{dyn} = \frac{k_1 + k_2 i}{L_1 + L_1 i} \tag{35'''}$$

which will be different from (32).

Frequency ratio is correspondingly:

$$R = \frac{\left(k_1 + k_2 i^2\right)\left(L_1 + L_2 i\right)}{\left(\frac{J}{l_1^2} + L_1 + L_2 i^2\right)\left(k_1 + k_2 i^2\right)}, \tag{35^{IV}}$$

whence it follows that R will always be less than one, in contrast to expression (33). A comparison of the two systems that incorporate the levers of the second and first kinds gives the following information:

1. In both systems, the natural frequency will be the same, as well as the frequency equations of the system. However, the latter cannot be asserted with respect to systems having a different configuration, because the transfer ratios of levers of the second and first kind are fundamentally different in signs. For the lever of the second kind, the transfer ratio has sign (–), and the expression (32) can be obtained accordingly from (5.35'') by introducing the sign (–) for the transfer ratio *i*.
2. The differences are that in the systems there will be different frequencies of dynamic absorbing and different dependencies of these frequencies on the parameters of the system.
   For levers of the second kind, if the condition $k_1=k_2 i$ is satisfied, the dynamic mode will be absent (provided that the friction forces are small), which corresponds to the value $R\to 0$.
   Such a combination of parameters changes the scheme of interaction of forces, as it follows from the transfer function (35) and changes the dynamic properties of the system as a whole. Similarly, when $L_1=L_2 i$ combined, the frequency ratio $R\to 0$ and the system works without implementing the dynamic absorbing mode.
3. A system with a lever of the second kind, has a mode in which, when combining parameters $k_1=k_2 i$ and $L_1=L_2 i$, a kind of "self-jamming" of the system is possible. In this case, an object with a mass-and-inertia parameter J will react to external disturbances y and $y'(y = y')$ in a different way than is presented from classical positions.
4. Since the ratio R varies within limits $0\leq R\leq\infty$, this makes it possible to obtain corresponding characteristics with dynamic absorbing frequencies before and after resonance modes. With $R=1$, as already noted, the emergence of a specific mode is possible, and the uncertainty 0/0 requires additional consideration.

5. With levers of the second and first kind, both types of vibration protection system are "locked" at high frequencies, but the locking levels will be different.
6. A system with a lever of the first kind behaves like a normal system with one degree of freedom, for which $K_{red}=k_1=k_2 i$, $L_{red}=L_1=L_2 i$ can be determined, with the corresponding dynamic properties for such systems.
   Summarizing such a view, it would also be noted that in a system with a lever of the first kind, essentially two branches of force interaction are parallel, therefore, it becomes possible to determine the reduced values of the corresponding parameters. At the same time, the lever of the second kind introduces into the system a different scheme for building power interaction.
7. The noted differences in the introduction of levers of the first and second kind emerge in schemes with kinematic perturbation. With force perturbations, the levers of the first kind and the second kind (this follows from the consideration of the transfer functions) will have the same frequency properties, although they will have different design versions of the system.

Leverages in the Transmission of Mechanical Effects

In mechanical oscillatory systems, lever linkages can have a noticeable effect on the expansion of the possibilities to change, in a certain way, the dynamic state of the system, which is of interest in approaches oriented to the use of structural interpretations (Eliseev & Eliseev, 2020). Consider the simplest oscillatory systems with levers of the second and first kind.

Levers of the Second Kind. We will write for the computational scheme presented in Figure 10, the coefficient of the reduced stiffness of the system:

$$K_p = \frac{K_p^1 \cdot K_p^2}{K_p^1 + K_p^2}, \quad (36)$$

where

$$K_p^1 = k_1 \quad (37)$$

Figure 10. The computational scheme with a lever of the second kind

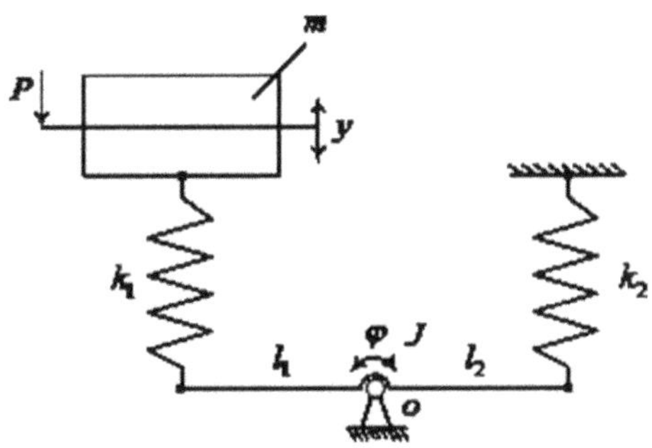

$$K_p^2 = k_2 \cdot \frac{l_2^2}{l_1^2} \tag{38}$$

We make the substitution of expressions (37) and (38) into (36), and after the transformations we get:

$$K_p = \frac{k_1 k_2 \cdot \frac{l_2^2}{l_1^2}}{k_1 + k_2 \cdot \frac{l_2^2}{l_1^2}}. \tag{39}$$

Let us prove the validity of expressions (36)–(39). Let us write the equations of kinetic and potential energies for the scheme shown in Figure 10:

$$\mathrm{T} = \frac{1}{2} m\dot{y}^2 + \frac{1}{2} J\dot{\varphi}^2, \Pi = \frac{1}{2} k_1 \left(y - \varphi \cdot l_1\right)^2 + \frac{1}{2} k_2 l_2^2 \varphi^2. \tag{40}$$

Using the Lagrange II equation in the absence of friction, we find the differential equations of motion of this system:

$$m\ddot{y} + k_1 y - k_1 l_1 \varphi = 0 \tag{41}$$

$$J\ddot{\varphi} - k_1 l_1 y + \varphi\left(k_1 l_1^2 + k_2 l_2^2\right) = 0 \tag{42}$$

The structural diagram of a dynamically equivalent automatic control system (ACS) for the computational scheme shown in Figure 10, has the form, as shown in Figure 11.

Figure 11. The structural diagram of the system (in Figure 10) with a lever linkage of the second kind

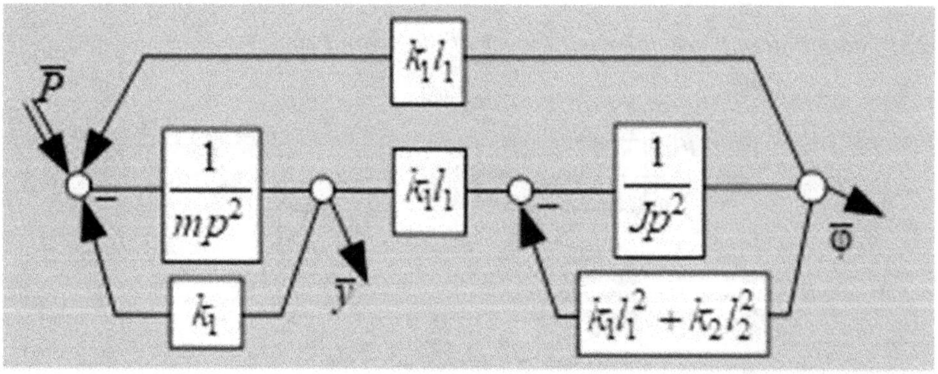

Another type of structural interpretation in the form of a dual mechanical chain is presented in Figure 12.

Figure 12. The mechanical chain of a system with a lever linkage of the second kind

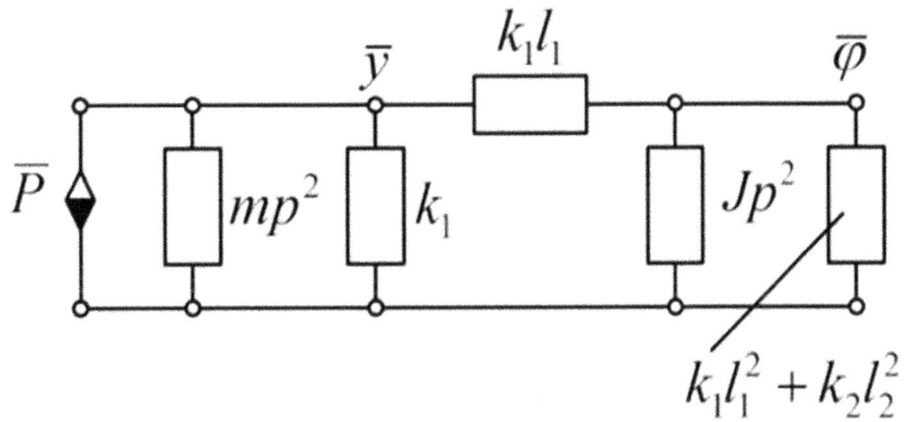

If we accept that $J\ddot{\varphi} = 0$, then from equation (42) we can find:

$$\varphi = y \frac{k_1 l_1}{k_1 l_1^2 + k_2 l_2^2}. \tag{43}$$

We substitute (43) into (41) and after transformation we get:

$$m\ddot{y} + K_p \cdot y = 0 \tag{44}$$

where

$$K_p = \frac{k_1 k_2 l_2^2}{k_1 l_1^2 + k_2 l_2^2}. \tag{45}$$

As you can see, expressions (39) and (44) are the same. Having assumed that $i = -\frac{l_2}{l_1}$, we get the final expression for the reduced stiffness coefficient:

$$K_p = \frac{k_1 k_2 i^2}{k_1 + k_2 i^2}. \tag{46}$$

A System With Two Levers of the Second Kind

Consider the system of levers presented in Figure 13.

Figure 13. The computational scheme with two levers of the second kind

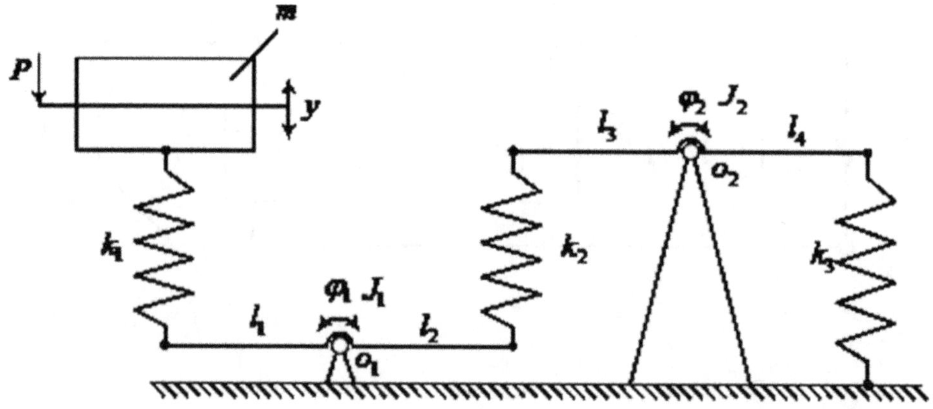

From which the coefficient of the reduced stiffness of the system is determined by the formula:

$$K_p = \frac{\left(\dfrac{K_p^1 \cdot K_p^2}{K_p^1 + K_p^2}\right) \cdot K_p^3}{\left(\dfrac{K_p^1 \cdot K_p^2}{K_p^1 + K_p^2}\right) + K_p^3}, \tag{47}$$

where

$$K_p^1 = k_1 \tag{48}$$

$$K_p^2 = k_2 \cdot \frac{l_2^2}{l_1^2} \tag{49}$$

$$K_p^3 = k_3 \cdot \frac{l_2^2}{l_1^2} \cdot \frac{l_4^2}{l_3^2} \tag{50}$$

We substitute expressions (48)–(40) into (47), and after some transformations we get:

$$K_p = \frac{k_1 k_2 k_3 \cdot \frac{l_2^2}{l_1^2} \cdot \frac{l_4^2}{l_3^2}}{k_1\left(k_2 + k_3 \cdot \frac{l_4^2}{l_3^2}\right) + k_2 k_3 \cdot \frac{l_2^2}{l_1^2} \cdot \frac{l_4^2}{l_3^2}}. \quad (51)$$

Let us prove the validity of expressions (47)–(51). Let us write the equations of kinetic and potential energies for the scheme shown in Figure 13:

$$\begin{aligned} \mathrm{T} &= \frac{1}{2} m \dot{y}^2 + \frac{1}{2} J_1 \dot{\varphi}_1^2 + \frac{1}{2} J_2 \dot{\varphi}_2^2, \\ \Pi &= \frac{1}{2} k_1 \left(y - \varphi_1 \cdot l_1\right)^2 + \frac{1}{2} k_2 \left(\varphi_2 \cdot l_3 - \varphi_1 \cdot l_2\right)^2 + \frac{1}{2} k_3 l_4^2 \varphi_2^2. \end{aligned} \quad (52)$$

Using the Lagrange equation of the second kind (with an insignificant effect of friction forces), we write the differential equations of motion of this system:

$$m\ddot{y} + k_1 y - k_1 l_1 \varphi_1 = 0 \quad (53)$$

$$J_1 \ddot{\varphi}_1 - k_1 l_1 y + \varphi_1 \left(k_1 l_1^2 + k_2 l_2^2\right) - k_2 l_2 l_3 \varphi_2 = 0 \quad (54)$$

$$J_2 \ddot{\varphi}_2 + \varphi_2 \left(k_2 l_3^2 + k_3 l_4^2\right) - k_2 l_2 l_3 \varphi_1 = 0 \quad (55)$$

The structural diagram for the system shown in Figure 13, will take the form as shown in Figure 14.

Figure 14. The structural diagram of the system containing two first-kind lever linkages

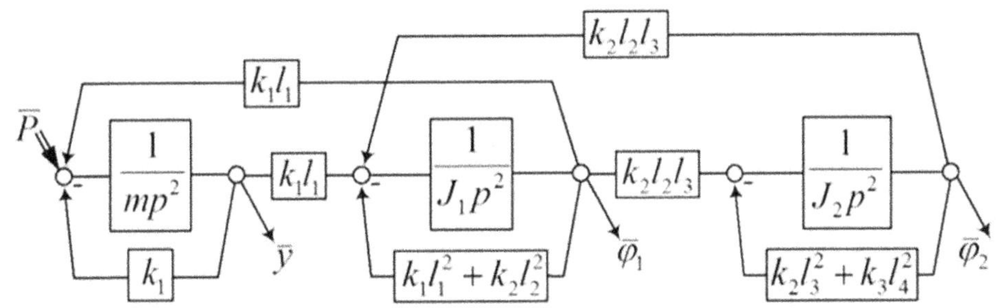

Figure 15. The mechanical chain of the system containing second-kind lever linkages

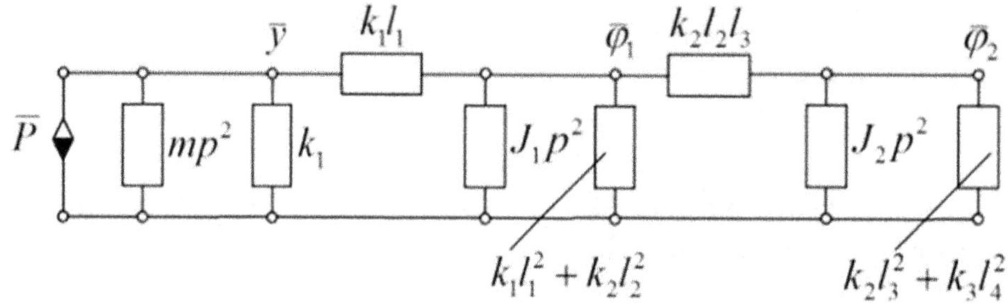

In this case, the structural interpretation of the same system in the form of a dual mechanical chain can be represented in Figure 15. Such schemes can be used for research based on the methods of electromechanical analogies (Druzhinsky, 1977; Eliseev et al., 2019).

If we suppose the moments of inertia of the levers are small, then $J_1\ddot{\phi}_1=0$, $J_2\ddot{\phi}_2=0$, then from equation (55) we express $\phi 2$.

$$\varphi_2=\varphi_1\frac{k_2l_2l_3}{k_2l_3^2+k_3l_4^2} \tag{56}$$

Substitute (56) into (54) and find $\phi 1$ –

$$\varphi_1=y\frac{k_1l_1\left(k_2l_3^2+k_3l_4^2\right)}{k_1l_1\left(k_2l_3^2+k_3l_4^2\right)+k_2k_3l_2^2l_4^2}. \tag{57}$$

We substitute (43) into (53) and after the transformation we get:

$$m\ddot{y}+K_p\cdot y=0 \tag{58}$$

where

$$k_p=\frac{k_2k_3\cdot\frac{l_2^2}{l_1^2}\cdot\frac{l_4^2}{l_3^2}}{k_1\left(k_2+k_3\cdot\frac{l_4^2}{l_3^2}\right)+k_2k_3\cdot\frac{l_2^2}{l_1^2}\cdot\frac{l_4^2}{l_3^2}} \tag{59}$$

As we see, expressions (40) and (59) coincide, which was to be proved. Having accepted $i_1 = -\frac{l_2}{l_1}$ and $i_2 = -\frac{l_4}{l_3}$, we obtain the final expression for the reduced stiffness coefficient of the system:

$$K_p = \frac{k_1 k_2 k_3 \cdot i_1^2 \cdot i_2^2}{k_1\left(k_2 + k_3 \cdot i_2^2\right) + k_2 k_3 \cdot i_1^2 \cdot i_2^2} \quad (60)$$

A System With Three Lever Linkages of the Second Kind

Complicating the task, we turn to the system presented in Figure 16.

Figure 16. The computational scheme with three levers of the second kind

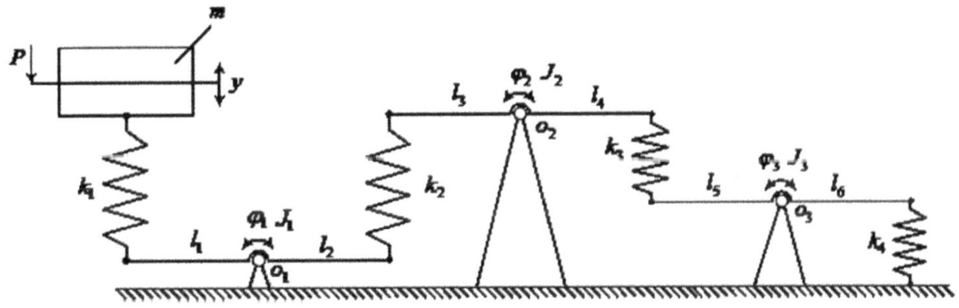

We assume that for the scheme presented in Figure 16, the coefficient of the reduced stiffness of the system is determined by the formula:

$$K_p = \frac{\left(\frac{\left(\frac{K_p^1 \cdot K_p^2}{K_p^1 + K_p^2}\right) \cdot K_p^3}{\left(\frac{K_p^1 \cdot K_p^2}{K_p^1 + K_p^2}\right) + K_p^3}\right) \cdot K_p^4}{\left(\frac{\left(\frac{K_p^1 \cdot K_p^2}{K_p^1 + K_p^2}\right) \cdot K_p^3}{\left(\frac{K_p^1 \cdot K_p^2}{K_p^1 + K_p^2}\right) + K_p^3}\right) + K_p^4} \quad (61)$$

where

$$K_p^1 = k_1 \tag{62}$$

$$K_p^2 = k_2 \cdot \frac{l_2^2}{l_1^2} \tag{63}$$

$$K_p^3 = k_3 \cdot \frac{l_2^2}{l_1^2} \cdot \frac{l_4^2}{l_3^2} \tag{64}$$

$$K_p^3 = k_3 \cdot \frac{l_2^2}{l_1^2} \cdot \frac{l_4^2}{l_3^2} \cdot \frac{l_6^2}{l_5^2} \tag{65}$$

Make the substitution of expressions (62–63) into (61) and after some transformations we get:

$$K_p = \frac{k_1 k_2 k_3 k_4 \cdot \frac{l_2^2}{l_1^2} \cdot \frac{l_4^2}{l_3^2} \cdot \frac{l_6^2}{l_5^2}}{k_1 \left[k_2 \left(k_3 + k_4 \cdot \frac{l_6^2}{l_5^2} \right) + k_3 k_4 \cdot \frac{l_4^2}{l_3^2} \cdot \frac{l_6^2}{l_5^2} \right] + k_2 k_3 k_4 \cdot \frac{l_2^2}{l_1^2} \cdot \frac{l_4^2}{l_3^2} \cdot \frac{l_6^2}{l_5^2}} \tag{66}$$

Let us prove the validity of expressions (61)–(66). Let us write the equations of kinetic and potential energy for the scheme shown in Figure 16:

$$\mathrm{T} = \frac{1}{2} m \dot{y}^2 + \frac{1}{2} J_1 \dot{\phi}_1^2 + \frac{1}{2} J_2 \dot{\phi}_2^2 + \frac{1}{2} J_3 \dot{\phi}_3^2 \tag{67}$$

$$\Pi = \frac{1}{2} k_1 \left(y - \phi_1 \cdot l_1 \right)^2 + \frac{1}{2} k_2 \left(\phi_2 \cdot l_3 - \phi_1 \cdot l_2 \right)^2 + \frac{1}{2} k_3 \left(\phi_3 \cdot l_5 - \phi_2 \cdot l_4 \right)^2 + \frac{1}{2} k_4 l_6^2 \phi_3^2 .$$

Using the Lagrange equation of the second kind with a slight effect of friction forces, we find the differential equations of motion of this system:

$$m \ddot{y} + k_1 y - k_1 l_1 \phi_1 = 0 \tag{68}$$

$$J_1\ddot{\phi}_1 - k_1 l_1 y + \phi_1\left(k_1 l_1^2 + k_2 l_2^2\right) - k_2 l_2 l_3 \phi_2 = 0 \tag{69}$$

$$J_2\ddot{\phi}_2 + \phi_2\left(k_2 l_3^2 + k_3 l_4^2\right) - k_2 l_2 l_3 \phi_1 - k_3 l_4 l_5 \phi_3 = 0 \tag{70}$$

$$J_3\ddot{\phi}_3 + \phi_3\left(k_3 l_5^2 + k_4 l_6^2\right) - k_3 l_4 l_5 \phi_2 = 0 \tag{71}$$

The structural diagram of the equivalent ACS for the computational scheme shown in Figure 16, will take the form as shown in Figure 17.

Figure 17. The structural diagram of a system containing three lever linkages of the second kind

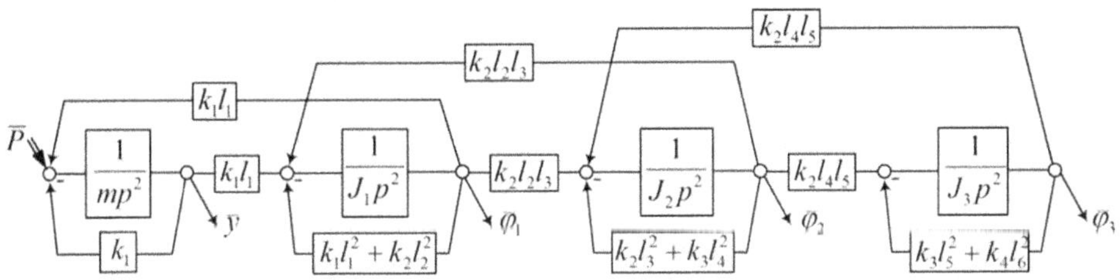

The corresponding structural interpretation in the form of a dual mechanical chain is shown in Figure 18.

Figure 18. The mechanical chain of a system containing three lever linkages of the second kind

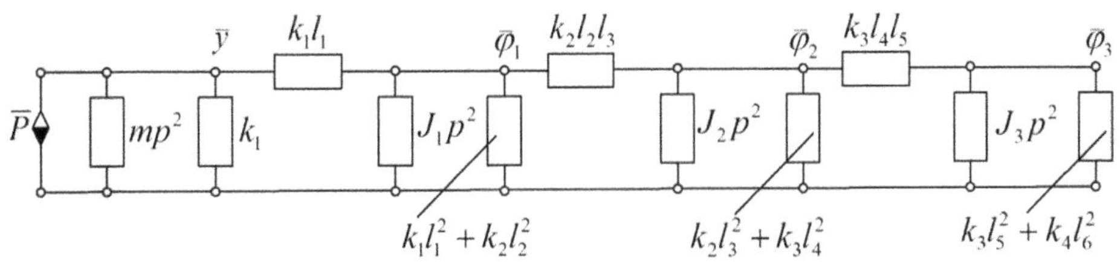

We accept that $J_1\ddot{\phi}_1 = 0,\ J_2\ddot{\phi}_2 = 0,\ J_3\ddot{\phi}_3 = 0$, and, producing successive transformations, select the coefficient of reduced stiffness for this system:

$$K_p = \frac{k_1k_2k_3k_4 \cdot \frac{l_2^2}{l_1^2} \cdot \frac{l_4^2}{l_3^2} \cdot \frac{l_6^2}{l_5^2}}{k_1\left[k_2\left(k_3 + k_4 \cdot \frac{l_6^2}{l_5^2}\right) + k_3k_4 \cdot \frac{l_4^2}{l_3^2} \cdot \frac{l_6^2}{l_5^2}\right] + k_2k_3k_4 \cdot \frac{l_2^2}{l_1^2} \cdot \frac{l_4^2}{l_3^2} \cdot \frac{l_6^2}{l_5^2}} \tag{72}$$

Having accepted that $i_1 = -\frac{l_2}{l_1}$, $i_2 = -\frac{l_4}{l_3}$, and $i_3 = -\frac{l_6}{l_5}$, we will find the final expression for the reduced stiffness coefficient of the system:

$$K_p = \frac{k_1k_2k_3k_4 \cdot i_1^2 \cdot i_2^2 \cdot i_3^2}{k_1\left[k_2\left(k_3 + k_4 \cdot i_3^2\right) + k_3k_4 \cdot i_2^2 \cdot i_3^2\right] + k_2k_3k_4 \cdot i_1^2 \cdot i_2^2 \cdot i_3^2} \tag{73}$$

All the results are summarized in Table 1. The obtained formula is also valid for mechanical oscillatory systems containing *n* lever linkages of the second kind. In this case, the expression for the coefficient of reduced stiffness takes the form:

$$K_p = \frac{k_1k_2...k_4 \cdot \left(i_1^2 \cdot i_2^2 ... i_n^2\right)}{k_1\left[k_2\left[....\left[k_n\left[k_{n-1}\left(k_n + k_{n+1} \cdot i_n^2\right) + k_n k_{n+1} \cdot i_{n-1}^2 \cdot i_n^2\right] + k_{n-1}k_n k_{n+1} \cdot i_{n-2}^2 \cdot i_{n-1}^2 \cdot i_n^2\right]...\right] + k_2k_3...k_{n+1} \cdot \left(i_1^2 \cdot i_2^2 ... i_n^2\right)} \tag{74}$$

Table 1. Values of reduced stiffness for systems with lever linkage of the second kind

Number of levers	Computational scheme	Reduced stiffness coefficient of the system
1	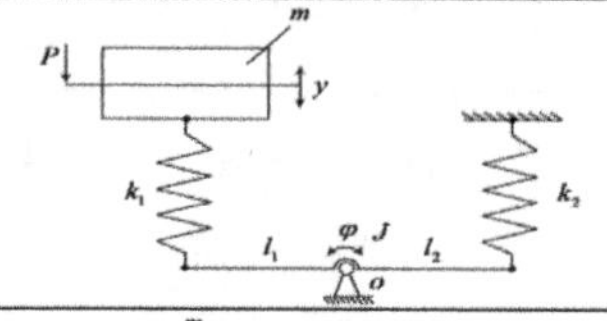	$K_p = \frac{k_1k_2i^2}{k_1 + k_2i^2}$
2	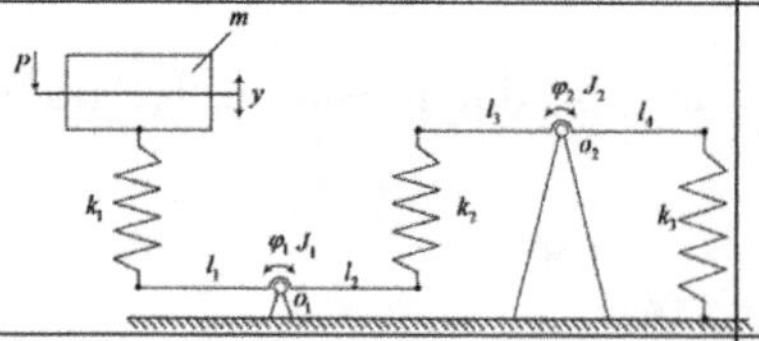	$K_p = \frac{k_1k_2k_3 \cdot i_1^2 \cdot i_2^2}{k_1\left(k_2 + k_3 \cdot i_2^2\right) + k_2k_3 \cdot i_1^2 \cdot i_2^2}$
3	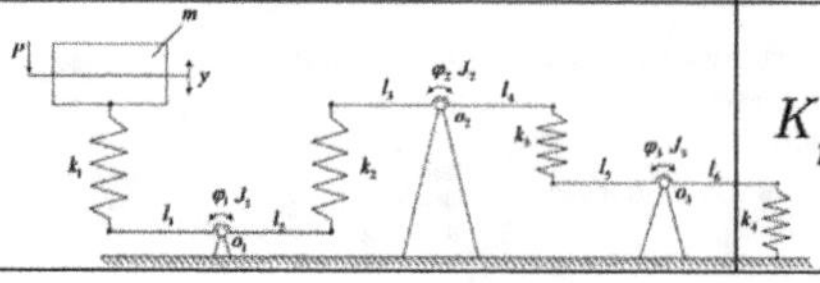	$K_p = \frac{k_1k_2k_3k_4 \cdot i_1^2 \cdot i_2^2 \cdot i_3^2}{k_1\left[k_2\left(k_3 + k_4 \cdot i_3^2\right) + k_3k_4 \cdot i_2^2 \cdot i_3^2\right] + k_2k_3k_4 \cdot}$

In conclusion, consideration of lever relationships of the second kind, we will make several preliminary conclusions.

1. According to the studies carried out above, it can be argued that in a system consisting of elements interconnected by the second-kind lever linkages, the connection takes the form of a consecutive connection, taking into account the features of the lever linkage of the n kind.
2. The coefficient of reduced stiffness for a system with n levers of the second kind, in the general case takes the form:

$$K_p = \frac{\left(\frac{\left(\left(\frac{K_p^1 \cdot K_p^2}{K_p^1 + K_p^2}\right) \cdot \ldots\right) \cdot K_p^{n-1}}{\left(\left(\frac{K_p^1 \cdot K_p^2}{K_p^1 + K_p^2}\right) + \ldots\right) + K_p^{n-1}}\right) \cdot K_p^n}{\left(\frac{\left(\left(\frac{K_p^1 \cdot K_p^2}{K_p^1 + K_p^2}\right) \cdot \ldots\right) \cdot K_p^{n-1}}{\left(\left(\frac{K_p^1 \cdot K_p^2}{K_p^1 + K_p^2}\right) + \ldots\right) + K_p^{n-1}}\right) + K_p^n} \tag{75}$$

where

$$K_p^1 = k_1,\ K_p^2 = k_2 \cdot \frac{l_2^2}{l_1^2}, \ldots,\ K_p^{n-1} = k_{n-1} \cdot \frac{l_2^2}{l_1^2} \cdot \ldots \cdot \frac{l_n^2}{l_{n-1}^2},\ K_p^n = k_n \cdot \frac{l_2^2}{l_1^2} \cdot \ldots \cdot \frac{l_n^2}{l_{n-1}^2} \cdot \frac{l_{n+2}^2}{l_{n+1}^2}.$$

A System Containing a Lever of the First Kind

1. Consider the scheme presented in Figure 19. We assume that for the scheme shown in Figure 19, the coefficient of the reduced stiffness of the system is determined by the formula:

$$K_p = \frac{K_p^1 \cdot K_p^2}{K_p^1 + K_p^2} \tag{76}$$

where

$$K_p^1 = k_1 \tag{77}$$

$$K_p^2 = k_2 \cdot \frac{l_2^2}{\left(l_1 + l_2\right)^2} \qquad (78)$$

Figure 19. The computational scheme with a first-kind lever

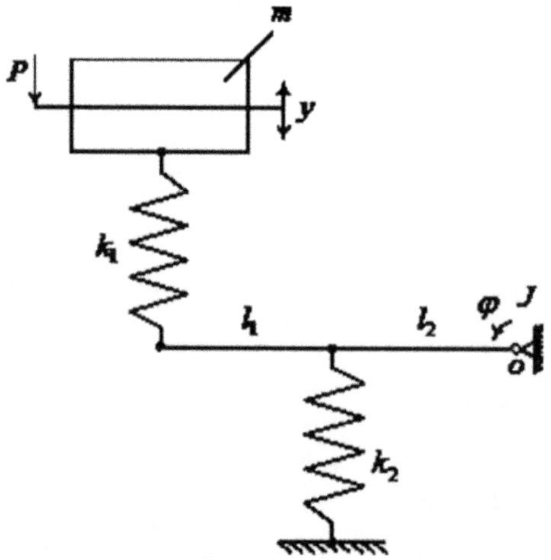

Make the substitution of expressions (77) and (78) into (76) and after the transformations we get:

$$K_p = \frac{k_1 k_2 \cdot \dfrac{l_2^2}{\left(l_1 + l_2\right)^2}}{k_1 + k_2 \cdot \dfrac{l_2^2}{\left(l_1 + l_2\right)^2}} \qquad (79)$$

To prove the validity of expressions (76–79), we write the equations of kinetic and potential energy for the scheme shown in Figure 19:

$$\mathrm{T} = \frac{1}{2} m\dot{y}^2 + \frac{1}{2} J\dot{\phi}^2, \ \Pi = \frac{1}{2} k_1 \left(y - \phi \cdot \left(l_1 + l_2\right)\right)^2 + \frac{1}{2} k_2 l_2^2 \phi^2 \qquad (80)$$

Using the Lagrange equation of the second kind (with a slight effect of friction forces), we find the differential equations of motion of this system:

$$m\ddot{y} + k_1 y - k_1 (l_1 + l_2)\varphi = 0 \qquad (81)$$

$$J\ddot{\phi} - k_1\left(l_1 + l_2\right)y + \phi\left(k_1\left(l_1 + l_2\right)^2 + k_2 l_2^2\right) = 0 \tag{82}$$

The structural diagram of the equivalent ACS for the computational scheme is shown in Figure 20, and the corresponding structural interpretation in the form of a dual mechanical chain is shown in Figure 21.

Figure 20. The structural diagram of the system with a first-kind lever linkage

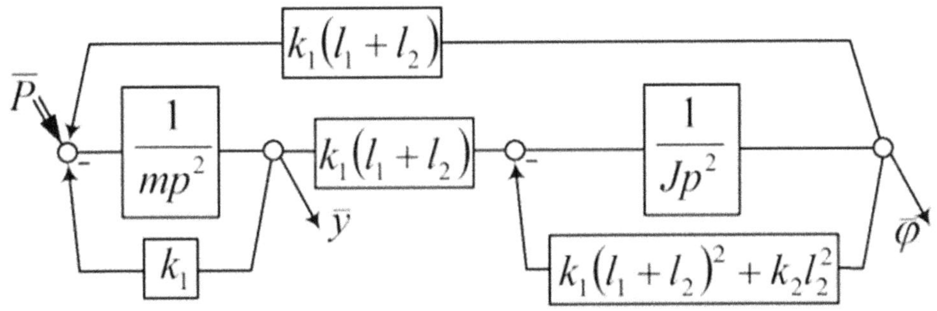

Figure 21. The mechanical chain of the system with a first-kind lever linkage

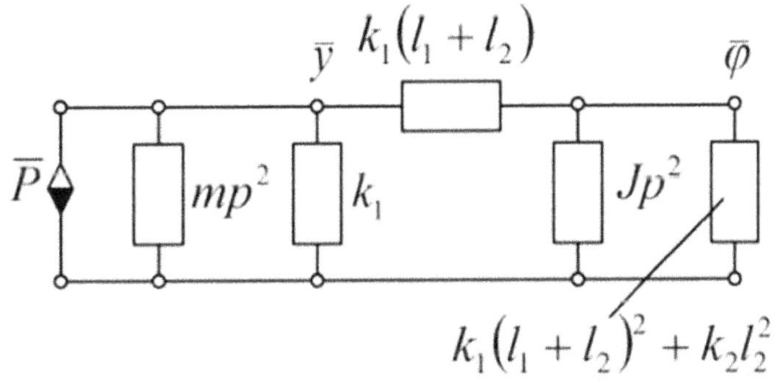

If we accept that $J\ddot{\phi} = 0$, then from equation (68) we can express:

$$\phi = y\frac{k_1\left(l_1 + l_2\right)}{k_1\left(l_1 + l_2\right)^2 + k_2 l_2^2} \tag{83}$$

We substitute (43) into (41) and after conversion we get:

$$m\ddot{y} + K_p \cdot y = 0 \tag{84}$$

where

$$K_p = \frac{k_1 k_2 \left(l_1 + l_2\right)^2}{k_1 + k_2 \dfrac{l_2^2}{\left(l_1 + l_2\right)^2}} \tag{85}$$

As you can see, expressions (79) and (85) are the same. Having assumed that $i = \frac{l_2}{\left(l_1 + l_2\right)}$, we get the final expression for the reduced stiffness coefficient:

$$K_p = \frac{k_1 k_2 i^2}{k_1 + k_2 i^2} \tag{86}$$

2. Consider the system of levers presented in Figure 22. For the scheme presented in Figure 22, let the coefficient of the reduced stiffness of the system be determined by the formula:

$$K_p = \frac{\left(\dfrac{K_p^1 \cdot K_p^2}{K_p^1 + K_p^2}\right) \cdot K_p^3}{\left(\dfrac{K_p^1 \cdot K_p^2}{K_p^1 + K_p^2}\right) + K_p^3} \tag{87}$$

where

$$K_p^1 = k_1 \tag{88}$$

$$K_p^2 = k_2 \cdot \frac{l_2^2}{\left(l_1 + l_2\right)^2} \tag{89}$$

$$K_p^3 = k_3 \cdot \frac{l_2^2}{\left(l_1 + l_2\right)^2} \cdot \frac{l_4^2}{\left(l_3 + l_4\right)^2} \tag{90}$$

We substitute expressions (88), (89) and (90) into (87) and, after some transformations, we obtain:

$$K_p = \frac{k_1 k_2 k_3 \cdot \dfrac{l_2^2}{\left(l_1 + l_2\right)^2} \cdot \dfrac{l_4^2}{\left(l_3 + l_4\right)^2}}{k_1 \left(k_2 + k_3 \cdot \dfrac{l_4^2}{\left(l_3 + l_4\right)^2}\right) + k_2 k_3 \cdot \dfrac{l_2^2}{\left(l_1 + l_2\right)^2} \cdot \dfrac{l_4^2}{\left(l_3 + l_4\right)^2}} \tag{91}$$

Let us prove the validity of expressions (87)–(91). To do this, we write down the equations of kinetic and potential energy for the scheme shown in Figure 16:

$$\mathrm{T} = \frac{1}{2} m \dot{y}^2 + \frac{1}{2} J_1 \dot{\varphi}_1^2 + \frac{1}{2} J_2 \dot{\varphi}_2^2, \tag{92}$$

$$\Pi = \frac{1}{2} k_1 \left(y - \varphi_1 \cdot (l_1 + l_2)\right)^2 + \frac{1}{2} k_2 \left(\varphi_2 \cdot (l_3 + l_4) - \varphi_1 \cdot l_2\right)^2 + \frac{1}{2} k_3 l_4^2 \varphi_2^2 .$$

Using the Lagrange equation of the second kind (with a slight effect of friction forces), we find the differential equations of motion of this system:

$$m\ddot{y} + k_1 y - k_1 \left(l_1 + l_2\right) \phi_1 = 0 \tag{93}$$

$$J_1 \ddot{\phi}_1 - k_1 \left(l_1 + l_2\right) y + \phi_1 \left(k_1 \left(l_1 + l_2\right)^2 + k_2 l_2^2\right) - k_2 l_2 \left(l_3 + l_4\right) \phi_2 = 0 \tag{94}$$

$$J_2 \ddot{\phi}_2 + \phi_2 \left(k_2 \left(l_3 + l_4\right)^2 + k_3 l_4^2\right) - k_2 l_2 \left(l_3 + l_4\right) \phi_1 = 0 \tag{95}$$

The structural diagram of the equivalent ACS for the computational scheme shown in Figure 22, will take the form as shown in Figure 23, and the corresponding structural interpretation in the form of a dual mechanical chain is shown in Figure 24.

If we accept that $J_1 \ddot{\phi}_1 = 0,\ J_2 \ddot{\phi}_2 = 0$, then applying the algorithm used above, we express the coefficient of the reduced stiffness of the system:

Figure 22. The computational scheme with two levers of the first kind

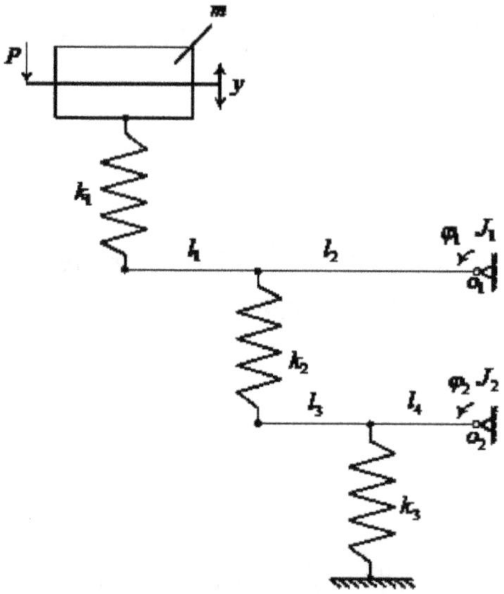

Figure 23. The structural diagram of a system containing two lever linkages of the first kind

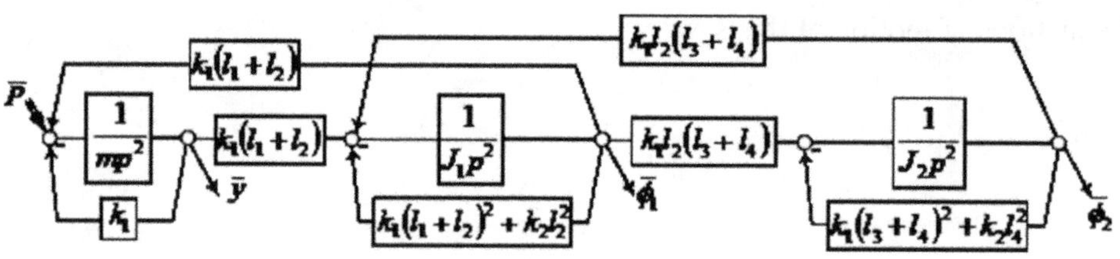

Figure 24. The mechanical chain system containing two lever linkages of the first kind

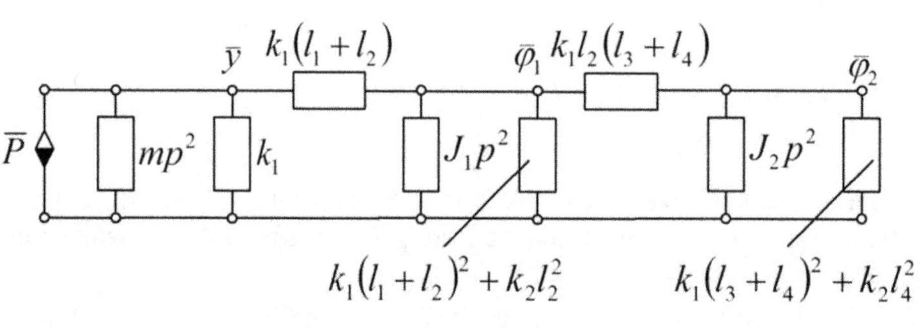

$$K_p = \frac{k_1 k_2 k_3 \cdot \frac{l_2^2}{\left(l_1 + l_2\right)^2} \cdot \frac{l_4^2}{\left(l_3 + l_4\right)^2}}{k_1 \left(k_2 + k_3 \cdot \frac{l_4^2}{\left(l_3 + l_4\right)^2}\right) + k_2 k_3 \cdot \frac{l_2^2}{\left(l_1 + l_2\right)^2} \cdot \frac{l_4^2}{\left(l_3 + l_4\right)^2}} \tag{96}$$

Having accepted that $i_1 = \frac{l_2}{l_1 + l_2}$ and $i_2 = \frac{l_4}{l_3 + l_4}$, we get the final expression for the reduced stiffness coefficient of the system:

$$K_p = \frac{k_1 k_2 k_3 \cdot i_1^2 \cdot i_2^2}{k_1 \left(k_2 + k_3 \cdot i_2^2\right) + k_2 k_3 \cdot i_1^2 \cdot i_2^2} \tag{97}$$

The approach can be extended to a mechanical oscillatory system containing n lever linkages of the first kind. Then the expression for the coefficient of reduced stiffness for a system with n lever linkages of the second kind, takes the form:

$$K_p = \frac{k_1 k_2 \ldots k_4 \cdot \left(i_1^2 \cdot i_2^2 \ldots i_n^2\right)}{\left[\begin{array}{l} k_1 \left[k_2 \left[\ldots \left[k_n \left[k_{n-1} \left(k_n + k_{n+1} \cdot i_n^2\right) + k_n k_{n+1} \cdot i_{n-1}^2 \cdot i_n^2\right] + k_{n-1} k_n k_{n+1} \cdot i_{n-2}^2 \cdot i_{n-1}^2 \cdot i_n^2\right] \ldots\right] + \\ \qquad + k_2 k_3 \ldots k_{n+1} \cdot \left(i_1^2 \cdot i_2^2 \ldots i_n^2\right) \end{array}\right]} \tag{98}$$

3. Consider a mechanical oscillatory system with two levers of the second or first kind. The expression for determining the natural frequency of the system, shown in Figure 10 and 19 in view of (45), can be written in the form of:

$$\omega_c = \sqrt{\frac{k_1}{m} \cdot \frac{i^2}{\frac{k_1}{k_2} + i^2}} = \sqrt{\omega_{c_1} \cdot A} \tag{99}$$

where $A = \frac{i^2}{\frac{k_1}{k_2} + i^2} = \frac{i^2}{a_1 + i^2}$, and *i* may take values:

- for levers of the second kind $i = -\frac{l_2}{l_1}$;

- for levers of the first kind $i = \dfrac{l_2}{l_1 + l_2}$.

Given a series of numerical values l_1, l_2, k_1, k_2, m we construct a dependence for the possible values of the coefficient *A*, as shown in Figures 25 and 26.

Figure 25. The range of possible values of A, with the constant a and the change of i

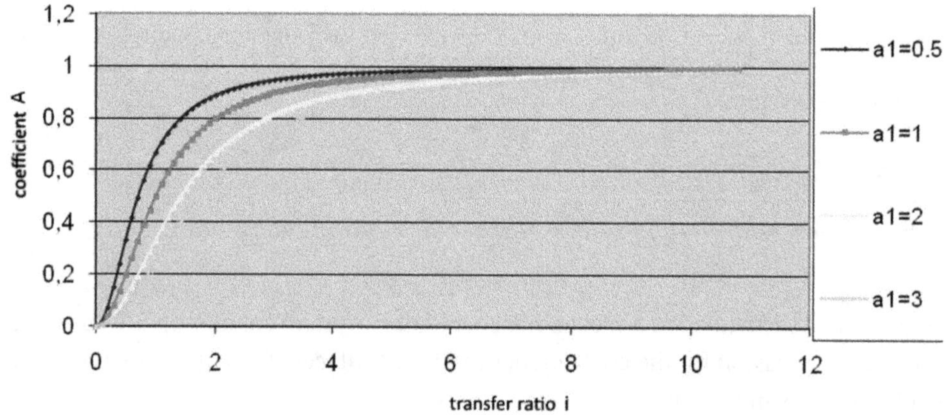

Figure 26. The range of possible values of A, with the constant i and the change of a

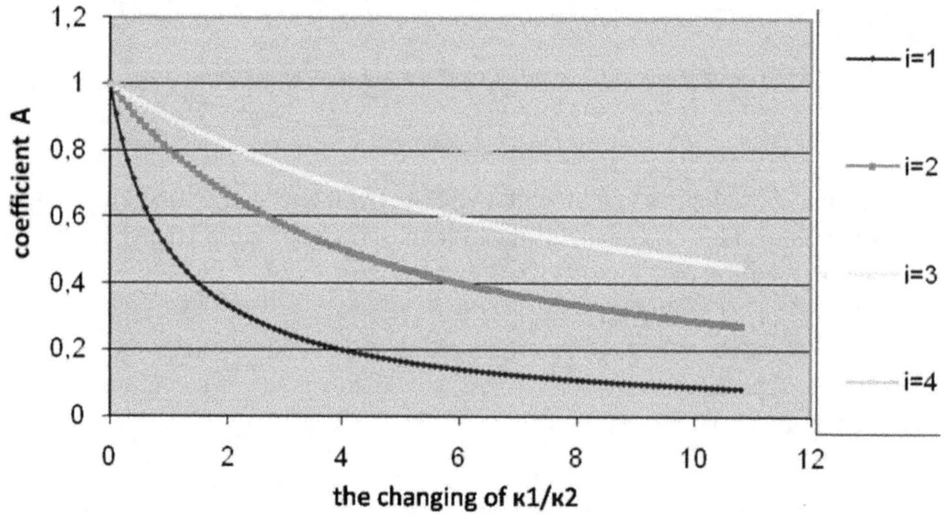

The dependencies shown in Figure 25, have obvious non-linear properties, and with an increase of *a* non-linear properties decrease. The construction area can be divided into three zones:

1. Zone of a sharp increase in *A* with small changes of *i*;
2. Zone of uniform change *A* with corresponding changes of *i*;
3. A asymptotically approaches unity.

With the constant i (Figure 26) there is a noticeable change in A, especially for small a. Later, as a changes, a decrease of A occurs. And, with an increase of i, the nonlinear properties decrease. The expression for the period of natural oscillations is obtained in the form of:

$$T = \frac{2\pi}{\sqrt{\frac{k_1}{m} \cdot \frac{i^2}{\frac{k_1}{k_2} + i^2}}} \tag{100}$$

4. The expression for determining the frequency of natural oscillations of the system shown in Figures 13 and 22 in view of (60), can be written in the form of:

$$\omega_c = \sqrt{\frac{k_1}{m} \cdot \frac{i_1^2 \cdot i_2^2}{\left[\frac{k_1}{k_3} + \frac{k_1}{k_2} \cdot i_2^2\right] + i_1^2 \cdot i_2^2}} = \sqrt{\omega_{c_1} \cdot A} \tag{101}$$

where

$$A = \frac{i_1^2 \cdot i_2^2}{\left[\frac{k_1}{k_3} + \frac{k_1}{k_2} \cdot i_2^2\right] + i_1^2 \cdot i_2^2} = \frac{i_1^2 \cdot i_2^2}{\left[a_2 + a_1 \cdot i_2^2\right] + i_1^2 \cdot i_2^2}$$

i_1 and i_2 can take values:

- for levers of the second kind рода $i_1 = -\frac{l_2}{l_1}$, $i_2 = -\frac{l_4}{l_3}$;
- for levers of the first kind $i_1 = \frac{l_2}{l_1 + l_2}$, $i_2 = \frac{l_4}{l_3 + l_4}$.

The expression for the period of natural oscillations is obtained in the form of:

$$T = \frac{2\pi}{\sqrt{\frac{k_1}{m} \cdot \frac{i_1^2 \cdot i_2^2}{\left[\frac{k_1}{k_3} + \frac{k_1}{k_2} \cdot i_2^2\right] + i_1^2 \cdot i_2^2}}} \tag{102}$$

5. The expression to determine the frequency of natural oscillations of the system, we write in the form of:

$$\omega_c = \sqrt{\frac{k_1}{m} \cdot \frac{i_1^2 \cdot i_2^2 \cdot i_3^2}{\left[\frac{k_1}{k_4} + \frac{k_1}{k_3} i_3^2 + \frac{k_1}{k_2} \cdot i_2^2 \cdot i_3^2\right] + i_1^2 \cdot i_2^2 \cdot i_3^2}} = \sqrt{\omega_{c_1} \cdot A} \tag{103}$$

where

$$A = \frac{i_1^2 \cdot i_2^2 \cdot i_3^2}{\left[\frac{k_1}{k_4} + \frac{k_1}{k_3} \cdot i_3^2 + \frac{k_1}{k_2} \cdot i_2^2 \cdot i_3^2\right] + i_1^2 \cdot i_2^2 \cdot i_3^2} = \frac{i_1^2 \cdot i_2^2 \cdot i_3^2}{\left[a_3 + a_2 \cdot i_3^2 + a_1 \cdot i_2^2 \cdot i_3^2\right] + i_1^2 \cdot i_2^2 \cdot i_3^2},$$

i_1, i_2 and i_3 can take values:

- for levers of the second kind, $i_1 = -\frac{l_2}{l_1}$, $i_2 = -\frac{l_4}{l_3}$, and $i_3 = -\frac{l_6}{l_5}$;
- for levers of the first kind $i_1 = \frac{l_2}{l_1 + l_2}$, $i_2 = \frac{l_4}{l_3 + l_4}$, and $i_3 = \frac{l_6}{l_5 + l_6}$.

We write the expression for the period of natural oscillations using (103):

$$T = \frac{2\pi}{\sqrt{\frac{k_1}{m} \cdot \frac{i_1^2 \cdot i_2^2 \cdot i_3^2}{\left[\frac{k_1}{k_4} + \frac{k_1}{k_3} i_3^2 + \frac{k_1}{k_2} \cdot i_2^2 \cdot i_3^2\right] + i_1^2 \cdot i_2^2 \cdot i_3^2}}} \tag{104}$$

The results can be extended to systems with n lever linkages. Then, the expression, allowing us to determine the natural frequency, for mechanical systems with n lever linkages of the second or first kind, will be written in the following form:

$$\omega_c = \sqrt{\frac{k_1}{m} \cdot \frac{i_1^2 \cdot i_2^2 \cdot \ldots \cdot i_n^2}{\left[\frac{k_1}{k_{n+1}} + \frac{k_1}{k_n} i_n^2 + \frac{k_1}{k_{n-1}} \cdot i_{n-1}^2 \cdot i_n^2 + \ldots + \frac{k_1}{k_2} \cdot i_2^2 \cdot i_3^2 \cdot \ldots \cdot i_n^2\right] + i_1^2 \cdot i_2^2 \cdot \ldots \cdot i_n^2}} \tag{105}$$

$$= \sqrt{\omega_{c_1} \cdot A}$$

where

$$A = \frac{i_1^2 \cdot i_2^2 \cdot \ldots \cdot i_n^2}{\left[\frac{k_1}{k_{n+1}} + \frac{k_1}{k_n} i_n^2 + \frac{k_1}{k_{n-1}} \cdot i_{n-1}^2 \cdot i_n^2 + \ldots + \frac{k_1}{k_2} \cdot i_2^2 \cdot i_3^2 \cdot \ldots \cdot i_n^2\right] + i_1^2 \cdot i_2^2 \cdot \ldots \cdot i_n^2} =$$

$$= \frac{i_1^2 \cdot i_2^2 \cdot \ldots \cdot i_n^2}{\left[a_n + a_{n-1} \cdot i_n^2 + a_{n-2} \cdot i_{n-1}^2 \cdot i_n^2 + \ldots + a_1 \cdot i_2^2 \cdot i_3^2 \cdot \ldots \cdot i_n^2\right] + i_1^2 \cdot i_2^2 \cdot \ldots \cdot i_n^2},$$

and $i_1, i_2 \ldots i_n$ are transfer coefficients for levers of the second and first kind.

Using (5.105), we obtain the expression for the period of natural oscillations of a system having n lever linkages of the second or first kind, in the form of:

$$T = \frac{2\pi}{\sqrt{\frac{k_1}{m} \cdot \frac{i_1^2 \cdot i_2^2 \cdot \ldots \cdot i_n^2}{\left[\frac{k_1}{k_{n+1}} + \frac{k_1}{k_n} i_n^2 + \frac{k_1}{k_{n-1}} \cdot i_{n-1}^2 \cdot i_n^2 + \ldots + \frac{k_1}{k_2} \cdot i_2^2 \cdot i_3^2 \cdot \ldots \cdot i_n^2\right] + i_1^2 \cdot i_2^2 \cdot \ldots \cdot i_n^2}}} \quad (106)$$

Thus,

- With the introduction of a leverage, there is a decrease in the resonant frequency and, accordingly, a decrease in the oscillation period;
- Lever relationships have the same shape for both levers of second and first kind;
- Based on the introduction of lever linkages, it becomes possible to obtain a controlled oscillation damper acting by changing the number of lever linkages and the arm lengths of each of them.

A NEW APPROACH IN ASSESSING THE POSSIBILITIES OF CONSECUTIVE CONNECTIONS OF TYPICAL ELEMENTS IN STRUCTURAL INTERPRETATIONS OF VIBRATION PROTECTION SYSTEMS

In the works Eliseev et al. (2007) and Belokobylsky and Sitov (2007) the consecutive connection of units (Figure 27) W_1 and W_2, for which the following relations exist:

$$W = \frac{W_1 W_2}{W_1 + W_2} \quad (107)$$

where W_1 and W_2 are transfer functions of typical units of the structure or complex resistances (Eliseev et al., 2008).

If we consider point (2) (Figure 27) in such a way that elements of two units (points (2′) and (2″)) are connected in it, then between points (2′) and (2″) you can introduce a new unit, a lever, as shown in Figure 28.

Figure 27. The scheme of the consecutive connection of units

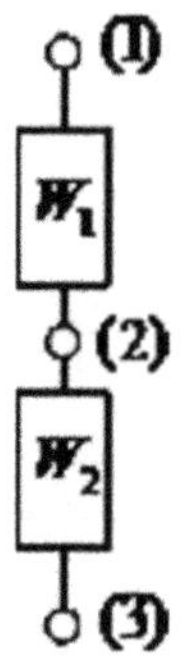

Figure 28. The consecutive connection with spaced-apart points of contact (2′) and (2″): l-the lever length

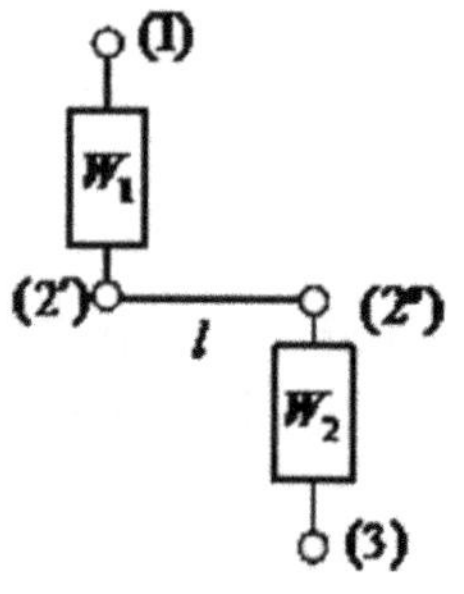

The lever may be, as has been shown, a non-inertial unit, and in some cases its mass-and-inertia properties can be taken into account. For the lever, it is possible to introduce points of support, assuming that a lever of either the second or the first kind is inserted into the interaction scheme by this means. The point of support of the lever can be associated with a fixed reference system (Figure 29).

Figure 29. The lever of the second kind (a) and the lever of the first kind (b) in the connection of typical units

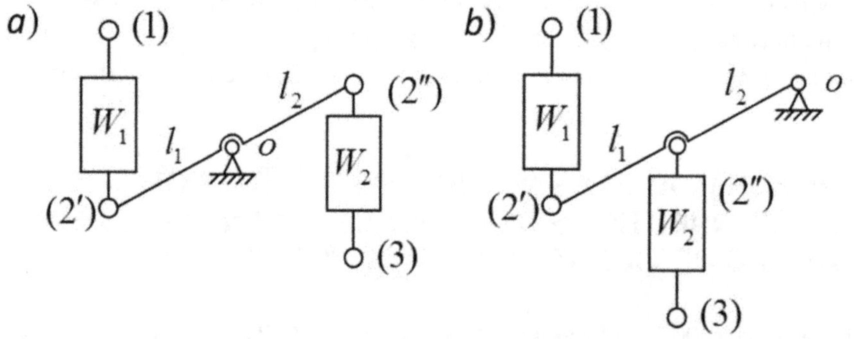

If we compare the compounds in Figures 27 and 29, it can be noted that they will be different. Continuing consideration of the possibilities of introducing a leverage, one can make an assumption about the "fixing" of a support point on one of the elements of the system, which is included in the general scheme. And, finally, you can consider the connection by a lever, in which the "point of support" will be sliding. Such situations, in essence, arise when a solid body oscillates in the form of a beam located on two supports. Special coordinate systems can be used in this case (Eliseev et al., 2008).

The Method for Determining Reduced Stiffness

Figure 30a presents one of the options for introducing a lever unit (lever of the second kind), and Figure 30b shows a computational scheme, including elastic and mass elements ensuring the inclusion of a bundle of W_1,W_2 into the oscillatory system.

Figure 30. The computational schemes of the consecutive connection of W_1 and W_2 through the lever unit: a) the general case; b) W_1 and W_2 are elastic elements

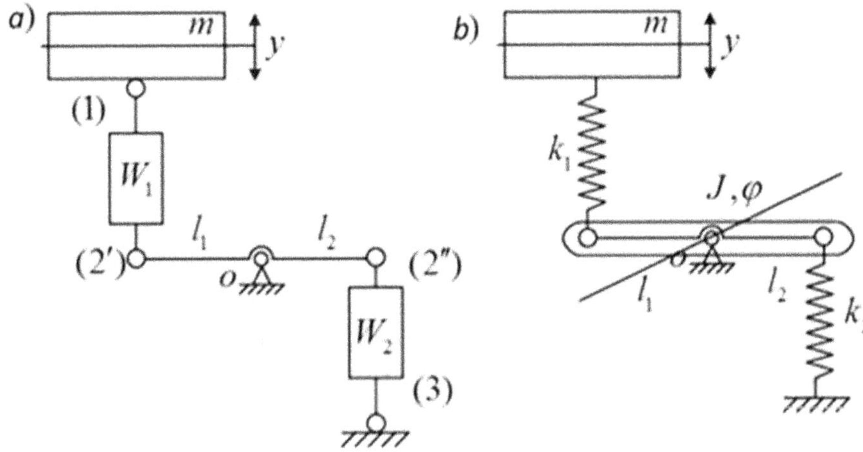

Developing a generalized way of taking into account the properties of a leverage, we will assume that the lever has a moment of inertia and can make angular motions φ. Using the Lagrange equation of the second kind and assuming that the force *P* is applied to the mass, we obtain the equations of motion in the form of:

$$\begin{cases} m\ddot{y} + k_1 y - k_1 l_1 \varphi = P \\ J\ddot{\varphi} + \left(k_1 l_1^2 + k_2 l_2^2\right)\varphi - k_1 l_1 y = k_2 l_2 \end{cases} \tag{108}$$

The structural diagram of the system is shown in Figure 31. It consists of two partial subsystems, marked by contours I and II, connected by elastic constraints. If you accept that *J*=0 (i.e., the lever will be weightless), then when you take into account:

$$\phi = y \frac{k_1 k_2 l_1^2}{k_1 l_1^2 + k_2 l_2^2}, i = \frac{l_2}{l_1} \tag{109}$$

we will obtain:

Figure 31. The structural diagram of the system in Figure 30b

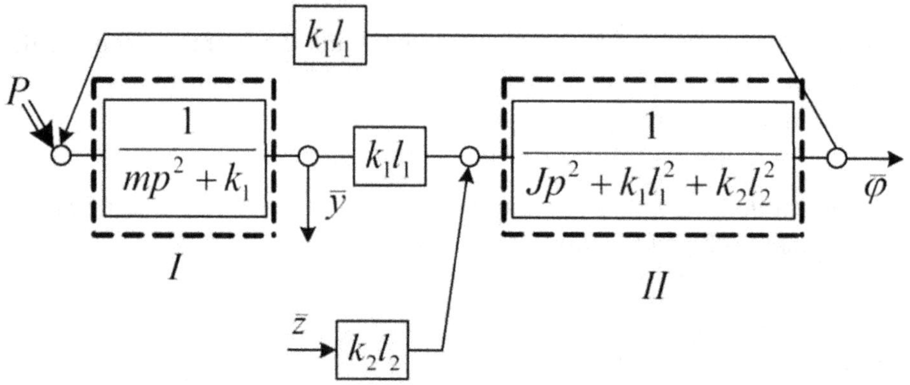

$$m\ddot{y} + y\frac{k_1k_2i^2}{k_1 + k_2i^2} = 0 \tag{110}$$

From (110) it follows that the additional constraint in the form of a lever of the second kind, having a transfer ratio i, is equivalent to the introduction of a new spring into the system, but with a different elasticity k_2i^2. The corresponding computational scheme is shown in Figure 32, by which one can see that the introduction of the lever does not change the nature of the initial connection, the elements interact in series in accordance with (107), but the lever changes the reduced stiffness.

Figure 32. The computational scheme for the system in Figure 30a: a) the consecutive connection with the introduction of the lever (i); b) the reduced computational scheme with the introduction of the lever (i); c) the computational scheme of the regular consecutive connection

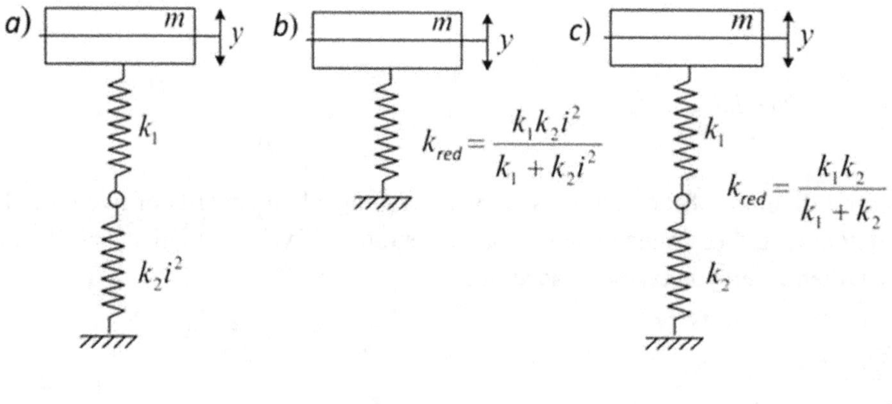

If $i = 1$, then the introduction of a lever does not change the reduced stiffness, but allows spacing apart the point of connection of two elements, k_1 and k_2. When $i = 0$ (l_2=0), the original system degenerates, just as with i=∞ (l_1=0). Thus, the limits of the change of i are determined by the conditions $0<i<\infty$ (but also $i\neq 0$ and $i\neq\infty$). Using the diagram in Figure 30b, we will make some transformations (J=0) and we will obtain the structure (Figure 33), allowing us to find the reduced stiffness.

Figure 33. The structural diagram of the system in Figure 30a: a) taking into account lever properties and m≠0; b) taking into account lever properties and m=0

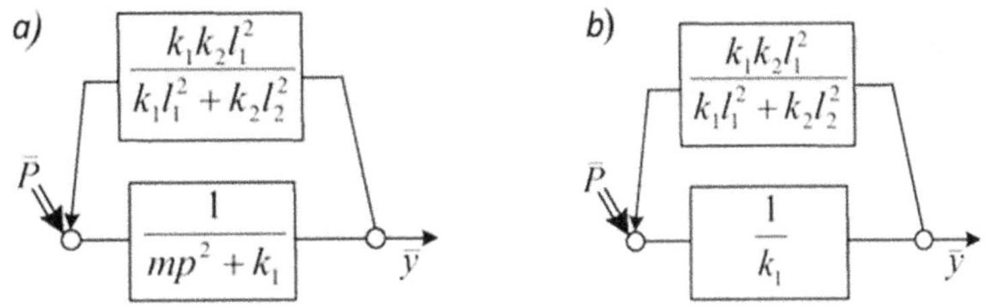

For the case of Figure 33a we find the transfer function of the system at the output *y* and the input *P*:

$$\bar{W}(p)=\frac{\bar{y}}{\bar{P}}=\frac{k_1l_1^2+k_2l_2^2}{m_1p^2+\dfrac{k_1k_2i^2}{k_1+k_2i^2}} \tag{111}$$

from where it follows that:

$$K_{red}=\frac{k_1k_2i^2}{k_1+k_2i^2}. \tag{112}$$

If to use the structural diagram in Figure 33b, it is also possible to find the transfer function corresponding to the ratio of "output and input," which has the form:

$$W=\frac{\bar{y}}{\bar{P}}=\frac{k_1l_1^2+k_2l_2^2}{k_1k_2l_2^2}=\frac{1}{K_{red}} \tag{113}$$

where $K_{red}=\dfrac{k_1k_2i^2}{k_1+k_2i^2}$.

Note that the structural diagram in Figure 33a is (in the terminology introduced in Druzhinsky, 1977), a "kinetostatic model", and the structural diagram in Figure 33b corresponds to a static scheme, that is, is a static model. The results obtained in the two approaches coincide, and the properties of the lever, therefore, do not depend on the frequency properties of the system.

The reduced stiffness can be obtained on the basis of conventional approaches, as is done, for example, in Eliseev et al. (2008), but we must take into account that when a lever is introduced, a transition to another class of systems occurs where it is necessary to take into account the moment ratios, turning angles, moments of forces.

Accounting for the Introduction of Motion Transformation Devices

If we assume that $W_1=Lp^2$, and $W_2=k$; which corresponds to the diagram in Figure 34a, the computational scheme with the introduction of mass m and kinematic perturbation z has the form, as shown in Figure 34b. Assuming that the kinetic and potential energy of the system is determined by the expressions

Figure 34. The consecutive connection of a motion transformation device and a spring

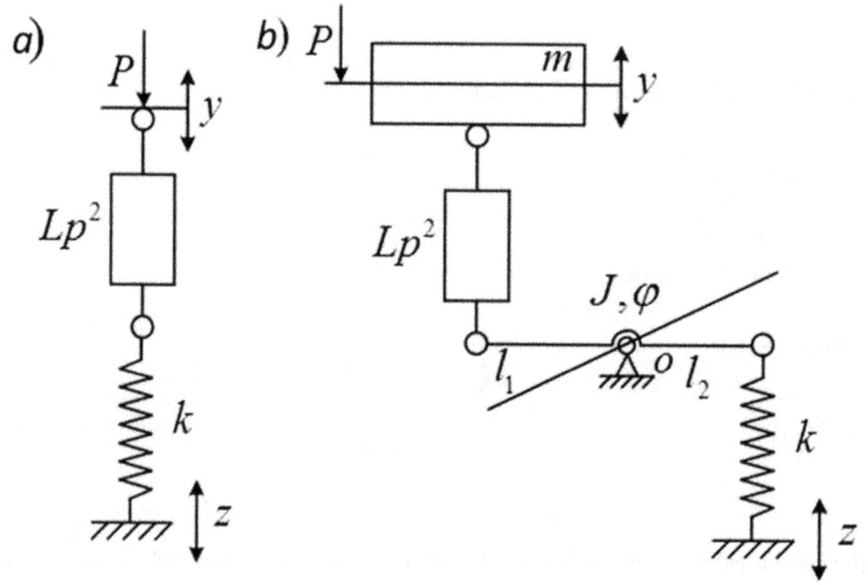

$$\begin{aligned} T &= \frac{1}{2}m\dot{y}^2 + \frac{1}{2}L\left(\dot{y} - l_1\dot{\phi}\right)^2 + \frac{1}{2}J\dot{\phi}^2, \\ \Pi &= \frac{1}{2}k\left(\phi l_2 - z\right)^2, \end{aligned} \tag{114}$$

we obtain a system of differential equations:

$$\begin{cases} m\ddot{y} + L\ddot{y} - Ll_1\ddot{\phi} = P, \\ J\ddot{\phi} + Ll_1^2\ddot{\phi} - Ll_1\ddot{y} + kl_2^2\phi = kl_2 z \end{cases} \tag{115}$$

and build a structural diagram (Figure 35).

Figure 35. The structural diagram of the system in Figure 34b

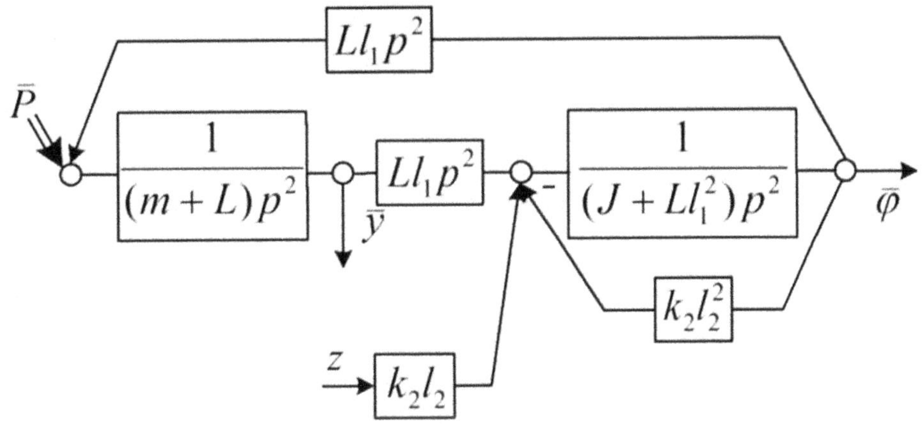

Using which, we find the transfer functions of the system:

$$W(p)=\frac{\bar{y}}{\bar{P}}=\frac{\left(J+Ll_1^2\right)p^2+kl_2^2}{(m+L)p^2\left[\left(J+Ll_1^2\right)p^2+kl_2^2\right]-L^2l_1^2p^4} \tag{116}$$

If J=0, and then m=0, we obtain accordingly:

$$W'(p)=\frac{\bar{y}}{\bar{P}}=\frac{Ll_1^2p^2+kl_2^2}{(m+L)p^2\left[Ll_1^2p^2+kl_2^2\right]-L^2l_1^2p^4} \tag{117}$$

$$\bar{W}(p)=\frac{\bar{y}}{\bar{P}}=\frac{Ll_1^2p^2+kl_2^2}{(m+L)p^2\left[Ll_1^2p^2+kl_2^2\right]-L^2l_1^2p^4}, \tag{118}$$

Assuming, as before, that $i=-\frac{l_2}{l_1}$, we find:

$$W_{red}=\frac{\bar{y}}{\bar{P}}=\frac{Lp^2ki^2}{Lp^2+ki^2} \tag{119}$$

Note that this result could be obtained immediately, if we assume that $W_1=Lp^2$, $W_2=k$, then:

$$W_{red}=\frac{\bar{y}}{\bar{P}}=\frac{Lp^2k}{Lp^2+k} \tag{120}$$

but in this case the lever linkage would not be taken into account.

The transfer function of the "input to output", which corresponds not to the force, but to the kinematic effect, has the form:

$$W(p) = \frac{\overline{y}}{\overline{Z}} = \frac{Ll_1^2 p^2 + kl_2^2}{(m+L)p^2[Ll_1^2 p^2 + kl_2^2] - L^2 l_1^2 p^4} \tag{121}$$

The denominator (120) is the frequency equation, coincides with the denominator (118), which is to be expected, therefore, W_{red}, determined from (121), will also correspond to the expression (119). Similarly, the interactions of any other sets of typical units can be tested.

Introducing First-Kind Lever Into the Connection Structure

The spacing-apart of points (2′) and (2″) with the help of a lever of the first kind is shown in Figure 36.

Figure 36. The computational (a) scheme and the structural diagram of the oscillatory system (b) with a first-kind lever

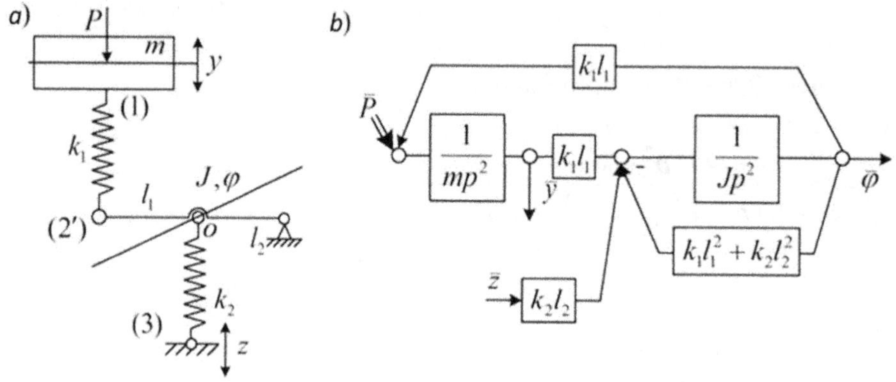

The system of differential equations in this case is:

$$\begin{cases} m\ddot{y} + k_1 y - k_1 l\phi = P, \\ J\ddot{\phi} + \phi\left(k_1 l_1^2 + k_2 l_2^2\right) - k_1 l y = k_2 l_2 z. \end{cases} \tag{122}$$

Using the structural diagram (Figure 36b), we find that:

$$W(p) = \frac{\overline{y}}{\overline{P}} = \frac{Jp^2 + k_1 l_1^2 + k_2 l_2^2}{(mp^2 + k_1)(Jp^2 + k_1 l_1^2 + k_2 l_2^2) - k_1 l_1^2} \tag{123}$$

$$W(p) = \frac{\bar{\varphi}}{\bar{P}} = \frac{k_1 l_1}{A} \tag{124}$$

$$W(p) = \frac{\bar{\varphi}}{\bar{Z}} = \frac{(mp^2 + k_1)k_2 l_2}{A} \tag{125}$$

$$W(p) = \frac{\bar{y}}{\bar{Z}} = \frac{k_1 k_2 l_1 l_2}{A} \tag{126}$$

where $A = \left(mp^2 + k_1\right)\left(Jp^2 + k_1 l_1^2 + k_2 l_2^2\right) - k_1^2 l^2$

If J=0 and m=0, then:

$$K_{red} = \frac{k_1 k_2 i^2}{k_1 + k_2 i^2} \tag{127}$$

where $i = \frac{l_2}{l} = \frac{l_2}{l_1 + l_2}$

Thus, the lever of the first kind, determining the reduced stiffness, which is formed as a consecutive connection of the spring k_1 and k_2 with the point (2′) and (2″), separated by the first-kind lever inertialess lever, gives an expression of the same type as the expression (113).

Moreover, if i=1, the lever does not contribute anything special compared to a conventional consecutive connection. If i=0, then l_2=0, which corresponds to the connection k_2 with a fixed point (such a connection becomes impossible). If i>0, then physically it means moving the point (2″) to the left, which will not change the qualitative picture of the results.

Thus, the use of the inertia-free lever of the second kind or first kind for separation of the points of connection of elements and (points (2′) and (2″)) does not change the type of connection, i.e. the consecutive connection type remains, however, the expression for the reduced connection characteristic will include the lever transfer ratios. Obtaining the above characteristics obtained from a combination of typical units can be obtained from the transfer functions, the so-called static systems. However, similar results can be obtained using frequency equations (the denominator of the transfer function), making the appropriate transformations. Accounting for the inertial properties of the leverage is a separate task and requires appropriate physical interpretation.

The Generalized Approach When Taking Into Account the Support of the Lever on the Unit

A lever linkage can also be implemented when the support point belongs to one of the elements of the general scheme, for example, it can be an element of mass or the support point that belongs to the element. In this case, the second point may be adjacent to a fixed base or connect with other elements. Examples of such a system are shown in Figure 37.

Figure 37. The schemes of oscillatory systems with levers, the support points of which belong to other elements (in all variants, the support point placed on the element of mass): a) the support is on the element with mass m_3; b) the support is on the element with mass m_3; c) the support is on the element with mass m_2

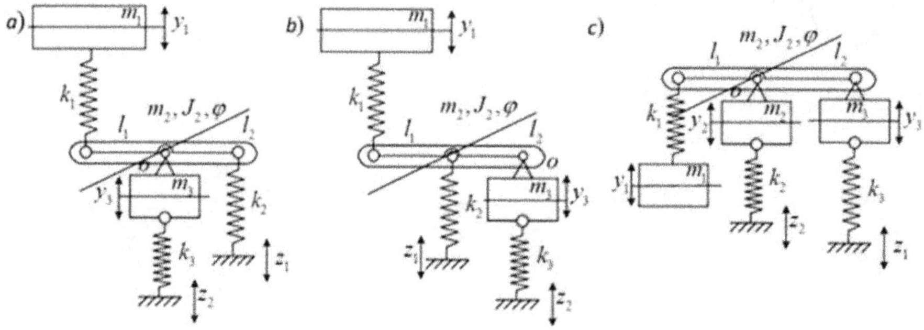

For the case shown in Figure 37a, we write the expressions for the kinetic and potential energy:

$$T = \frac{1}{2} m_1 \dot{y}_1^2 + \frac{1}{2} J_2 \dot{\phi}^2 + \frac{1}{2} J_2 \left(\frac{\dot{y}_3}{l_1} \right)^2 + \left(m_2 + m_3 \right) \dot{y}_3^2$$

$$\Pi = \frac{1}{2} k_1 \left(y_1 - l_1 \phi - y_3 \right)^2 + \frac{1}{2} k_2 \left(l_2 \phi + y_3 \right)^2 + \frac{1}{2} k_3 \left(y_3 - Z_2 \right)^2 \tag{128}$$

and the corresponding system of differential equations of motion:

$$m \ddot{y}_1 + k_1 y_1 - k_1 l \phi + k_1 y_3 = P_1,$$

$$J_2 \ddot{\phi} + \phi \left(k_1 l_1^2 + k_2 l_2^2 \right) + k_1 l_1 y_1 - y_3 \left(k_1 l_1 - k_2 l_2 \right) = k_2 l_2 Z_1,$$

$$\left(m_2 + m_3 \right) \ddot{y}_3 + y_3 \left(k_1 + k_2 + k_3 \right) + k_1 y_1 - \phi \left(k_1 l_1 - k_2 l_2 \right) = k_2 l_2 Z_1 + k_3 Z_2 + P_3. \tag{129}$$

The structural diagram of the system is shown in Figure 38. The system of equations (129) takes into account possible types of disturbances (force and kinematic ones), which is also reflected in the structural scheme (see. Figure 38a). In Figure 38b is represented by a dual mechanical chain, built according to the rules set forth in Eliseev et al. (2008).

Figure 38. The structural diagram (a) corresponding to Figure 37a: P_1, P_3 – force perturbations; Z_1, Z_2 – kinematic perturbations (a) and its dual mechanical chain (b)

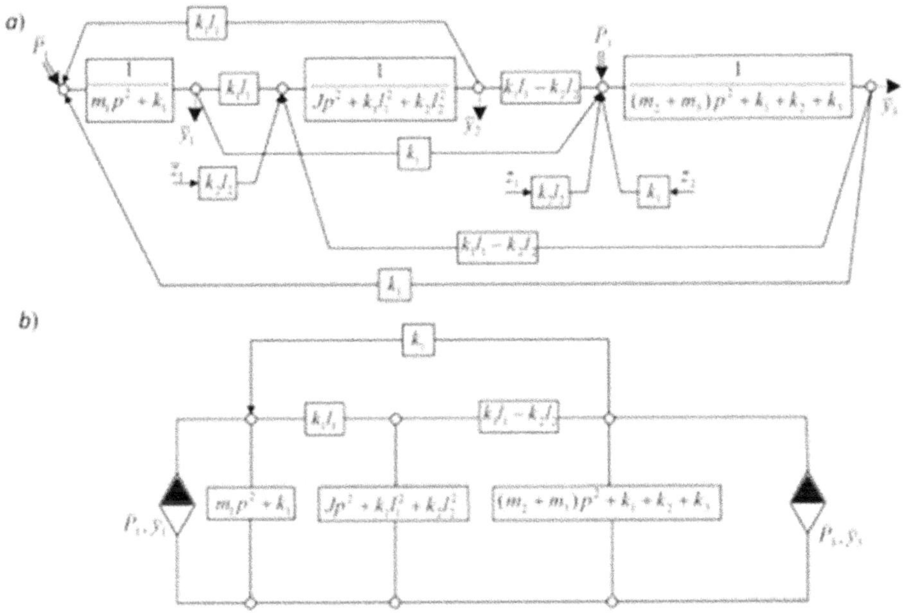

Note that in this system there are constraints k_1 that turn the system into non-planar. Having accepted that $\varphi=y2$ we will rewrite the system (129) in the form:

$$\begin{aligned} &a_{11}y_1 + a_{12}y_2 + a_{13}y_3 = b_1, \\ &a_{21}y_1 + a_{22}y_2 + a_{23}y_3 = b_2, \\ &a_{31}y_1 + a_{32}y_2 + a_{33}y_3 = b_3 \end{aligned} \tag{130}$$

and determine the corresponding coefficients:

$$\begin{aligned} &a_{11} = m_1 p^2 + k_1, a_{12} = -k_1 l_1, a_{13} = k_1, \\ &a_{21} = -k_1 l_1, a_{22} = Jp^2 + k_1 l_1^2 + k_2 l_2^2, a_{23} = -(k_1 l_1 - k_2 l_2), \\ &a_{31} = k_1, a_{32} = -(k_1 l_1 - k_2 l_2), a_{33} = (m_2 + m_3) p^2 + k_1 + k_2 + k_3, \\ &b_1 = P_1, b_2 = k_2 l_2 z_1, b_3 = k_2 l_2 z_1 + k_3 z_2 + P_3. \end{aligned} \tag{131}$$

Without resorting to the construction of transfer functions, we write in accordance with the method of determinants (Eliseev et al., 2008) the frequency equation of the system in the form:

$$B = a_{11}a_{22}a_{33} - a_{11}a_{21}a_{32} + a_{12}a_{21}a_{32} - a_{12}a_{21}a_{33} + a_{11}a_{23}a_{31} - a_{13}a_{22}a_{31}. \quad (132)$$

We will assume that the lever is inertia-free (J_2=0) and m_2+m_3=0, then:

$$\begin{aligned} B = m_1p^2\Big[&\left(k_{11}^2 + k_2l_2^2\right)\left(k_1 + k_2 + k_3\right) - \left(k_1l_1 - k_2l_2\right)^2 - \left(k_1l_1 - k_2l_2\right)\Big] + \\ &+k_1\left(k_1l_1^2 + k_2l_2^2\right)\left(k_1 + k_2 + k_3\right) - k_1\left(k_1l_1 - k_2l_2\right)^2 + k_1^2l_1\left(k_1l_1 - k_2l_2\right) - \\ &-k_1^2l_1^2\left(k_1 + k_2 + k_3\right) - k_1^2\left(k_1l_1 - k_2l_2\right) - k_1^2\left(k_1l_1^2 + k_2l_2^2\right). \end{aligned} \quad (133)$$

When k_1l_1–k_2l_2=0, the expression (132) is simplified and takes the form:

$$\begin{aligned} B = m_1p^2\Big[&\left(k_{11}^2 + k_2l_2^2\right)\left(k_1 + k_2 + k_3\right) + k_1\left(k_1l_1^2 + k_2l_2^2\right)\left(k_1 + k_2 + k_3\right) - \\ &-k_1^2l_1^2\left(k_1 + k_2 + k_3\right) - k_1^2\left(k_1l_1^2 + k_2l_2^2\right). \end{aligned} \quad (134)$$

We divide (133) by $\left(k_{11}^2 + k_2l_2^2\right)$ (k_1+k_2+k_3), which corresponds to the translation of this term into the structure of the numerator of the transfer function, then we get:

$$B = m_1p^2 + k_1 - \frac{k_1^2l_1^2}{k_1l_1^2 + k_2l_2^2} - \frac{k_1^2}{k_1 + k_2 + k_3} \quad (135)$$

or

$$B = m_1p^2 + \frac{k_1k_2l_1^2}{k_1l_1^2 + k_2l_2^2} - \frac{k_1}{k_1 + k_2 + k_3} \quad (136)$$

After the introduction $i = \frac{l_2}{l_1}$, we find that:

$$B = m_1p^2 + K_{red} - \frac{k_1^2}{k_1 + k_2 + k_3} \quad (137)$$

Where $K_{red} = \frac{k_1k_2i^2}{k_1 + k_2i^2}$ (the sign i is taken into account when creating expressions for potential energy)

Considering that is had been accepted by us that $k_1l_1-k_2l_2=0$, then $k_1 = k_2 \frac{l_2}{l_1} = k_2 i$, we write:

$$\frac{k_1k_2i^2}{k_1+k_2i^2} = \frac{k_1^2}{k_1+k_2+k_3} \tag{138}$$

From relation (137) it can be obtained that, under the condition of $k_1l_1-k_2l_2=0$, a simplified system ($J_2=0$, $m_2+m_3=0$) with a lever having $i>1$, in which a quasi-zero stiffness regime arises if the following relation is fulfilled:

$$k_3 = k_2(i-1) \tag{139}$$

Thus, the inertia-free lever in the circuit with an intermediate support on the elastic element implements the effect of the formation of the reduced characteristic according to the formulas (111), (121). If we assume that $k_3=\infty$ in the diagram (see Figure 37a), the result in this particular case coincides with the calculations according to the diagram in Figure 30b.

A Unit as a Beam-Type Lever

When taking into account the inertial properties of a lever that does not have a fixed turning point, we consider a beam-type scheme (Figure 39).

Figure 39. The computational scheme of the consecutive connection of $W_1=k_1$ and $W_2=k_2$ with the lever given its inertial properties

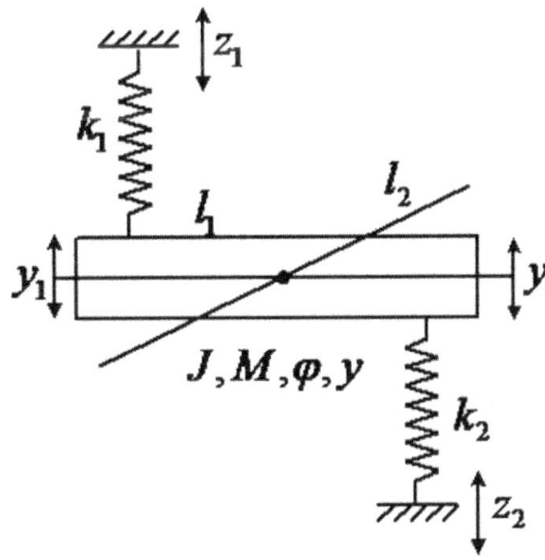

Let us write for the system, shown in Figure 39, expressions for kinetic and potential energy:

$$T = \frac{1}{2}M\dot{y}^2 + \frac{1}{2}J\dot{\phi}^2,$$
$$\Pi = \frac{1}{2}k_1\left(y_1 - z_1\right)^2 + \frac{1}{2}k_2\left(y_2 - z_2\right)^2. \tag{140}$$

and introduce a number of necessary relations:

$$\varphi = \frac{y_2 - y_1}{l_1 + l_2}, y = \frac{y_2 l_1 - y_1 l_2}{l_1 + l_2}, y_1 = y - l_1\varphi, y_2 = y + l_2\phi.$$

Then

$$T = \frac{1}{2}M\left(a_1\dot{y}_1 + a_2\dot{y}_2\right)^2 + \frac{1}{2}Jb_1^2\left(\dot{y}_2 - \dot{y}_1\right)^2$$

or

$$T = \frac{1}{2}M\left(a_1^2\dot{y}_1^2 + 2a_1a_2a_2\dot{y}_1\dot{y}_2 + a_2^2\dot{y}_2^2\right) + \frac{1}{2}Jb_1^2\left(\dot{y}_2^2 - \dot{y}_2\dot{y}_1 + \dot{y}_1^2\right).$$

To obtain differential equations of motion of systems, we use the Lagrange second-kind equations:

$$Ma_1\ddot{y}_1 + Jb_1^2\ddot{y}_1 + Ma_1a_2\ddot{y}_1 - Jb_1^2\ddot{y}_1 + k_1y_1 = k_1z_1$$
$$Ma_2\ddot{y}_2 + Jb_1^2\ddot{y}_2\phi + Ma_1a_2\ddot{y}_1 - Jb_1^2\ddot{y}_1 + k_2y_2 = k_2z_2. \tag{141}$$

The structural diagram corresponding to the computational scheme in Figure 39, has the form in accordance with Figure 40, from where we can obtain the transfer function of the system shown in Figure 40:

$$W(p) = \frac{\bar{y}_2}{\bar{z}_1} = \frac{k_1}{[(Ma_1^2 + Jb_1^2)p^2 + k_1][(Ma_2^2 + Jb_2^2)p^2 + k_2] - (Ma_1a_2 + Jb_1^2)p^4} \tag{142}$$

Let us make the assumption that $M \neq 0$ and $J=0$,, but first we find the frequency equation:

$$B = \left(Ma_1^2 + Jb_1^2\right)\left(Ma_2^2 + Jb_1^2\right)p^4 + p^2k_1\left(Ma_2^2 + Jb_1^2\right) +$$
$$+ p^2k_2\left(Ma_1^2 + Jb_1^2\right) + k_1k_2 - p^4\left(Ma_1a_2 - Jb_1^2\right)^2$$

which after some transformations will take the form:

$$B = p^4 MJb_1 \left(a_1 + a_2\right)^2 + p^2 \left[k_1 \left(Ma_2^2 + Jb_1^2\right) + k_2 \left(Ma_1^2 + Jb_1^2\right)\right] + k_1 k_2. \tag{143}$$

If J=0, then

$$B = p^2 M\left[k_1 a_2^2 + k_2 a_1^2\right] + k_1 k_2 \tag{144}$$

or

$$B = p^2 M + \frac{k_1 k_2}{k_1 a_2^2 + k_2 a_1^2} \tag{145}$$

where

$$a_1 = \frac{l_2}{l_1 + l_2}, a_2 = \frac{l_1}{l_1 + l_2}, k_1 a_2^2 = \frac{k_1 l_1^2}{\left(l_1 + l_2\right)^2}, k_2 a_1^2 = \frac{k_2 l_2^2}{\left(l_1 + l_2\right)^2} \tag{146}$$

We substitute (144) into (143), and after some transformations, by assuming that $i = \frac{l_2}{l_1 + l_2}$, we find:

$$B = p^2 M + \frac{k_1 k_2}{k_1 + k_2 i^2} + \frac{2k_1 k_2 i}{k_1 + k_2 i^2} + \frac{k_1 k_2 i^2}{k_1 + k_2 i^2} = p^2 M + \frac{k_1 k_2 (1+i)^2}{k_1 + k_2 i^2}. \tag{147}$$

Analyzing (5.147) we can speak about the reduced characteristic for a weighty lever with J=0, but $M \neq 0$. In this case, we obtain:

$$k_{red} = \frac{k_1 k_2}{k_1 + k_2 i^2} + \frac{2k_1 k_2 i}{k_1 + k_2 i^2} + \frac{k_1 k_2 i^2}{k_1 + k_2 i^2} = \frac{k_1 k_2 i^2}{k_1 + k_2 i^2} + \frac{k_1 k_2 (1 + 2i)}{k_1 + k_2 i^2} \tag{148}$$

whence it follows that another component occurs in the expression for reduced stiffness:

$$\frac{k_1 k_2 \left(1 + 2i\right)}{k_1 + k_2 i^2} \tag{149}$$

If $k_1 = k_2 = k, a_1 = a_2 = 1/2, i = \frac{l_2}{l_1} = 1$, then

$$k_{red} = 2k \tag{150}$$

In this particular case, the consecutive connection of elastic elements degenerates into a parallel one.

We will assume that M=0 and J≠0, and then we obtain that:

$$B = p^2 J\left[k_1 b_1^2 + k_2 b_1^2\right] + k_1 k_2 \tag{151}$$

where

$$b_1 = \frac{1}{l_1 + l_2}, k_1 b_1^2 = \frac{k_1}{(l_1 + l_2)^2}, k_2 b_1^2 = \frac{k_2}{(l_1 + l_2)^2}. \tag{152}$$

We substitute (151) into (150) and after some transformations, by assuming that $i = \frac{l_2}{l_1}$, we get:

$$B = p^2 J + \frac{k_1 k_2}{k_1 + k_2}(l_1 + l_2)^2 = p^2 J + k_{seq}(l_1 + l_2)^2 \tag{153}$$

$$k_{red} = k_{seq}(l_1 + l_2)^2 \tag{154}$$

The system of differential equations of motion in the generalized coordinates y, φ for the system shown in Figure 39, has the form:

$$\begin{aligned} &M\ddot{y} + y(k_1 + k_2) - \phi(k_1 l_1 - k_2 l_2) = k_1 z_1 + k_2 z_2, \\ &J\ddot{\phi} + y(k_1 + k_2) - \phi(k_1 l_1^2 - k_2 l_2^2) = k_1 l_1 z_1 + k_2 l_2 z_2. \end{aligned} \tag{155}$$

The corresponding structural diagram of the system is shown in Figure 41, where the transfer function can be determined

$$W(p) = \frac{\bar{\phi}}{\bar{z}_1} = \frac{(Mp^2 + k_1 + k_2)c_1 l_1 + k_2(k_1 l_1 - k_2 l_2)}{(Mp^2 + k_1 + k_2)(Jp^2 + k_1 l_1^2 + k_2 l_2^2) - (k_1 l_1 - k_2 l_2)^2} \tag{156}$$

for which the frequency equation has the form:

$$B = \left(Mp^2 + k_1 + k_2\right)\left(Jp^2 + k_1 l_1^2 + k_2 l_2^2\right) - \left(k_1 l_1 - k_2 l_2\right)^2 \tag{157}$$

If we accept that J=0, then the expression (157) will be simplified to:

$$B = Mp^2\left(k_1 l_1^2 + k_2 l_2^2\right) + k_1\left(k_1 l_1^2 + k_2 l_2^2\right) - k_2\left(k_1 l_1^2 + k_2 l_2^2\right) - k_1^2 l_1^2 + 2k_1 k_2 l_1 l_2 - k_2^2 l_2^2. \tag{158}$$

Figure 40. The structural diagram of the system in Figure 39

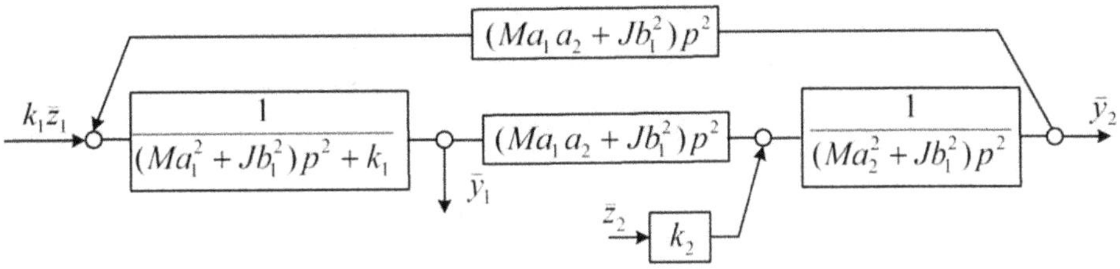

Figure 41. The structural diagram of the system in Figure 30

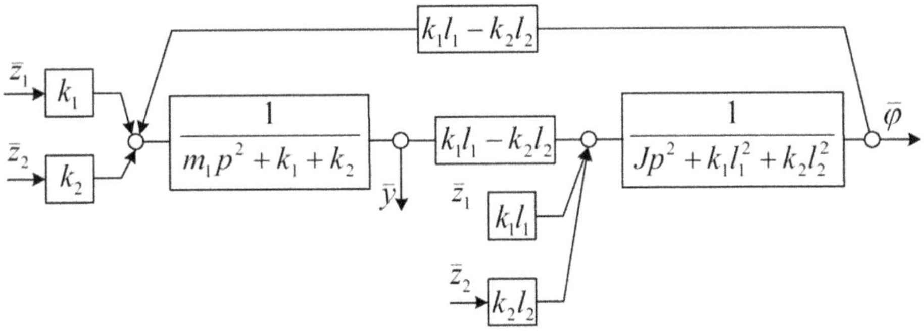

Having made some transformations, we will find the reduced coefficient:

$$k_{red} = \frac{k_1k_2l_2^2}{k_1l_1^2 + k_2l_2^2} + \frac{2k_1k_2l_1l_2}{k_1l_1^2 + k_2l_2^2} + \frac{k_1k_2l_1^2}{k_1l_1^2 + k_2l_2^2} \tag{159}$$

or

$$k_{red} = \frac{k_1k_2i^2}{k_1 + k_2i^2} + \frac{2k_1k_2i^2}{k_1 + k_2i^2} + \frac{k_1k_2}{k_1 + k_2i^2} \tag{160}$$

where $i = \frac{l_2}{l_1}$ (features of the lever of the second kind are taken into account in the expression for potential energy (140)).

From the analysis (160) follows:

- if l_2=0, then i=0 and $k_{red}=k_2$;
- if i=1, then $k_{red} = \frac{4k_1k_2}{k_1 + k_2i^2} = 4k_{seq}$;
- if $k_1=k_2=k$, then $i=1–k_{red}=2k$;
- if $i=\infty–k_{red}=k_1$.

Returning to the expression (5.43), we write that:

$$B = p^4 MJb_1 (a_1 + a_2)^2 + p^2 \left[k_1 \left(Ma_2^2 + Jb_1^2 \right) + k_2 \left(Ma_1^2 + Jb_1^2 \right) \right] + k_1 k_2 .$$

Whence, denoting $\frac{J}{M} = \gamma$ and dividing into b_1^2, we find:

$$B = p^4 MJb_1 + p^2 M \left[k_1 \frac{a_2^2}{b_1^2} + \gamma k_1 + k_2 \frac{a_1^2}{b_1^2} + \gamma k_2 \right] + k_1 k_2 \tag{161}$$

where

$$a_1 = \frac{l_2}{l_1 + l_2}, a_2 = \frac{l_1}{l_1 + l_2}, b_1 = \frac{1}{l_1 + l_2}. \tag{162}$$

Substituting (162) into (161) and making the transformations, we get:

$$B = p^4 MJb_1 + p^2 M \left[k_1 \left(l_1^2 + \gamma \right) + k_1 \left(l_1^2 + \gamma \right) \right] + k_1 k_2 \tag{163}$$

whence we find the full (taking into account the mass-and-inertia properties of the lever) value of k_{red}:

$$k_{red} = \frac{k_1 k_2 (1+i)^2}{k_1 + k_2 i^2 + \frac{J}{Ml_1^2}(k_1 + k_2)} \tag{164}$$

Let us denote that $\frac{J}{Ml_1^2} = \theta$, then the expression (164) takes the form:

$$k_{red} = \frac{k_1 k_2 (1+i)^2}{k_1 + k_2 i^2 + \theta (k_1 + k_2)} \tag{165}$$

Taking Account of the Possible "Sliding" Point of the Lever Support

To assess the properties of the lever with a sliding point of support, let us consider the system presented in Figure 42.

Figure 42. The computational scheme of the system (a beam on two supports)

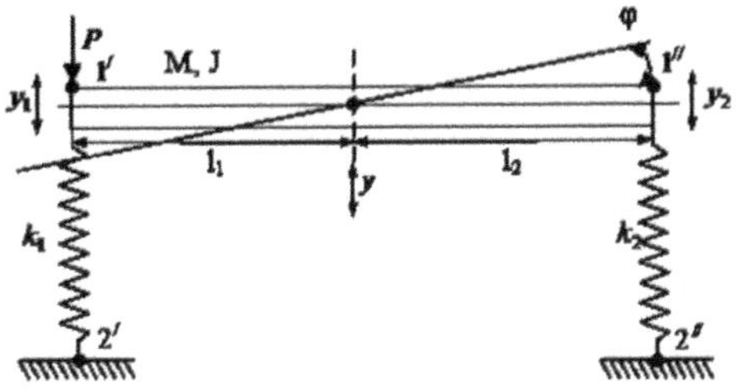

We write the equations for the kinetic and potential energies in the system of generalized coordinates y, φ:

$$T = \frac{1}{2}J\dot{\varphi}^2 + \frac{1}{2}M\dot{y}^2, \Pi = \frac{1}{2}k_1\left(y - l_1\varphi\right)^2 + \frac{1}{2}k_2\left(y + l_2\varphi\right)^2 \tag{166}$$

Using the Lagrange formalism of the second kind, we obtain the equations of motion of the system:

$$\begin{aligned} &J\ddot{\phi} + \phi\left(k_1 l_1^2 + k_2 l_2^2\right) - y\left(k_1 l_1 - k_2 l_2 = 0,\right. \\ &M\ddot{y} + y\left(k_1 + k_2\right) - \phi\left(k_1 l_1 - k_2 l_2\right) = 0, \end{aligned} \tag{167}$$

and build the structural diagram, appropriate to the computational scheme (Figure 42), as shown in Figure 43.

Figure 43. The structural diagram of the system in the y,φ coordinate system

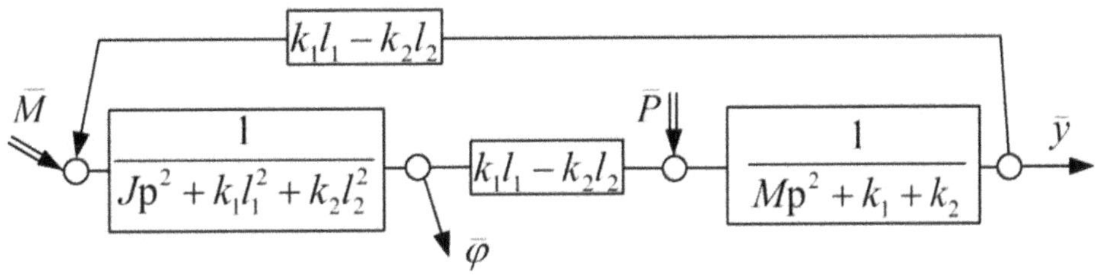

We also write expressions for the kinetic and potential energy in the y_1, y_2 coordinate system:

$$\begin{aligned} T &= \frac{1}{2}Ja^2\left(\dot{y}_2 - \dot{y}_1\right)^2 + \frac{1}{2}M\left(a_1\dot{y}_1 + a_2\dot{y}_2\right)^2, \\ \Pi &= \frac{1}{2}k_1 y_1^2 + \frac{1}{2}k_2 y_2^2, \end{aligned} \tag{168}$$

where

$$a^2 = \frac{1}{(l_1 + l_2)^2}, a_1 = \frac{l_2}{l_1 + l_2}, a_2 = \frac{l_1}{l_1 + l_2}.$$

Using similar techniques, we obtain the corresponding equations of motion of the system:

$$\begin{aligned} &Ja^2\ddot{y}_1 - Ja^2\ddot{y}_2 + Ma_1^2\ddot{y}_1 - Ma_1a_2\ddot{y}_2 + k_1y_1 = 0, \\ &Ja^2\ddot{y}_2 - Ja^2\ddot{y}_1 + Ma_2^2\ddot{y}_2 - Ma_1a_2\ddot{y}_1 + k_2y_2 = 0, \end{aligned} \tag{169}$$

and the structural diagram of the system will take the form shown in Figure 44.

Figure 44. The structural diagram of the system presented in Figure 42 in the y_1, y_2 coordinate system

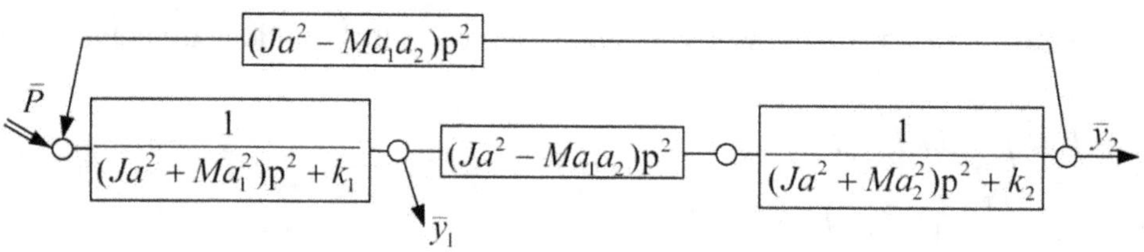

The transfer function of the system is written as:

$$W(p) = \frac{\bar{y}_1}{\bar{P}} = \frac{\left(Ja^2 + Ma_1^2\right)p^2 + k_2}{p^4\left(JMa^2a_1^2 + JMa^2a_2^2\right) + p^2\left[k_1\left(Ja^2 + Ma_1^2\right) + k_2\left(Ja^2 + Ma_1^2\right)\right] + k_1k_2}, \tag{170}$$

where will we find the frequency equation of the system:

$$p^4\left(JMa^2a_1^2 + JMa^2a_2^2\right) + p^2\left(k_1\left(Ja^2 + Ma_1^2\right) + k_2\left(Ja^2 + Ma_1^2\right) + k_1k_2. \tag{171}$$

If we assume that J=0, M=0, then applying force P at point (1′), we have only one spring k_1. We will get the same, if the force P is applied at the point (1″), then the parallel connection degenerates into spring k_2.

Thus, if a regular parallel connection is subject to the "spacing-apart" procedure and, instead of points (1) and (2), points (1′), (1″), (2′) and (2″) are introduced into consideration (Figure 42), then the interaction parameters between the units are determined by external factors correlated with actions at points (1′), (1″), (2′) and (2″). So, if the force P is applied only at the point (1′), then instead of the total stiffness k_1+k_2(as in the usual parallel connection), we obtain the value of the reduced stiffness k_{red}=k_1, and with a load at the point (1″) it follows that k_{red}=k_2, respectively (Figure 43). When the force P is applied at an intermediate point, for example, E (Figure 45) at the distance r from the center of gravity of a weighty lever, it is necessary to take into account that the force loading circuit will have its own characteristics (see Figure 45).

Figure 45. The computational scheme to determine power loads at point (1′) and (1″)

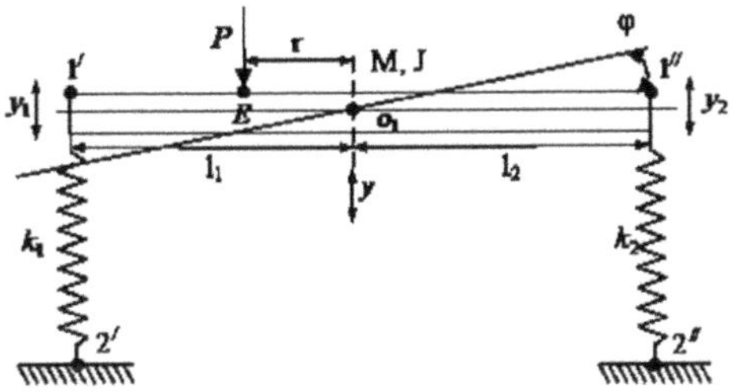

When the force P is transferred to the center of mass of the lever $o_{1'}$, a moment of forces $M_{ext}=P\cdot r$ appears in the system; therefore, when considering the motion in the y, φ coordinate system as external perturbation forces, the corresponding terms P and Pr emerge in the right-hand part of the system of differential equation (169).

If the y_1 and y_2 coordinate system is used, then the forces at the point (1′) and (1″) will have two components.

$$P_{1'} = P\frac{l_2}{l_1+l_2} + P\frac{r}{l_1+l_2} \tag{172}$$

$$P_{1''} = P\frac{l_2}{l_1+l_2} - P\frac{r}{l_1+l_2} \tag{173}$$

or after transformation:

$$P_{1'} = P\frac{l_2+r}{l_1+l_2} \tag{174}$$

$$P_{1''} = P\frac{l_2-r}{l_1+l_2} \tag{175}$$

If $r=l_1$, then we will have:

$$P_{1'} = P \tag{176}$$

$$P_{1'} = 0. \tag{177}$$

However, in the derivation of the transfer functions, it will be necessary to make appropriate adjustments to the structural diagrams, as shown in Figure 46.

Figure 46. The structural schemes with adjustments: a) the y_0 coordinate system, φ; b) the $y_{1'}$, y_2 coordinate system

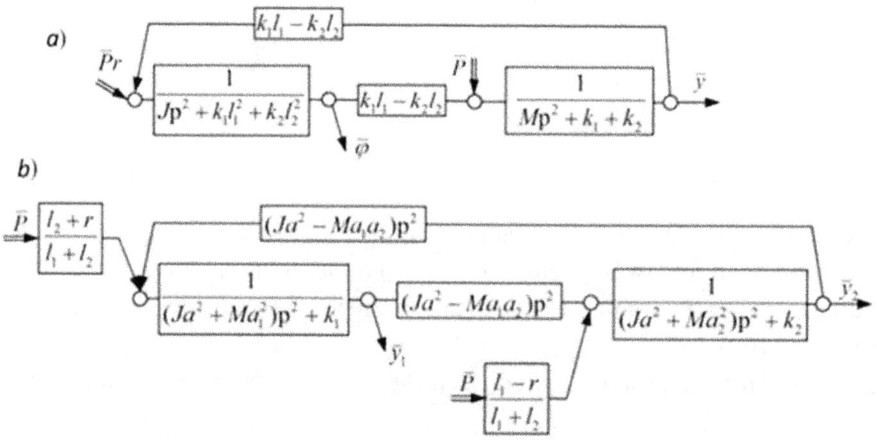

In connection with the correction made, it is obtained that:

$$W(p) = \frac{\bar{y}_{1'}}{\bar{P}_{1'}} = \left(\frac{l_2 + r}{l_1 + l_2}\right)\left[\frac{(Ja^2 + Ma_1^2)p^2 + k_2}{p^4(JMa^2a_1^2 + JMa^2a_2^2) + p^2[k_1(Ja^2 + Ma_1^2) + k_2(Ja^2 + Ma_1^2)] + k_1k_2}\right] + \\ + \left(\frac{l_1 - r}{l_1 + l_2}\right)\left[\frac{(Ja^2 + Ma_1^2)p^2 + k_2}{p^4(JMa^2a_1^2 + JMa^2a_2^2) + p^2[k_1(Ja^2 + Ma_1^2) + k_2(Ja^2 + Ma_1^2)] + k_1k_2}\right] \tag{178}$$

or after transformation:

$$W(p) = \frac{\bar{y}_{1'}}{\bar{P}_{1'}} = \frac{(l_2 + r)[(Ja^2 + Ma_1^2)p^2 + k_2] + (l_1 - r)[(Ja^2 + Ma_1^2)p^2 + k_2]}{(l_1 + l_2)p^4(JMa^2a_1^2 + JMa^2a_2^2) + p^2[k_1(Ja^2 + Ma_1^2) + k_2(Ja^2 + Ma_1^2)] + k_1k_2} \tag{179}$$

When J=0 and M=0, the expression (179) takes the form:

$$W = \frac{y_{1'}}{P_{1'}} = \frac{(l_2 + r)k_2 + (l_1 - r)k_1}{(l_1 + l_2)[k_1k_2]} \tag{180}$$

and when $l_1=l_2=l$,

$$W = \frac{y_{1'}}{P_{1'}} = \frac{(l+r)k_2 + (l-r)k_1}{2l[k_1k_2]} \tag{181}$$

respectively.

In a more compact form (181) is written as:

$$W = \frac{y_{1'}}{P_{1'}} = \frac{(l+r)k_2 + (l-r)k_1}{2l[k_1k_2]} \tag{182}$$

Summarizing some of the research, we can conclude that if the lever has mass and moment of inertia, then in this form it represents the usual computational scheme of a mechanical oscillatory system in the form of a solid body on two elastic supports (a beam on two elastic supports). In this case, as it were, the line between the consecutive and parallel connection is erased, if the application of force corresponds to its action on the lever arm between the spacing-apart points ($1'$), ($1''$), ($2'$) and ($2''$) (point E in Figure 45).

However, the results obtained are of practical importance, since the developed approach makes it possible to introduce into consideration the details of the interaction, determined by the nature of the superimposed connections implemented through mechanical units, mechanical chains. The latter is important for a more complete account of the constructive properties and characteristics of dynamic oscillatory systems, when their dynamic characteristics are determined for practical purposes.

Note also that for all the similarity of the loading schemes, there are quite definite differences. With a consecutive connection, an external force is applied to point (1) and "spaced apart" with a lever to point 2 ($2'$ and $2''$), while point 3 can be connected with a fixed reference system (fixed surface) or come into contact with the corresponding elements (units) of the system, in which the initial formation is included as a constituent part. Therefore, a consecutive connection essentially defines a connection between points 1 and 3. In this plan, a parallel connection is considered as a connection between points 1 and 2, which can be distributed using levers to points ($1'$), ($1''$), ($2'$) и ($2''$).

If a lever is considered, having inertial properties as shown in Figure 47a and b, then external influences are applied to various points:

- With a consecutive connection – to the point 1 (Figure 47a);
- With parallel connection – to the point $1'$ or $1''$ (Figure 47b).

Not excluding the possibility of coincidence of the results in cases *a*) and *b*) (Figure 47), we should note that the difference between the two types of compounds is still rather fundamental. This refers to the fact that the external influence is applied to different points of the mechanical system. The case presented in Figure 47a and b, is of a specific nature, reflecting the blurring of the difference between a purely consecutive and a purely parallel connection. As studies show, combined and more complex variants are also possible. When the parameters of the lever are zeroed, they are reduced to the original known rules for connecting elements.

Figure 47. The schemes of the consecutive (a) and parallel (b) connection of two springs k_1 and k_2 using a lever

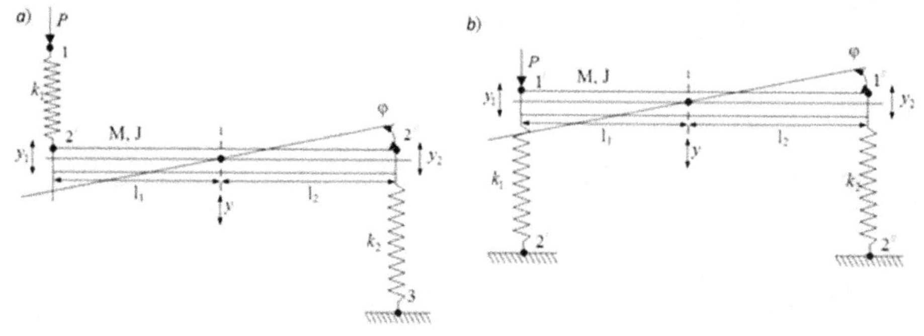

The above makes it difficult to use a lever in structural interpretations as another additional type unit of the first level into the expanded set. But it definitely introduces a special method or a new type of interconnection of two units that takes into account an important property in specific constructive and technical solutions, when several simultaneously operating units take part in elastic interactions, the line of action of the efforts in which is not on the same line.

Developing this approach, one can imagine the connection of a certain set of links (these can be springs, dampers, etc.), which are connected through a complex lever, as shown in Figure 48a and b.

Figure 48. Connection diagrams using a three-point planar lever

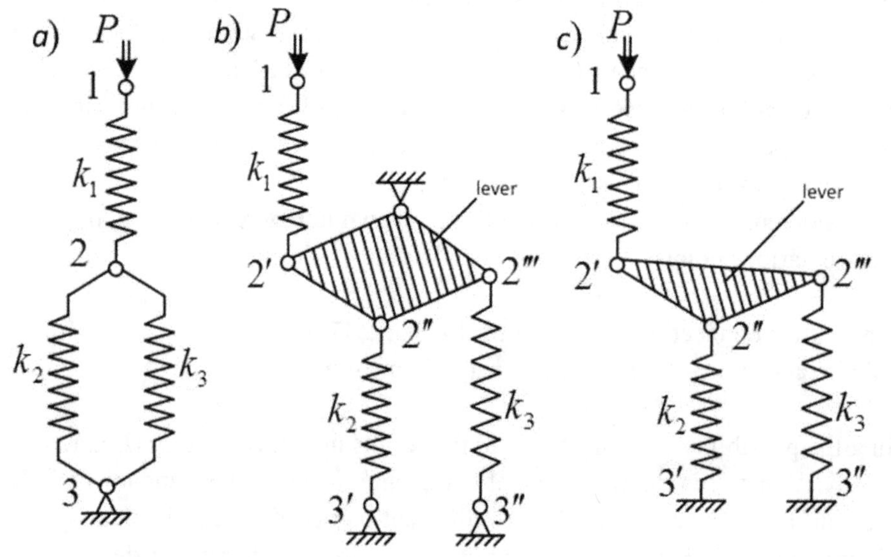

If Figure 48a presents the scheme of consecutive-parallel connection of three springs k_1, k_2 and k_3, the reduced stiffness of which has the form:

$$K_{red} = \frac{(k_2 + k_3)k_1}{k_1 + k_2 + k_3} \tag{183}$$

then Figure 48b and c shows possible connection options that can be obtained by inserting a lever with a fixed point of support or a weighty lever at which the center of rotation is not fixed.

CONCLUSION

Thus, there is an opportunity to implement new ways of connecting typical elementary units through a lever, to create some basis for finding patterns for constructing complex oscillatory structures that various applications can find. It is not excluded that the construction of molecular structures, in addition to the classical ideas about elastic interactions, can be based on the use of special types (or types) of compounds.

REFERENCES

Belokobylsky, S. V., Eliseev, S. V., & Kashuba, V. B. (2013). *Applied Problems of the Structural Theory of Vibration Protection Systems*. Polytekhnika Publisher.

Belokobylsky, S. V., Eliseev, S. V., & Sitov, I. S. (2013). *Dynamics of Mechanical Systems: Lever Linkages and Inertial-Elastic Constraints*. Polytekhnika Publisher.

Belokobylsky, S. V., & Sitov, I. S. (2007). Ways and means of changing the dynamic state of the technological complex for vibration smoothing. Modern technologies. System analysis. *Modeling*, (2), 46–52.

Druzhinsky, I. A. (1977). *The Mechanical Chains*. Mashinostroenie Publ.

Eliseev, A.V., Kuznetsov, N.K. & Moskovskikh, A.O. (2019). The Dynamics of Machines: System Conception, Structural Schemes and Connections of Elements. *Innovative Mechanical Engineering*.

Eliseev, S. V., & Eliseev, A. V. (2020). *Theory of Oscillations: Structural Mathematical Modeling in Problems of Dynamics of Technical Objects. Studies in Systems, Decision and Control* (Vol. 252). Springer International Publishing. doi:10.1007/978-3-030-31295-4

Eliseev, S. V., & Khomenko, A. P. (2009). Oscillation stability in two-dimensional systems with non-traditional constraints. Modern technologies. System analysis. *Modeling*, (2), 8–12.

Eliseev, S. V., Khomenko, A. P., & Upir', R. Yu. (2009). Mechatronics of vibration protection systems with lever linkages. Modern technologies. System analysis. *Modeling*, (3), 8–16.

Eliseev, S. V., Lukyanov, A. V., Reznik, Yu. N., & Khomenko, A. P. (2006). *Dynamics of Mechanical Systems with Additional Ties*. Irkutsk State University.

Eliseev, S. V., Reznik, Yu. N., Khomenko, A. P., & Zasyadko, A. A. (2008). *Dynamic synthesis in generalized problems of vibration protection and vibration isolation of technical objects*. ISU Publ.

Eliseev, S. V., & Upir', R. Yu. (2008). Mechatronic approaches in the tasks of vibration protection of machines and equipment. Modern technologies. System analysis. *Modeling*, (4), 8–16.

Eliseev, S. V., Upir', R. Yu., Nasnikov, D. N., & Logunov, A. S. (2007). The system of new model links in the structural representations of mechanical oscillatory systems. In *Innovative technologies of transport and industry: Proc. of the 45th International scientific and practical conf.* (*Vol. 3*, pp. 173-182). Khabarovsk: FEDUPS Publishing House.

Ermoshenko, Yu. V., & Fomina, I. V. (2009). Dynamic absorbing in vibration protection systems using L-shaped lever linkages. Modern technologies. System analysis. *Modeling*, (2), 85–88.

Genkin, M. D., & Ryaboy, V. M. (1988). *Inert-vibration isolation systems: Ultimate opportunities, optimal structures*. Nauka Publ.

Ivanov, B. G. (2007). *Development of methods for calculating the dynamics and strength of transport equipment aggregates with articulated kinematic relations* [PhD thesis]. Samara.

Upir', R. Yu. (2009). *Dynamics of mechanical oscillatory systems taking into account the spatial forms of the connection of elementary units* [PhD thesis]. IrGUPS Publ.

ADDITIONAL READING

Belokobylytskyi, S. V., Eliseev, A. V., Sitov, I. S., & Artyunin, A. I. (2017). Mathematical modeling of chain mechanical systems. Systems. Methods. *Technologies*, *1*(33), 7–18.

Eliseev, A. V., & Bol'shakov, R. S. (2022). System approaches to generalized assessment of dynamic properties of technical objects under connected loads. Modern Technologies. System Analysis. *Modeling*, *1*(73), 39–52.

Eliseev, A. V., Kargapol'tsev, S. K., Bol'shakov, R. S., & Eliseev, S. V. (2020). Mechanical chains in problems of correcting dynamic states of vibration machines. *Bulletin of Irkutsk State Technical University*, *24*(4), 718–727. doi:10.21285/1814-3520-2020-4-718-727

Eliseev, A. V., & Kopylov, Y. R. (2014). Features of technical implementation of vibration strengthening technology. *Modern Science and Technology*, *12*(2), 152–155.

Eliseev, A. V., Mironov, A. S., & Eliseev, S. V. (2022). Formation of mathematical models of vibrational interactions between elements of transport and technological equipment. Bulletin of Voronezh State University. Series. *System Analysis and Information Technologies*, *1*, 32–42.

Eliseev, A. V., Sitov, I. S., & Kuznetsov, N. K. (2022). System approach to assessing dynamic states of transport and technological objects: Structural schemes, transfer functions, dynamic invariants. Systems. Methods. *Technologies*, *2*(54), 7–19.

Eliseev, S. V., Kargapol'tsev, S. K., Bol'shakov, R. S., & Eliseev, A. V. (2020). Mechanical chains in problems of correcting dynamic states of vibration machines. *Bulletin of Irkutsk State Technical University*, *24*(4), 718–727. doi:10.21285/1814-3520-2020-4-718-727

Orlenko, A. I., Eliseev, A. V., & Sitov, I. S. (2017). Mathematical models of contact interactions during movement with detachment stages. Systems. Methods. *Technologies*, *4*(36), 18–24.

Chapter 6
Forms and Features of Implementation of Lever Linkages in Mechanical Vibrational Systems:
Features of Dynamic Effects

ABSTRACT

The chapter discusses mathematical models of lever mechanisms in mechanical systems and features of their modeling. Mathematical models are considered taking into account the options for the placement of reference points: the installation of a pore on a movable element, the connection of two elements with an intermediate link in the form of a lever of the second kind, a variant of a non-supporting lever. Lever connections and dynamic effects are evaluated taking into account the introduction of additional connections. The features of the structure and mathematical model of a system with lever mechanisms under kinematic disturbances are considered. The dynamic properties of coordinate blocking and the influence of the choice of the coordinate system are described.

INTRODUCTION

In the theory of mechanical chains, the lever mechanism (or lever) is used as a form of connection of typical elements (Druzhinsky, 1977; Eliseev et al., 2008), however, the lever is not included in the set of standard units, which is associated with its specific features. In Eliseev et al. (2012b) and Khomenko and Eliseev (2010), it was noted that lever linkages result lead to changes in the topological integrity of a mechanical chain, just as it happens by analogy in the theory of electrical circuits when a transformer is introduced. In the general case, if the transfer ratio of the lever is not equal to one, the mechanical chain takes on a special form (or "metric"), which is associated with the emergence of parallel forces acting on the lever or intermediate solid body, which can have various forms of placing the fulcrum.

DOI: 10.4018/978-1-6684-7237-8.ch006

CONSTRUCTION OF MATHEMATICAL MODELS: FEATURES OF LEVER LINKAGES IN MECHANICAL OSCILLATORY SYSTEMS

Consider the computational scheme of a technological machine in the form of a chain mechanical system with two degrees of freedom, as shown in Figure 1a-f, where the following notation is used: m_1 and m_2 are the mass of the elements of the chain system; $k_1 \div k_4$ are stiffnesses of the elastic elements; z_1 and z_2 are kinematic external influences; point (1) is the junction point of the elements k_2 and k_3; points (1′) and (1″) are the ends of the lever with lengths l_1 and l_2; point (1) in Figure 1a is the junction point of the two elements k_1 and k_2; in Figure 1b-e point (1) is the lever support; in Figure 1f it is the center of mass of the system. Input lever in Figure 1b-e is weightless; in Figure 1f the lever has a mass m and a moment of inertia J. To describe the state of the system, coordinates $y_1 \div y_4$ in the fixed frame are used; y and φ (in Figure 1f) are the coordinates of the lever as a solid body.

Figure 1. Variants of the location of the contact point of two typical elements of k_2 and k_3 using lever linkages

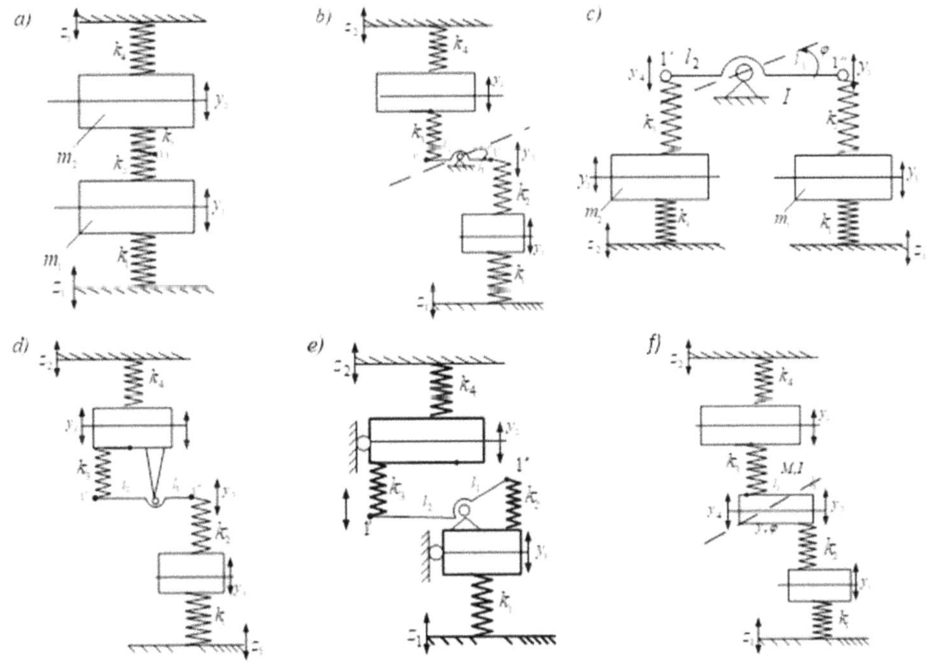

The problem is to evaluate the possible forms of joining elements with parameters k_2 and k_3 (Figure 1a), if the point of connection point (1) can be separated by a lever (l_1 and l_2) by points (1′) and (1″). The first option is related to the consideration of the pivot point (point (1)) of the lever associated with the fixed rack, as shown in Figure 1b and c.

To simplify the subsequent calculations, we will assume that the lever also has a moment of inertia J and has the ability to rotate through an angle φ. Further, after the derivation of the equations, it will be assumed that $J = 0$. Thus, the initial computational scheme will have three degrees of freedom, y_1, y_2 and φ. In this case, φ is determined unambiguously through y_3 and y_4, when using lever linkages.

Comparative Evaluation of Placement Options for Control Points

As a basic model, we consider the computational scheme in Figure 1a where the elements k_2 and k_3 are connected in series, which can be reflected through the reduced stiffness:

$$k_{red} = \frac{k_2 k_3}{k_2 + k_3} \tag{61}$$

The expressions for the kinematic and potential energies of the base model (Figure 1) are:

$$T = \frac{1}{2} m_1 \dot{y}_1^2 + \frac{1}{2} m_2 \dot{y}_2^2, \tag{2}$$

$$\Pi = \frac{1}{2} k_1 (y_1 - z_1)^2 + \frac{1}{2} k_{red} (y_2 - y_1)^2 + \frac{1}{2} k_4 (y_2 - z_2)^2. \tag{2'}$$

We write the system of differential equations of motion:

$$m_1 \ddot{y}_1 + y_1 (k_1 + k_{red}) - y_2 k_{red} = k_1 z_1, \tag{3}$$

$$m_2 \ddot{y}_2 + y_2 (k_{red} + k_4) - y_1 k_{red} = k_4 z_2. \tag{4}$$

The structural diagram of a dynamically equivalent automatic control system is shown in Figure 2.

Figure 2. The structural diagram of the system corresponding to Figure 1a

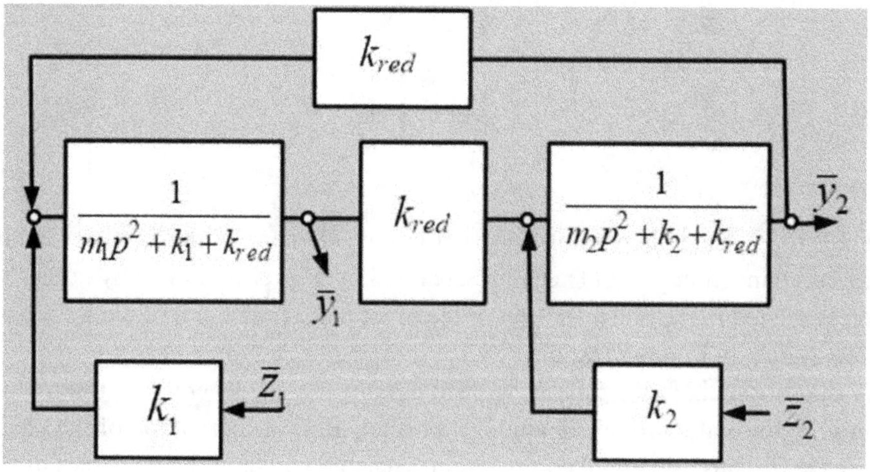

The system under consideration, as follows from Figure 2, has two partial systems with partial oscillation frequencies:

$$\omega_1^2 = \frac{k_1 + k_{red}}{m_1}, \tag{5}$$

$$\omega_2^2 = \frac{k_2 + k_{red}}{m_2} \tag{6}$$

Note that the structural diagram is a graphical analogue of equations (3) and (4) in the field of Laplace transformations ($p = j\omega, j = \sqrt{-1}$). In the system (Figure 2), the connections between the partial structures are elastic. In this case, the resistance forces are considered small. The system has two "input signals" (kinematic perturbation $\bar{z}_1$ and $\bar{z}_2$), which allows us to find a number of transfer functions. We will assume that in all the cases considered, the perturbation has the same frequency. If, a, then the transfer function from on y_1 and y_2 will take the form:

$$\bar{W}_1 = \frac{\bar{y}_1}{\bar{z}_1} = \frac{k_1(m_2p^2 + k_2 + k_{red})}{(m_1p^2 + k_1 + k_{red})(m_2p^2 + k_2 + k_{red}) - k_{red}^2}, \tag{7}$$

$$\bar{W}_2 = \frac{\bar{y}_2}{\bar{z}_1} = \frac{k_1k_{red}}{(m_1p^2 + k_1 + k_{red})(m_2p^2 + k_2 + k_{red}) - k_{red}^2}, \tag{8}$$

Similarly, other necessary transfer functions can be found.

Let us find the expression for the kinetic and potential energy for the system in Figure 1b.

$$T = \frac{1}{2}m_1\dot{y}_1^2 + \frac{1}{2}m_2\dot{y}_2^2 + \frac{1}{2}J\dot{\varphi}^2, \tag{9}$$

$$\Pi = \frac{1}{2}k_1(y_1 - z_1)^2 + k_2(y_3 - y_1)^2 + \frac{1}{2}k_3(y_4 - y_2)^2 + \frac{1}{2}k_4(y_2 - z_2)^2. \tag{10}$$

In this case, the relations are satisfied, or then, taking into account the type of a lever linkage, it can be assumed that:

$$y_3 = -\frac{l_1}{l_2}y_4 \text{ or } y_3 = -iy_4 \tag{11}$$

where $i = \frac{l_1}{l_2}$ is called the transfer ratio of the lever. Note that when a lever of a different type the sign i is taken into account, that is, in the direction of the velocity points of the lever. Expressions (9)–(10) can be converted to the form:

$$T = \frac{1}{2} m_1 \dot{y}_1^2 + \frac{1}{2} m_2 \dot{y}_2^2 + \frac{1}{2} J \frac{\dot{y}_4^2}{l_2^2}, \tag{12}$$

$$\Pi = \frac{1}{2} k_1 (y_1 - z_1)^2 + \frac{1}{2} k_2 (-y_1 - i y_4)^2 + \frac{1}{2} k_3 (y_4 - y_2)^2 + \frac{1}{2} k_4 (y_2 - z_2)^2. \tag{13}$$

Let us assume that $\frac{J}{l_2^2} = m_0$, then the system of differential equations of motion is written as:

$$m_1 \ddot{y}_1 + y_1 (k_1 + k_2) + k_2 i y_4 = k_1 z_1, \tag{14}$$

$$m_2 \ddot{y}_2 + y_2 (k_3 + k_4) - y_4 k_3 = k_2 z_2, \tag{15}$$

$$m_0 \ddot{y}_4 + y_4 (k_2 i^2 + k_3) + y_1 k_2 i - k_3 y_2 = 0. \tag{16}$$

The structural diagram of the system in Figure 1b in coordinates y_1, y_2 and y_4 is presented in Figure 3.

Figure 3. The structural diagram of the system corresponding to Figure 1b

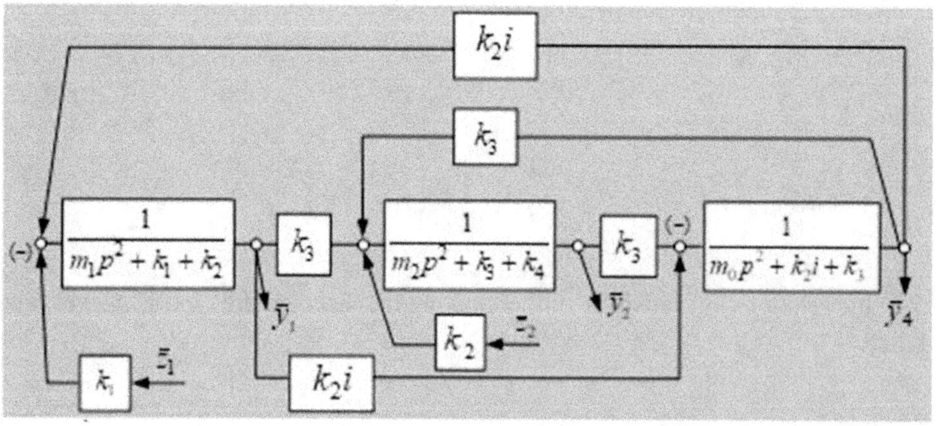

From Figure 1b it follows that the introduction of a mass-and-inertia lever added, compared with the scheme in Figure 1a, an additional degree of freedom. In this case, the cross-couplings of the system are elastic; the frequencies of the partial systems are determined accordingly:

$$\omega_1^2 = \frac{k_1 + k_2}{m_1}, \tag{17}$$

$$\omega_2^2 = \frac{k_3 + k_4}{m_2}, \tag{18}$$

$$\omega_3^2 = \frac{k_2 i}{m_0}. \tag{19}$$

To take into account the dynamic effects of the introduction of a weightless lever (rod), we assume that $m_0 = 0$. From (16) we find:

$$y_4 = -y_1 \frac{k_2 i}{k_2 i^2 + k_3} + y_2 \frac{k_3}{k_2 i^2 + k_3} \tag{20}$$

$$a = \frac{k_2 i}{k_2 i^2 + k_3}, \quad b = \frac{k_3}{k_2 i^2 + k_3} \tag{21}$$

then

$$y_4 = -y_1 a + y_2 b. \tag{22}$$

We substitute (22) into (14) and (15), we obtain the system of equations

$$m_1 \ddot{y}_1 + y_1 (k_1 + k_2 - k_2 a i) + k_2 i b y_2 = k_1 z_1, \tag{23}$$

$$m_2 \ddot{y}_2 + y_2 (k_3 + k_4 - k_3 b) + k_3 a y_1 = k_2 z_2. \tag{24}$$

Taking into account relations (21), these equations will take the form:

$$m_1 \ddot{y}_1 + y_1 \left(\frac{k_1 k_2 i^2 + k_1 k_3 + k_2 k_3}{k_2 i^2 + k_3} \right) + y_2 \frac{k_2 k_3 i}{k_2 i^2 + k_3} = k_1 z_1, \tag{25}$$

$$m_2\ddot{y}_2 + y_2\left(\frac{k_2k_3i^2 + k_4k_2i^2 + k_4k_3}{k_2i^2 + k_3}\right) + y_1\frac{k_2k_3i}{k_2i^2 + k_3} = k_2z_2. \tag{26}$$

When $i = 1$ (25), (26) reduces to a system of equations that differs from the system of equations (3), (4) in that the signs in front of the members $k_{red}y_1$ in (3) and $k_{red}y_2$ in (4) will be not a minus, but a plus. This means that the introduction of a lever of the first kind with $i = 1$ produces a change in the sign of the connection between partial systems. If in the initial system (without introducing a lever of the first kind), such a connection was positive (Figure 2), then, taking into account the introduction of a lever linkage of the specified type, the nature of the cross-coupling between partial systems changes, it becomes negative. This leads to a change in the denominator of the transfer functions (7), (8), i.e., to a change in the characteristic equation of a system. As for the choice of i, in which the system of equations (25), (26) would coincide with the system of equations (3), (4), this situation is not implemented.

If $i = 0$, then this corresponds to $l_1 = 0$, then (25), (26) will take the form of:

$$m_1\ddot{y}_1 + y_1(k_1 + k_2) = k_1z_1, \tag{27}$$

$$m_2\ddot{y}_2 + y_2k_4 = k_2z_2. \tag{28}$$

The presence of a support for the lever in this case "destroys" the original system. It turns into a system of two oscillatory circuits that do not interact with each other. The remaining part of the lever $l_2 \neq 0$ with $l_1 = 0$ does not affect the motion. However, if $J \neq 0$, then the contour m_2, k_2 forms together with the fragment of the lever $l_2 \neq 0$ a system with two degrees of freedom, but the interaction with the contour m_1, k_1, k_2 is destroyed. If $i = \infty$, then $l_2 = 0$; in this case, there is a symmetric, in relation to the above situation.

If the connection between elements k_1 and k_2 is implemented with the help of a lever of the second kind, the center of rotation of which coincides with one of the ends of the lever, then the system of equations of motion (25), (26) is written as follows:

$$m_1\ddot{y}_1 + y_1\left(\frac{k_1k_2i^2 + k_1k_3 + k_2k_3}{k_2i^2 + k_3}\right) - y_2\frac{k_2k_3i}{k_2i^2 + k_3} = k_1z_1, \tag{29}$$

$$m_2\ddot{y}_2 + y_2\left(\frac{k_2k_3i^2 + k_4k_2i^2 + k_4k_3}{k_2i^2 + k_3}\right) - y_1\frac{k_2k_3i}{k_2i^2 + k_3} = k_2z_2. \tag{30}$$

With such a transition, it is necessary to take into account that the expression (11) takes the form:

$$y_3 = iy_4, \tag{31}$$

which leads to significant changes in the properties of the system.

If i does not take extremal values, then for $i < 1$ and $i > 1$, it will be possible to observe a change in the dynamic properties of the system. Compare the two systems of equations of motion (3), (4) and (29), (30):

$$\left.\begin{aligned} m_1\ddot{y}_1 + y_1(k_1 + k_{red}) - y_2k_{red} = k_1z_1, \\ m_2\ddot{y}_2 + y_2(k_{red} + k_4) - y_1k_{red} = k_4z_2, \end{aligned}\right\} \tag{32}$$

$$\left.\begin{aligned} m_1\ddot{y}_1 + y_1\left(\frac{k_1k_2i^2 + k_1k_3 + k_2k_3}{k_2i^2 + k_3}\right) - y_2\frac{k_2k_3i}{k_2i^2 + k_3} = k_1z_1, \\ m_2\ddot{y}_2 + y_2\left(\frac{k_2k_3i^2 + k_4k_2i^2 + k_4k_3}{k_2i^2 + k_3}\right) - y_1\frac{k_2k_3i}{k_2i^2 + k_3} = k_2z_2. \end{aligned}\right\} \tag{32'}$$

From the system of equations (32), (32'), we can make the relations:

$$k_2+k_3i = k_2i^2+k_3, \tag{33}$$

from where

$$i^2k_2-i(k_2+k_3)+k_3 = 0. \tag{34}$$

The solution (34) with respect to i is:

$$i = \frac{k_2 + k_3}{2k_2} \pm \frac{(k_2 - k_3)}{2k_2}. \tag{35}$$

Thus, it is possible to find the values of i for which the two systems of equations (3), (4) and (29), (30) coincide:

$$i_1 = 1,\ i_2 = \frac{k_3}{k_2}. \tag{36}$$

Consideration of the computational scheme shown in Figure 1c gives the results similar to those obtained for the scheme in Figure 1b. However, the use of the diagram in Figure 1c gives a clearer picture of dynamic interactions. In Eliseev et al. (2012b), the features of systems with lever linkages of the first and second kinds were considered, where it was shown that the choice of i is also determined by the stability conditions.

The Case of the Location of the Support on the Movable Element

Consider a system whose computational scheme (Figure 1d) assumes the presence of a support on a mass of m_2. In this case, a lever of the first kind is considered, which has an inertia moment J with respect to the support point and a mass m. The center of gravity of the lever coincides with the fulcrum. Since there is a joint at the point (1), according to the diagram in Figure 1d, can be converted to a form, as shown in Figure 4.

Figure 4. The transformed computational scheme corresponding to the diagram in Figure 1d

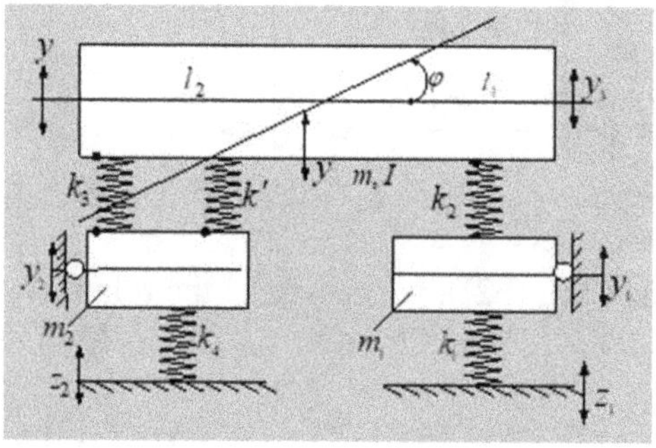

Note that when converting the fulcrum of the lever of the first kind, the point (1) is divided by a spring with stiffness k_0' that connects the lever and the mass m_2. In further calculations, this spring will be removed assuming that $k_0' \to \infty$. We introduce the variable $y - y_2 = y_0$, then we write the expression for kinetic and potential energies using the coordinates:

$$T = \frac{1}{2}J\dot{\varphi}^2 + \frac{1}{2}m\dot{y}^2 + \frac{1}{2}m_1\dot{y}_1^2 + \frac{1}{2}m_2\dot{y}_2^2, \tag{37}$$

$$\Pi = \frac{1}{2}k_1(y_1 - z_1)^2 + \frac{1}{2}k_2(y_3 - y_1)^2 + \frac{1}{2}k_4(y_2 - z_2)^2 + \frac{1}{2}k_3(y_4 - y_2)^2 + \frac{1}{2}k_0(y_0)^2. \tag{38}$$

For calculations, we introduce a series of relationships:

$$y = ay_3 + by_4,\ a = \frac{l_1}{l_1 + l_2},\ b = \frac{l_2}{l_1 + l_2},\ \text{Æ}= c(y_4 - y_3),\ c = \frac{1}{l_1 + l_2}. \tag{39}$$

Transform (37), (38) using relations (39) and get:

$$T = \frac{1}{2}m_1\dot{y}_1^2 + \frac{1}{2}m_2\dot{y}_2^2 + \frac{1}{2}Jc^2(\dot{y}_4 - \dot{y}_3)^2 + \frac{1}{2}m(a\dot{y}_3 + b\dot{y}_4)^2; \tag{40}$$

$$\Pi = \frac{1}{2}k_1(y_1 - z_1)^2 + \frac{1}{2}k_2(y_3 - y_1)^2 + \frac{1}{2}k_4(y_2 - z_2)^2 + \frac{1}{2}k_3(y_4 - y_2)^2 + \frac{1}{2}k_0(y_0)^2, \tag{41}$$

where $y_0 = y - y_2$.

Then assuming that $ay_3 + by_4 - y_2 = y_0$, we write the potential energy in the form:

$$\Pi = \frac{1}{2}k_1(y_1 - z_1)^2 + \frac{1}{2}k_2(y_3 - y_1)^2 + \frac{1}{2}k_4(y_2 - z_2)^2 + \frac{1}{2}k_3(y_4 - y_2)^2 + \frac{1}{2}k_0(ay_3 + by_4 - y_2)^2. \tag{42}$$

Note that $\dot{y} = y_0 + y_2 = ay_3 + by_4$, then in the expression of kinetic energy:

$$T = \frac{1}{2}m_1\dot{y}_1^2 + \frac{1}{2}m_2\dot{y}_2^2 + \frac{1}{2}Jc^2(\dot{y}_4 - \dot{y}_3)^2 + \frac{1}{2}m(\dot{y}_0 + \dot{y}_2)^2$$

you can consider the ratio:

$$y_3 = \frac{y_0 + y_2 - by_4}{a} = a_1 y_0 + a_1 y_2 - b_1 y_4.$$

We introduce a number of symbols:

$$y_3 = a_1 y_0 + a_1 y_2 - b_1 y_4, \quad a_1 = \frac{1}{a}, \quad b_1 = \frac{b}{a}.$$

We write the kinetic energy in the form:

$$'' = \frac{1}{2}m_1\dot{y}_1^2 + \frac{1}{2}m_2\dot{y}_2^2 + \frac{1}{2}Jc^2\left(\dot{y}_4 - a_1\dot{y}_0 + a_1\dot{y}_2 - b_1\dot{y}_4\right)^2 + \frac{1}{2}m\left(aa_1\dot{y}_0 + aa_1\dot{y}_2 - b_1a\dot{y}_4 + b\dot{y}_4\right)^2 \tag{43}$$

or

$$'' = \frac{1}{2}m_1\dot{y}_1^2 + \frac{1}{2}m_2\dot{y}_2^2 + \frac{1}{2}Jc^2\left(b_2y_4 - a_1y_0 - a_1y_2\right)^2 + \frac{1}{2}m\left(\dot{y}_0 + \dot{y}_2\right)^2, \tag{43'}$$

where the (y_1, y_2, y_4, y_0) coordinates are present. The potential energy of the system in this case takes the form:

$$\Pi = \frac{1}{2}k_1(y_1 - z_1)^2 + \frac{1}{2}k_2(a_1y_0 + a_1y_2 - b_1y_4 - y_1)^2 + \frac{1}{2}k_4(y_2 - z_2)^2 + \frac{1}{2}k_3(y_4 - y_2)^2 + \frac{1}{2}k_0(y_0)^2. \tag{44}$$

We make a number of auxiliary calculations and write the system of equations of motion:

$$m_1\ddot{y}_1 + y_1(k_1 + k_2) - y_2(-k_2a_1) + y_4(k_2b_1) + y_0(-k_2a_1) = k_1z_1, \tag{45}$$

$$\begin{aligned}&\ddot{y}_2(m_2 + Jc^2a_1^2 + m) + y_2(k_2a_1^2 + k_4 + k_3) + y_1(-k_2a_1) + \ddot{y}_4(-Jc^2a_1b_2) + \\ &\quad + y_4(-k_2a_1b_1 - k_3) + \ddot{y}_0(Jc^2a_1^2 + m) + y_0(k_2a_1^2) = k_4z_2,\end{aligned} \tag{46}$$

$$y_1(k_2b_1) + \ddot{y}_2(-Jc^2a_1b_2) + y_2(-k_2a_1b_1 - k_3) + \ddot{y}_4(Jc^2b_2^2) + y_4(k_2b_1^2 + k_3) + \ddot{y}_0(-Jc^2a_1b_2) + y_0(-k_2a_1b_1) = 0, \tag{47}$$

$$\ddot{y}_2(Jc^2a_1^2) + y_2(k_2a_1^2) + y_1(-k_2a_1) + \ddot{y}_4(-Jc^2a_1b_2) + y_4(-k_2a_1b_1 + k_3) + \ddot{y}_0(Jc^2a_1^2) + y_0(k_0 + k_2a_1^2) = 0. \tag{48}$$

The coefficients of the equations of motion in the (y_1, y_2, y_4, y_0) coordinates are presented in Table 1.

To go to the (y_1, y_2, y_4) coordinate system, we use the method described in Khomenko and Eliseev (2010) and perform the "zeroings of y_0". As a result, we obtain a system of equations in the (y_1, y_2, y_4) coordinates.

Table 1. The coefficients of the equations of motion (45)–(48)

a_{11} y_1	a_{12} y_2	a_{13} y_4	a_{14} y_0
$m_1p^2 + k_1 + k$	$-k_2a_1$	k_2b_1	$-k_2a_1$
a_{21}	a_{22}	a_{23}	a_{24}
$-k_2a_1$	$(m_2 + m + Jc^2a_1^2)p^2 + k_2a_1^2 + k_3 + k_4$	$-Jc^2a_1b_2p^2 - k_3 - k_2a_1b_1$	$(m + Jc^2a_1^2)p^2 + k_2a_1^2$
a_{31}	a_{32}	a_{33}	a_{34}
k_2b_1	$-Jc^2a_1b_2 - k_2a_1b_1 - k_3$	$Jc^2b_2p^2 + k_2b_1^2 + k_3$	$-Jc^2a_1b_2 - k_2a_1b_1$
a_{41}	a_{42}	a_{43}	a_{44}
$-k_2a_1$	$Jc^2a_1p^2 + k_2a_1^2$	$-Jc^2a_1b_2 - k_2a_1b_1$	$Jc^2a_1^2p^2 + k_0 + k_2a_1$
Q_1	Q_2	Q_3	Q_4
k_1z_1	k_4z_2	0	0

Note: Q_1–Q_4 are generalized forces.

Assume that $J = 0$ and $m = 0$, then we first get the equations:

$$\begin{cases} m_1\ddot{y}_1 + y_1(k_1 + k_2) - y_2k_2a_1 + y_4k_2b_1 = k_1z_1, \\ m_2\ddot{y}_2 + y_2(k_2a_1^2 + k_4 + k_3) - y_1k_2a_1 - y_4(k_2a_1b_1 + k_3) = k_4z_2, \\ y_1k_2b_1 - y_2(k_2a_1b_1 + k_3) + y_4(k_2b_1^2 + k_3) = 0. \end{cases} \tag{49}$$

The third equation from the system can be written as

$$y_1a_3 - a_4y_2 + a_5y_4 = 0, \tag{50}$$

where $a_3{=}k_2b_1$, $a_4{=}k_2a_1b_1{+}k_3$, $a_5 = k_2b_1^2 + k_3$, which is used to get the coordinates:

$$y_4 = \frac{a_4y_2 - a_3y_1}{a_5} = b_3y_2 - b_4y_1, \tag{51}$$

where, in turn, $b_3 = \frac{a_4}{a_5}; b_4 = \frac{a_3}{a_5}$.

Perform the substitution (51) into the system of equations (49) and get:

$$m_1\ddot{y}_1 + y_1(k_1 + k_2 - k_2b_1b_4) - y_2(k_2a_1 - k_2b_1b_3) = k_1z_1, \tag{52}$$

$$m_2\ddot{y}_2 + y_2(k_2a_1^2 + k_4 + k_3 - k_2a_1b_1b_3 - k_3b_3) - y_1(k_2a_1 - k_2a_1b_1b_4 - k_2b_1b_4) = k_4z_2. \tag{53}$$

For further use we will make the list of ratios:

$$\begin{gathered} a = \frac{l_1}{l_1 + l_2}; b = \frac{l_2}{l_1 + l_2}; c = \frac{1}{l_1 + l_2}; a_1 = \frac{1}{a}; b_1 = \frac{b}{a}; b_2 = 1 + b_1 = \frac{a+b}{a} \\ a_3 = k_2b_1; a_4 = k_2a_1b_1 + k_3; a_5 = k_2b_1^2 + k_3; b_3 = \frac{a_4}{a_5}; b_4 = \frac{a_3}{a_5}. \end{gathered} \tag{54}$$

Using (54) and taking the values of l_1 and l_2, equal to zero simultaneously and alternately, one can simplify the generalized computational schemes in Fig. 6.1, *b–f*, which are ultimately reduced to the basic or source diagram presented in Figure 1a.

We assume that $i = \frac{l_2}{l_1}$, then, relations (54) can be transformed:

$$\begin{gathered} a = \frac{l_1}{l_1 + l_2} = 1 + i; b = \frac{l_2}{l_1 + l_2} = \frac{i}{1+i}; b_1 = i; b_2 = 1 + i; a_3 = k_2i; \\ a_4 = k_2(i+1)i + k_3; a_5 = k_2i^2 + k_3; b_3 = \frac{k_2(i+1)i + k_3}{k_2i^2 + k_3}; b_4 = \frac{k_2i}{k_2i^2 + k_3}. \end{gathered} \tag{55}$$

Simplifications lead to comparable results.

The Case of the Connection of Two Elements With an Intermediate Unit in the Form of a Lever of the Second Kind

Consider the variant for the computational scheme in Figure 1d when using the lever of the second kind. In this case, the scheme of interaction of the elements will be different, as shown in Figure 5.

In the computational scheme (Figure 5), the following notation is adopted: the segment (1)(1″) corresponds to l_2, the segment (1)(1′) corresponds to l_1, the lever has the inertia moment J relative to the support (point (1)), φ is the angle of deflection of the lever from the mass element m_2, y_3 is the coordinate of the attachment point of the spring k_3 – point (1′), $i = \frac{l_2}{l_1}$. The expression for the kinetic and potential energies have the form:

$$T = \frac{1}{2} m_1 \dot{y}_1^2 + \frac{1}{2} m_2 \dot{y}_2^2 + \frac{1}{2} m_0 \dot{y}_3^2, \tag{56}$$

$$\Pi = \frac{1}{2} k_1 (y_1 - z_1)^2 + \frac{1}{2} k_2 (y_3' - y_1)^2 + \frac{1}{2} k_3 (y_3 - y_2)^2 + \frac{1}{2} k_4 (y_4 - z_2)^2. \tag{57}$$

We introduce a series of relationships: the angle of rotation φ in the relative motion of the lever

$$\varphi = \frac{y_2 - y_4}{l_2};\ y_4' = y_2 - y_4;\ y_3' = \frac{y_4'}{i};\ y_4' = \varphi l_2$$

y_4', y_3' – the coordinates of the points of attachment of the springs in the relative motion; let us express y_4' through $y_3' - y_4' = y_3' i$.

Figure 5. The computational scheme of the system in Figure 1d, but with a lever of the second kind

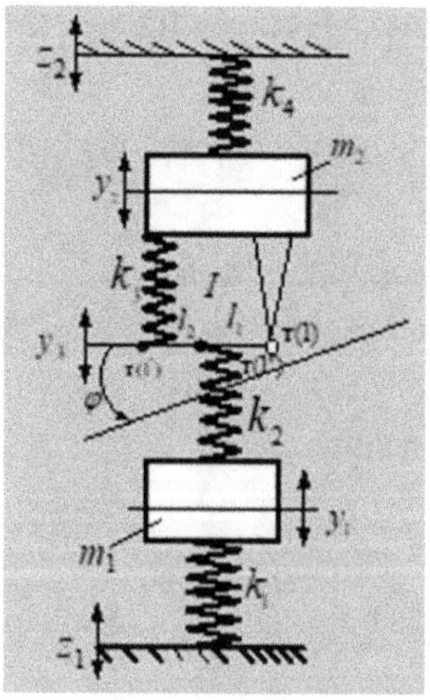

Then the absolute speed of motion of the reduced mass ($m_0 = \frac{J}{l_2^2}$) is determined. The offset of the point of attachment of the spring k_2 will take the form $y_3 = y_2 + y_3' = y_2 + y_4'$. Write (56), (57) in the form:

$$T = \frac{1}{2}m_1\dot{y}_1^2 + \frac{1}{2}m_2\dot{y}_2^2 + \frac{1}{2}m_0\dot{y}_4^2 = \frac{1}{2}m_1\dot{y}_1^2 + \frac{1}{2}m_2\dot{y}_2^2 + \frac{1}{2}m_0(\dot{y}_2 - \dot{y}_4')^2, \tag{58}$$

$$\Pi = \frac{1}{2}k_1(y_1 - z_1)^2 + \frac{1}{2}k_2(y_2 - y_3' - y_1)^2 + \frac{1}{2}k_3(y_4')^2 + \frac{1}{2}k_4(y_2 - z_2)^2. \tag{59}$$

Let us assume that $y_4' = y_2 - y_1$, $y_3' = \frac{y_4'}{i} = \frac{y_2 - y_4}{i}$. We make a series of calculations and obtain the system of equations:

$$m_1\ddot{y}_1 + y_1(k_1 + k_2) + y_2(-k_2) + \frac{k_2 y_4'}{i} = k_1 z_1, \tag{60}$$

$$\ddot{y}_2(m_0 + m_2) + y_2(k_2 + k_4) - y_1 k_2 - y_4' k_2 - \ddot{y}' m_0 = k_2 z_2, \tag{61}$$

$$\ddot{y}_4' m_0 + y_4'(k_3 + \frac{k_2}{i}) + y_1\frac{k_2}{i} - \ddot{y}_2'' m_0 - \frac{k_2}{i}y_2'' = 0. \tag{62}$$

If $m_0 = 0$ $(m_0 = \frac{J}{l_2^2})$, then the system of equations is written as:

$$m_1\ddot{y}_1 + y_1(k_1 + k_2) - y_2 k_2 + k_2 y_4' = k_1 z_1, \tag{63}$$

$$m_2\ddot{y}_2 + y_2(k_2 + k_4) - y_1 k_2 - \frac{y_4' k_2}{i} = k_2 z_2, \tag{64}$$

$$k_3 y_4' + \frac{y_4' k_2}{i} - \frac{y_2 k_2}{i} + \frac{y_1 k_2}{i} = 0 \tag{65}$$

From (65) we find the value of the coordinate:

$$y_4' = y_2 \frac{k_2 i}{k_3 i^2 + k_2} - \frac{k_2 i}{k_3 i^2 + k_2} y_1. \quad (66)$$

After substituting (66) into (63) and (64) we get:

$$m_1 \ddot{y}_1 + y_1 (k_1 + k_2 - \frac{k_2^2 i}{k_3 i^2 + k_2}) - y_2 (k_2 - \frac{k_2^2 i}{k_3 i^2 + k_2}) = k_1 z_1, \quad (67)$$

$$m_2 \ddot{y}_2 + y_2 (k_2 + k_4 - \frac{k_2^2 i}{k_3 i^2 + k_2}) - y_1 (k_2 - \frac{k_2^2 i}{k_3 i^2 + k_2}) = k_2 z_2, \quad (68)$$

or

$$m_1 \ddot{y}_1 + y_1 \left[\frac{(k_1 + k_2) k_3 i^2 + k_1 k_2}{k_3 i^2 + k_2} \right] - y_2 \left[\frac{k_2 k_3 i^2}{k_3 i^2 + k_2} \right] = k_1 z_1, \quad (69)$$

$$m_2 \ddot{y}_2 + y_2 \left[\frac{(k_2 + k_4) k_3 i^2 + k_2 k_4}{k_3 i^2 + k_2} \right] - y_1 \left[\frac{-k_3 i^2 k_2}{k_3 i^2 + k_2} \right] = k_2 z_2, \quad (70)$$

If we accept that $l_2 = 0$, then (69), (70) will be transformed into a system of equations:

$$m_1 \ddot{y}_1 + y_1 k_1 = k_1 z_1, \quad (71)$$

$$m_2 \ddot{y}_2 + y_2 k_4 = k_2 z_2, \quad (72)$$

When $i = 1$ ($l_2 = l_1$), we obtain, respectively:

$$m_1 \ddot{y}_1 + y_1 \left(\frac{k_1 k_3 + k_2 k_3 + k_1 k_2}{k_2 + k_3} \right) - y_2 \frac{k_3 k_2}{k_3 + k_2} = k_1 z_1, \quad (73)$$

$$m_2 \ddot{y}_2 + \frac{k_2 k_3 + k_3 k_4 + k_2 k_4}{k_3 + k_2} - y_2 \frac{k_3 k_2}{k_2 + k_3} = k_2 z_2, \quad (74)$$

We transform (73), (74) to the form:

$$m_1\ddot{y}_1 + y_1\left(k_1 + k_{?@}\right) - y_2k_{?@} = k_1z_1, \tag{75}$$

$$m_2\ddot{y}_2 + (k_4 + k_{red}) - y_2k_{red} = k_2z_2, \tag{76}$$

where $k_{red} = \dfrac{k_3k_2}{k_3 + k_2}$

Note that the system of equations (75), (76) corresponds to the configuration of the computational scheme in Figure 1a. Such a situation can only be for $i = 1$. If $l_1 = 0$, then $i \rightarrow \infty$, in this case, the elastic element k_3 "drops out" of the interaction.

When considering the computational scheme in Figure 5 note that when: $l_1 = 0$ the spring k_2 connects m_1 and m_2 directly, and the elastic element k_3 "leaves" the interaction, since $l_2 \neq 0$. The system in this case completely coincides with the chain system of the usual form. A similar situation occurs when $i = 1$. If $l_2 = 0$ when $l_1 = 0$, then the situation remains the same ($i = 1$). If $l_2 = 0$, and $l_1 \neq 0$, then the system splits into two blocks ($m_1p^2+k_1$) and ($m_2p^2+k_4$), and the connection of these blocks will occur only when $l_1 = 0$.

The Unsupported Lever Option

Consider a system of general form, when the lever has no support (Figure 6).

Figure 6. The computational scheme of the system with two degrees of freedom with a lever without a fixed point of support

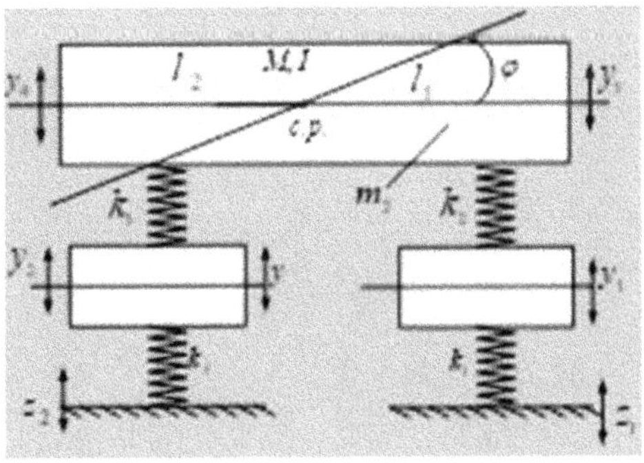

The expressions for the kinetic and potential energies are:

$$T = \frac{1}{2}m_1\dot{y}_1^2 + \frac{1}{2}m_2\dot{y}_2^2 + \frac{1}{2}M\dot{y}^2 + \frac{1}{2}J\dot{\varphi}^2, \tag{77}$$

$$\Pi = \frac{1}{2}k_1(y_1 - z_1)^2 + \frac{1}{2}k_2(y_3 - y_1)^2 + \frac{1}{2}k_3(y_4 - y_2)^2 + \frac{1}{2}k_4(y_2 - z_2)^2. \tag{78}$$

We introduce a series of relationships:

$$y_3 = y + l_2\varphi,\ y_4 = y - l_2\varphi,\ \varphi = \frac{y_3 - y_4}{l_1 + l_2},\ y = ay_4 + by_3,$$
$$a = \frac{l_1}{l_1 + l_2},\ b = \frac{l_2}{l_1 + l_2},\ c = \frac{1}{l_1 + l_2}. \tag{79}$$

We transform (77), (78) to the form:

$$T = \frac{1}{2}m_1\dot{y}_1^2 + \frac{1}{2}m_2\dot{y}_2^2 + \frac{1}{2}M(a\dot{y}_4 + b\dot{y}_3)^2 + \frac{1}{2}Jc^2(\dot{y}_3 - \dot{y}_4)^2, \tag{80}$$

$$\Pi = \frac{1}{2}k_1(y_1 - z_1)^2 + \frac{1}{2}k_2(y_3 - y_1)^2 + \frac{1}{2}k_3(y_4 - y_2)^2 + \frac{1}{2}k_4(y_2 - z_2)^2. \tag{81}$$

We make a series of intermediate calculations and write down the system of equations of motion in the (y_1, y_2, y_3, y_4) coordinates:

$$m_1\ddot{y}_1 + y_1(k_1 + k_2) + y_3(-k_2) = k_1z_1, \tag{82}$$

$$m_2\ddot{y}_2 + y_2(k_3 + k_4) - y_4k_3 = k_4z_2, \tag{83}$$

$$\ddot{y}_3(Mb^2 + Jc^2) + y_3(k_2) - \ddot{y}_4(Jc^2 - Mab) + y_1(-k_2) = 0 \tag{84}$$

$$\ddot{y}_4(Ma^2 + Jc^2) + y_4(k_3) - \ddot{y}_3(Jc^2 - Mab) - y_2k_3 = 0. \tag{85}$$

Let us assume that $M = 0, J = 0$, i.e. the lever is in this case weightless, but it has a turning point, then:

$$m_1\ddot{y}_1 + y_1(k_1 + k_2) - y_3k_2 = k_1z_1, \tag{86}$$

$$m_2\ddot{y}_2 + y_2(k_2 + k_4) - y_4k_3 = k_4z_2, \tag{87}$$

$$y_3k_2 - y_1k_2 = 0, \tag{88}$$

$$y_4k_3 - y_2k_3 = 0, \tag{89}$$

from where $y_4 = y_2$.

After substitutions in (86), (87) we get:

$$m_1\ddot{y}_1 + y_1(k_1 + k_2) - y_1k_2 = k_1z_1, \tag{90}$$

$$m_2\ddot{y}_2 + y_2k_3 + k_4y_2 - y_2k_3 = k_4z_2, \tag{91}$$

In this case, the system (82) –(85) is reduced to the form:

$$m_1\ddot{y}_1 + y_1k_1 = k_1z_1, \tag{92}$$

$$m_2\ddot{y}_2 + y_2k_4 = k_4z_2, \tag{93}$$

If the lever does not have mass-and-inertia properties, then it does not play any role in the scheme according to Fig. 6.1, *f*; the system of equations splits into two independent equations (92), (93).

Let us use Laplace transforms and write:

$$\left.\begin{aligned} &\bar{y}_3(A) - \bar{y}_4(B) - \bar{y}_1k_2 = 0, \\ &\bar{y}_3B + y_4C - k_3y_2 = 0. \end{aligned}\right\} \tag{94}$$

If J and M are not zero, then from the system of equations it follows that with

$$A = (Mb^2 + Jc^2)p^2 + k_2,\ B = (Mab - Jc^2)p^2,\ C = (Ma^2 + Jc^2)p^2 + k_3,$$

one can transform the system of equations (82)–(85). Using (94), we find that:

$$y_4 = \frac{y_2Ak_3 + y_1k_2B}{AC - B^2}; \tag{95}$$

$$y_3 = \frac{y_1k_2C + y_2k_3B}{AC - B^2}. \tag{96}$$

Let y_4=$k_2a_1y_1$+$k_3b_1y_2$, where $a_1=\dfrac{B}{AC-B^2}$; $b_1=\dfrac{A}{AC-B^2}$; in turn, y_3=$a_2k_2y_1$+$b_2k_3y_2$ where $a_2=\dfrac{C}{AC-B^2}$; $b_2=\dfrac{B}{AC-B^2}$; that is, b_2=a_1.

Perform the substitution (94), (95) into (82) and (83) and we obtain:

$$\bar{y}_1(m_1p^2+k_1+k_2)-k_2(a_2k_2y_1+b_2k_3y_2)=k_1z_1, \tag{97}$$

$$\bar{y}_2(m_2p^2+k_3+k_4)-k_3(a_1k_2y_1+b_1k_3y_2)=k_2z_2, \tag{98}$$

or

$$\bar{y}_1(m_1p^2+k_1+k_2-a_2k_2^2)-y_2k_2k_3b_2=k_1\bar{z}_1, \tag{99}$$

$$\bar{y}_2(m_2p^2+k_3+k_4-b_1k_3^2)-y_1k_3k_2a_1y_1=k_2z_2. \tag{100}$$

Since a_1=b_2, then in the system of equations (99), (100) there is some symmetry. Consider

$$k_2-a_2k_2^2=k_2-\frac{k_2^2C}{AC-B^2}=\frac{ACk_2-k_2B^2-k_2^2C}{AC-B^2};$$

$$k_2-a_2k_2^2=\frac{\left[(Mb^2+Jc^2)p^2+k_2\right]\left[(Ma^2+Jc^2)p^2+k_3\right]k_2}{AC-B^2}-\frac{k_2\left[(Mab-Jc^2)p^2\right]^2-k_2\left[(Ma^2+Jc^2)p^2+k_3\right]}{AC-B^2}.$$

If $p=0$, then $k_2-a_2k_2^2=\dfrac{k_2^2k_3-k_2^2k_3}{k_2k_3}=0$. If $p^2\rightarrow\infty$, then

$$k_2-a_2k_2^2=\frac{k_2(Mb^2+Jc^2)(Ma^2+Jc^2)-(Mab-Jc^2)^2k_2}{(Mb^2+Jc^2)(Ma^2+Jc^2)-(Mab-Jc^2)^2}=k_2.$$

Similar result will be regarding $k_3-k_3^2b_1$.

That is, when $p\rightarrow\infty$ and $M\neq 0, J\neq 0$, the system splits into two independent blocks (the motions are not related along y_1, y_2). If p = 0, then the interaction effect is absent, as well as when, but the reasons are different. If $p\neq 0$ and $p\neq\infty$, then we will have an intermediate case. As for the assessment of the influence of the intermediate body between m_1 and m_2 through point (1), the introduction of a weighty lever with a floating support gives the following features.

In the static case, the connection between m_1 and m_2 at $p = 0$ was provided through an elastic element with a reduced stiffness defined by the expression $k_{red} = \frac{k_2 k_3}{k_2 + k_3}$. With the introduction of the lever at $p = 0$, we must take into account the coefficients a_1=b_2 in the cross-coupling. When $p = 0$, they are equal to zero, since $b_2 = \frac{B}{AC - B^2}$, and B=(Mab–Jc^2)p^2. We estimate the values of the coefficients a_2 and b_1:

$$a_2 = \frac{C}{AC - B^2} = \frac{k_3}{k_2 k_3} = \frac{1}{k_2},\ b_1 = \frac{1}{k_3}.$$

If we make the substitution of such values into equation (99), (100), then the equations break up into autonomous ones; in this case, the connections between the elements m_1 and m_2, through the elastic elements k_2 and k_3, are not implemented. If we take into account the effect of forces at different frequencies, then there is one symmetry relation Mab=Jc^2 that makes the value $B = 0$, and therefore b_1 and a_2. In this case, at any frequency, the elements m_1 and m_2 do not interact with each other.

Thus, in the general case, the system, the computational scheme of which is shown in Figure 1f has, taking into account a weighty lever, 4 degrees of freedom. Therefore, it can be expected that the interaction between m_1 and m_2 will undergo significant changes at resonant frequencies four times. The value of the two frequencies associated with the lever is determined from the characteristic equation:

$$AC - B^2 = 0. \tag{101}$$

When resonant frequencies are reached, the coefficients b_1,b_2,a_2 reach infinite values. If we take each of the coefficients a_2 and b_1, then they take, at certain frequencies, zero values determined by the formulas:

$$\omega_{1dyn}^2 = \frac{k_2}{Mb^2 + Jc^2} \tag{102}$$

$$\omega_{2dyn}^2 = \frac{k_3}{Ma^2 + Jc^2} \tag{103}$$

At these frequencies, the system of elements m_1 and m_2, initially having a chain structure and a cross-coupling of an elastic type, will behave like a kind of solid body on the elastic supports $k_1 + k_2$ and $k_3 + k_4$ with a cross-coupling of inertial type. A characteristic feature of the connection of two elements according to the scheme shown in Figure 1a, with the introduction of a weighty lever with a floating support, is the expansion of the range of dynamic properties. At the same time, the parameters of the lever are significant, since the dynamic stiffness of the interaction gives a large variant of changes at different frequencies.

LEVER LINKAGES AND DYNAMIC EFFECTS FROM THE INTRODUCTION OF ADDITIONAL CONSTRAINTS

The elements of mechanical oscillatory systems, considered in the form of computational schemes or physical models of engineering objects, are most often mass-and-inertia elements, elastic units (springs), springs, suspensions, shock absorbers, dampers, dampers of various types and more complex devices having control systems (De Silva, 2000; Sun & Jing, 2015; Harris & Piersol, 2002; Eliseev et al., 2011a). In recent years, theoretical and applied research has focused on the search for and development of new technical means using various mechanisms and devices for motion transformation (Eliseev et al., 2011b; Eliseev & Artyunin, 2016; Karnovsky & Lebed, 2016; Belokobylsky et al., 2013). The complication of the constructive technical forms of vibration protection systems initiates the use of the analytical apparatus, methods of circuit theory, the theory of automatic control and mechatronics, which is reflected in the work on the use of ideas based on dynamic and electro-mechanical analogies. In the works dealing with the development of structural mathematical modeling methods (Eliseev & Khomenko, 2014; Kolovsky, 1976; Fomichev, 2010), the possibilities of constructing mathematical models focused on detailing the ideas about the features of dynamic interactions of elements of mechanical systems, which include various kinds of mechanisms, devices for transforming motion and energy transmission, are shown (Eliseev et al., 2011b; Fomichev, 2010; Baranovsky, 2000). Great interest in the tasks of the dynamics of technological vibration machines and vehicles for various purposes is associated with the development of the ideas of dynamic oscillation oscillations, the search for rational forms of using new dynamic effects with applications in various branches of technology. Various versions of the construction of dynamic vibration dampers have been reflected in a number of publications focused on the use of new structural materials, including those with controlled elastic-dissipative properties (Kuznetsov, 2009; Eliseev et al., 2012a). Issues related to the expansion of methodological positions in problems of dynamic oscillation damping, taking into account the specifics of dynamic states of objects with several degrees of freedom, which have a spatial metric (this is especially characteristic of vibration technological machines, and the reduction of vibrations for large energy installations, road building machines and vehicles), have been developed to a lesser extent. We consider the possibility of constructing structural mathematical models for objects with two degrees of freedom, in which dynamic damping modes are provided by introducing and using additional constraints implemented by lever mechanisms.

Construction of Mathematical Models: Consideration of Structure Features

A technical object on a movable base is considered as a solid with mass-and-inertia parameters M and J, (M is the mass of the object, J is the moment of inertia with respect to the center of gravity at p. O). The solid body rests on elastic elements with stiffnesses k_1, k_2 (Figure 7) at pp. A and B and, at the same time, on levers in the form of rigid weightless rods with turning points (pp. A_1, A_2). At the ends of the rods there are additional concentrated masses m_1 and m_2. The object, as a solid body, rests on the levers at pp. A and B, where elastic elements with stiffnesses k_1, k_2 are also fixed. The center of gravity of an object is determined by the lengths of the shoulders l_{10}, l_{20} (Figure 7). Elastic-lever supports, as shown in Figure 7, have the lengths of the shoulders, respectively, as the notation l_1, l_2, l_3, l_4. It is assumed that the system has linear properties in the absence of resistance forces and performs small oscillations relative to the position of stable equilibrium in the coordinates (y_1, y_2 and y_0, φ), associated with a fixed basis.

The position of the additional masses at the ends of the levers (m_1, m_2) is determined by the y_1', y_2' coordinates, and the angles of rotation φ_1 and φ_2, respectively. External disturbances in the system in Figure 7 are considered in the form of vibrations of the base $z(t)$, as well as in the form of a concentrated force Q_1, applied at p. A, the motion of which is determined by the coordinate y_1.

Figure 7. The computational scheme of the mechanical oscillatory system

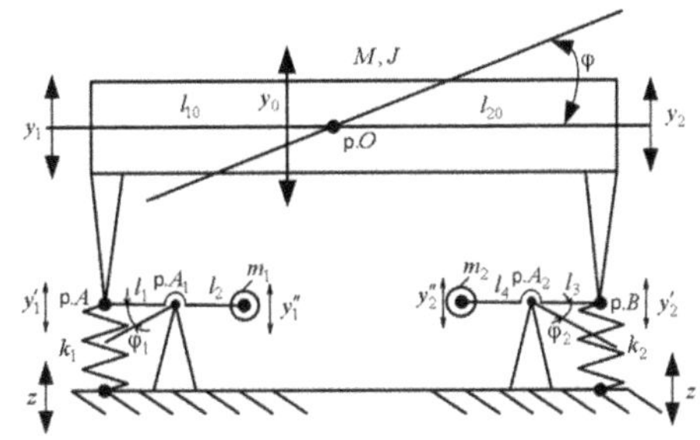

It is taken into account that there are relations between the coordinates of the system and its parameters:

$$y_0 = ay_1 + by_2,\ \phi = c\cdot(y2 - y1_),\ y1\_y0 - 11_{0\phi},\ y2 =_y 0 +_1 20_{\phi}, \tag{104}$$

where $a = \dfrac{l_{20}}{l_{10} + l_{20}}$, $b = \dfrac{l_{10}}{l_{10} + l_{20}}$, $c = \dfrac{1}{l_{10} + l_{20}}$.

We introduce transfer ratios of lever linkages, which can be represented as:

$$i_1 = \frac{l_2}{l_1},\ i_2 = \frac{l_4}{l_3},\ i_0 = \frac{l_{10}}{l_{20}}. \tag{105}$$

When considering kinematic perturbations from the side of the supporting surface $z(t)$, the object of protection (and the system as a whole with $Q_1 = 0$) participate in two motions: a transportation one – according to the law of $z(t)$ – and a relative one (of the object of protection around p. O, and levers with masses m_1, m_2 relative to points A_1, A_2, respectively). The motion parameters of mass-and-inertia elements for estimating the kinetic energy are determined in absolute motion, which is the sum of the transportation and relative motions. Thus, taking into account (104), we obtain that:

$$\begin{gathered} y_1 = \phi_1 l_1 + z, \phi_1 = \frac{y_1 - z}{l_1}, y_1'' = -\phi_1 l_2, y_1'' = -i_1\left(y_1 - z\right), \\ y_2 = \phi_2 l_3 + z, \phi_2 = \frac{y_2 - z}{l_3}, y_2'' = -\phi_2 l_4, y_2'' = -i_2\left(y_2 - z\right). \end{gathered} \tag{106}$$

To determine the kinetic and potential energies, we find the parameters of absolute motion:

$$\dot{y}_{m_1} = \dot{y}_1'' + \dot{z} = -(\dot{y}_1 - \dot{z})i_1 = -y_1 i_1 + z(1+i_1), \tag{107}$$

$$\dot{y}_{m_2} = \dot{y}_2'' + \dot{z} = -(\dot{y}_2 - \dot{z})i_2 = -y_2 i_2 + z(1+i_2). \tag{108}$$

Using (104)÷(108), one can write the expressions for the kinetic and potential energies necessary to obtain differential equations of motion.

The task of the research is to develop a method for constructing mathematical models for systems that have an object that interacts not only with ordinary elements, but also with lever mechanisms for various types of external disturbances. At the preliminary stage consider the case of a force perturbation along the coordinate координате y_1 ($Q_1 \neq 0$, $z(t) = 0$).

We write the expressions for the kinetic and potential energies in the y_1, y_2 coordinate system:

$$\text{T} = \frac{1}{2}M(a\dot{y}_1 + b\dot{y}_2)^2 + \frac{1}{2}Jc^2(\dot{y}_2 - \dot{y}_1)^2 + \frac{1}{2}m_1 i_1^2 \dot{y}_1'^2 + \cdot\frac{1}{2}m_2 i_2^2 \dot{y}_2'^2, \tag{109}$$

$$\Pi = \frac{1}{2}k_1 y_1^2 + \frac{1}{2}k_2 y_2^2. \tag{110}$$

To build the equations of motion, the technology is used using the Lagrange equation of the second kind (Harris & Piersol, 2002); will get:

$$\left(Ma^2 + Jc^2 + m_1 i_1^2\right)\cdot \ddot{y}_1 + y_1 k_1 + \left(Mab - Jc^2\right)\cdot \ddot{y}_2 = Q_1, \tag{111}$$

$$\left(Mb^2 + Jc^2 + m_2 i_2^2\right)\cdot \ddot{y}_2 + y_2 k_2 + \left(Mab - Jc^2\right)\cdot \ddot{y}_1 = 0. \tag{112}$$

Applying the Laplace transforms under zero initial conditions to (111) and (112) and assuming that $Q_1(t)$ is a harmonic function, it is possible to present (111) and (112) in operator form:

$$\bar{y}_1 \cdot \left[\left(Ma^2 + Jc^2 + m_1 i_1^2\right)\cdot p^2 + k_1\right] + \bar{y}_2 \cdot \left(Mab - Jc^2\right)\cdot p^2 = \bar{Q}_1, \tag{113}$$

$$\bar{y}_2 \cdot \left[\left(Mb^2 + Jc^2 + m_2 i_2^2\right)\cdot p^2 + k_2\right] + \bar{y}_1 \cdot \left(Mab - Jc^2\right)\cdot p^2 = 0, \tag{114}$$

where $p = j\omega$ is the complex variable ($j = \sqrt{-1}$); sign (–) corresponds to the transform image of the Laplace variable. Table 2 shows the coefficients of equations (113), (114).

Table 2. The coefficients of equations (113), (114) in the coordinates y_1, y_2 with force disturbance in operator form

a_{11}	a_{12}
$\left(Ma^2 + Jc^2 + m_1 i_1^2\right)\cdot p^2 + k_1$	$(Mab - Jc^2)p^2$
a_{21}	a_{22}
$(Mab - Jc^2)p^2$	$\left(Mb^2 + Jc^2 + m_2 i_2^2\right)\cdot p^2 + k_2$
Generalized forces	
$\bar{Q}_1$	0

The structural mathematical model of the initial system according to Figure 7 can be considered as a structural diagram of a dynamically equivalent automatic control system (Eliseev et al., 2011b; Khomenko & Eliseev, 2013). The structural diagram consists of two partial systems; interpartial constraints correspond to the inertial type of dynamic interactions.

Figure 8. The structural diagram of the initial system in Figure 7 in the y_1, y_2 coordinate system with a force external disturbance

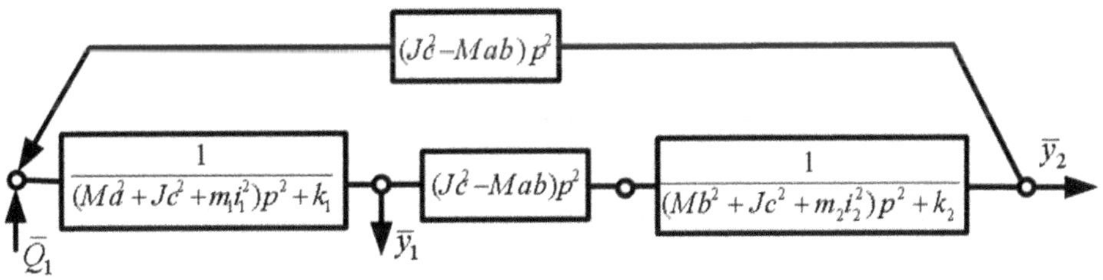

The transfer functions of the system, as follows from the structural diagram in Figure 8, are determined by the expressions:

$$W_1(p) = \frac{\bar{y}_1}{\bar{Q}_1} = \frac{\left(Mb^2 + Jc^2 + m_2 i_2^2\right)\cdot p^2 + k_2}{A_0(p)}, \tag{116}$$

$$W_2(p)=\frac{\bar{y}_2}{\bar{Q}_1}=\frac{\left(Jc^2-Mab\right)p^2}{A_0(p)}, \tag{117}$$

where $A_0(p)$ is the characteristic frequency equation of the system:

$$A_0(p)=\left[\left(Ma^2+Jc^2+m_1i_1^2\right)\cdot p^2+k_1\right]\cdot\left[\left(Mb^2+Jc^2+m_2i_2^2\right)\cdot p^2+k_2\right]-\left[\left(Jc^2-Mab\right)p^2\right]^2. \tag{118}$$

The transfer function of interpartial constraints under the force external action of Q_1 is:

$$W_{12}(p)=\frac{\bar{y}_2}{\bar{y}_1}=\frac{\left(Jc^2-Mab\right)p^2}{\left(Mb^2+Jc^2+m_2i_2^2\right)\cdot p^2+k_2}. \tag{119}$$

The transfer functions (116), (117), (119) are used to determine the frequency of the dynamic oscillation suppression and to estimate the dynamic interactions of the system in terms of the coordinates y_1, y_2. In this case, when the force perturbation Q_1 on the coordinate y_1 at the frequency:

$$\omega_{1dyn}^2=\frac{k_2}{Mb^2+Jc^2+m_2i_2^2} \tag{120}$$

the mode of dynamic oscillation damping along the coordinate $\bar{y}_1$ arises. The physical interpretation of dynamic oscillation damping is usually considered through the quasi-static condition, which is formed on the basis of the d'Alembert principle, although other approaches are possible related to structural ideas about mathematical models of dynamic interactions (Eliseev & Khomenko, 2014).

Features of the Mathematical Model of a System With Lever Mechanisms With Kinematic Perturbations

Using relations (104)÷(108), we write the expressions for the kinetic and potential energies at $z(t)\neq 0$ ($Q_1(t)=0$):

$$\mathrm{T}=\frac{1}{2}M\left(a\dot{y}_1+b\dot{y}_2\right)^2+\frac{1}{2}Jc^2\left(\dot{y}_2-\dot{y}_1\right)^2+\frac{1}{2}m_1\left[-\dot{y}_1i_1+\dot{z}\left(i_1+1\right)\right]^2+\frac{1}{2}m_2\left[-\dot{y}_2i_2+\dot{z}\left(i_2+1\right)\right]^2, \tag{121}$$

$$\Pi=\frac{1}{2}k_1\left(l_1\phi_1\right)^2+\frac{1}{2}k_2\left(l_3\phi_2\right)^2=\frac{1}{2}k_1l_1^2\left[\frac{\left(y_1-z\right)}{l_1}\right]^2+\frac{1}{2}k_2l_3^2\left[\frac{\left(y_2-z\right)}{l_3}\right]^2=\frac{1}{2}k_1\left(y_1-z\right)^2+\frac{1}{2}k_2\left(y_2-z\right)^2. \tag{122}$$

After the corresponding transformations Table 3 shows the coefficients of the equations of motion of the system in operator form in the $\bar{y}_1$, $\bar{y}_2$ coordinates with kinematic perturbation $\bar{z}$.

Table 3. The coefficients of the equations of motion of the system in coordinates y_1, y_2 with kinematic external disturbance

a_{11}	a_{12}
$\left(Ma^2 + Jc^2 + m_1 i_1^2\right) \cdot p^2 + k_1$	$(Mab - Jc^2)p^2$
a_{21}	a_{22}
$(Mab - Jc^2)p^2$	$\left(Mb^2 + Jc^2 + m_2 i_2^2\right) \cdot p^2 + k_2$
Generalized forces	
$\bar{Q}_1'$	$\bar{Q}_2'$
$\bar{z}\left[i_1\left(i_1 + 1\right)m_1 p^2 + k_1\right]$	$\bar{z}\left[i_2\left(i_2 + 1\right)m_2 p^2 + k_2\right]$

Kinematic perturbation in the system, as it follows from Table 3, forms a force action simultaneously in two coordinates, which predetermines the use of the principle of superposition to determine the transfer functions. The structural diagram of the system with kinematic perturbation is shown in Figure 9.

Figure 9. The structural diagram of the initial system in Figure 7 in the coordinate system y_1, y_2 with a kinematic external perturbation from the side of the support surface

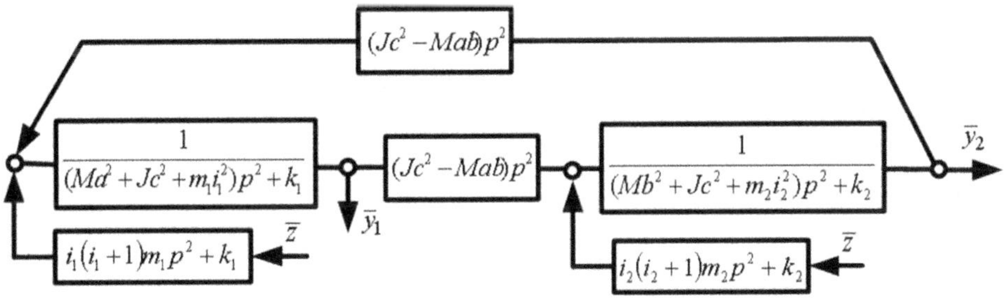

Using the structural diagram in Figure 9, we find the transfer functions of the system with kinematic perturbation:

$$W_1''(p)=\frac{\overline{y}_1}{\overline{z}}=\frac{\left[\left(Mb^2+Jc^2+m_2i_2^2\right)\cdot p^2+k_2\right]\cdot\left[i_1\left(i_1+1\right)m_1p^2+k_1\right]+ +\left[\left(Mab-Jc^2\right)p^2\right]\cdot\left[i_2\left(i_2+1\right)m_2p^2+k_2\right]}{A_0\left(p\right)}, \tag{123}$$

$$W_2''(p)=\frac{\overline{y}_2}{\overline{z}}=\frac{\left[\left(Ma^2+Jc^2+m_1i_1^2\right)\cdot p^2+k_1\right]\cdot\left[i_2\left(i_2+1\right)m_2p^2+k_2\right]+ +\left[\left(Mab-Jc^2\right)p^2\right]\cdot\left[i_1\left(i_1+1\right)m_1p^2+k_1\right]}{A_0\left(p\right)}, \tag{124}$$

where $A_0(p)$ is the characteristic frequency equation of the system and is defined by the expression (118).

3. The transfer function of interpartial relations is as follows:

$$W_{12}'\left(p\right)=\frac{\overline{y}_2}{\overline{y}_1}=\frac{a_{11}R_1+a_{12}R_2}{a_{22}R_2+a_{21}R_1}, \tag{125}$$

where a_{11}, a_{22}, a_{12}, a_{21} are determined from Table 3; R_1, R_2 respectively have the form:

$$R_1= i_1(i_1+1)m_1p^2+k_1, \tag{2126}$$

$$R_2= i_2(i_2+1)m_2p^2+k_2. \tag{127}$$

Comparison of Tables 2 and 3 shows that the coefficients of the equations of motion in the operator form do not change with variations in the form of external influences, but the right-hand sides of the equations (generalized forces) undergo significant changes.

Features of Dynamic Properties (Blocking)

When considering the transfer functions of the system $W_1''(p)$ и $W_2''(p)$, it can be concluded that kinematic perturbations along the inputs to the partial systems with coordinates y_1, y_2 can be "blocked". In this case, at the frequencies defined by the expressions:

$$\omega^2_{10dyn=}=\frac{k_1}{i_1\left(i_1+1\right)m_1}, \tag{128}$$

$$\omega_{20dyn}^2 = \frac{k_2}{i_2(i_2+1)m_2}, \tag{129}$$

and kinematic perturbing factors, it is possible to implement a mode of dynamic oscillation damping provided that the frequencies ω_{10dyn}^2 and ω_{20dyn}^2 coincide. In the general case, from (123), (124), which are fractional rational expressions, the numerator and denominator of which are fourth-order polynomials with even degrees, we can assume the possibility of the existence of motions when dynamic damping of oscillations both along the y_1 coordinate and along the y_2 coordinate will occur simultaneously.

If we assume that the condition $R_1 = R_2 = 0$ is fulfilled simultaneously, then the transfer function (125), which reflects the lever interactions between motions along the coordinates y_2 and y_1, gives an uncertainty of 0/0 type, the disclosure of which leads to some constant relation between the coordinates y_2 and y_1 that does not exclude the complete isolation of the object of protection in the form of a solid (M, J) from the vibrations of the supporting surface, since at this frequency the coordinates y_1, y_2 simultaneously pass through the position of dynamic equilibrium.

Possible forms of $\frac{y_2}{y_1}(\omega)$ dependency graphs are shown in Figure 10, which in a general form shows the possible options for graphs with various combinations of physical parameters of the system.

With an appropriate choice of parameters in the system, under the action of kinematic perturbations $z(t)$, motions with various forms of the distribution of oscillation amplitudes along the length $(l_{10}+l_{20})$ of a solid can be implemented, in particular with $y_2/y_1=1$ or $y_2/y_1= -1$, as well as dynamic damping modes along the coordinates y_1, y_2, as well as coordinate system y_0, φ.

Characteristic of dependence $\frac{y_2}{z}$ is the fact that when $\omega\to\infty$ the system enters the "locking" mode, with $N = W_1'(p) = \text{const}$. Dependencies of the same type $p\to\infty$ are also typical for graphs $\frac{y_2}{y_1}(\omega)$.

As for the dependencies of the transfer functions $\frac{y_2}{y_1}(\omega)$, they are quite complex in nature, which is also characterized by the presence of zero values of the numerator and denominator (125) when "locked" at high frequencies.

In the study of the influence of transfer ratios of lever linkages, defined by the expression (129) with equal external kinematic effects (129), it transforms to:

$$W_{12}'''(p) = \frac{\bar{y}_2}{\bar{y}_1} = \frac{[Ma(a-b)+2Jc^2]p^2 + k_1}{[Mb(b-a)+2Jc^2]p^2 + k_2} \tag{130}$$

In the general form, the graphs of $\frac{\bar{y}_2}{\bar{y}_1}(\omega)$ dependencies are shown in Figure 11.

When the zero is equal to the denominator (130), p. (2) is indicated on the abscissa axis. In turn, when the numerator (130) is equal to zero, the frequencies are indicated by p. (1) on the abscissa axis.

When considering dynamic effects in the coordinate system, features of motion can manifest themselves, which can be used to solve specific problems of vibration protection.

Figure 10. Variants of amplitude-frequency characteristics of $\frac{y_2}{z}(\omega)$ *with various combinations of parameters; pp. (1), (2) correspond to the dynamic damping of oscillations, pp.* ω_{1nat}, ω_{2nat} *correspond to the resonant modes*

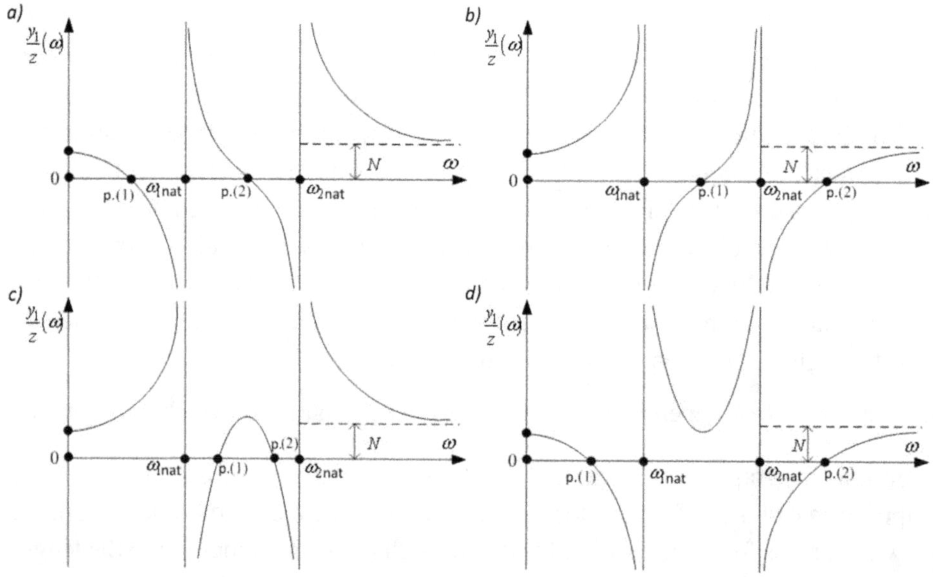

Figure 11. Variants of $\frac{y_2}{y_1}(\omega)$ *dependency graphs with equality of the "blocking" operators z(t) and various system parameters: a) the frequency of "zeroing"* y_2 *is less than the frequency of "zeroing"* y_1*; b) the frequency of "zeroing"* y_2 *is greater than the frequency of "zeroing"* y_1

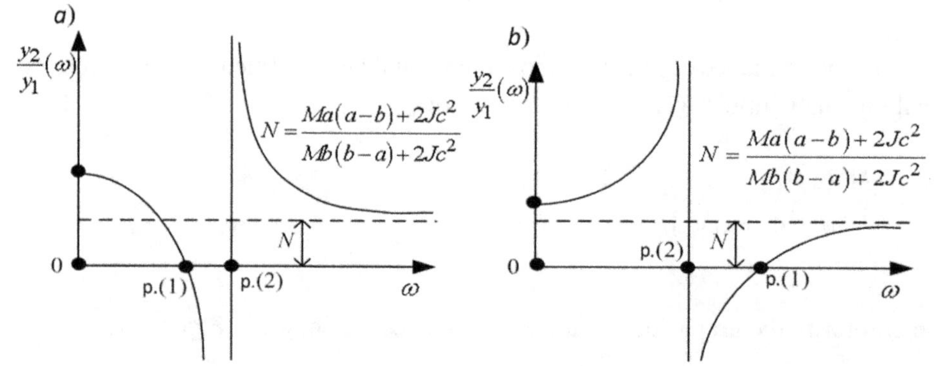

The Influence of the Choice of Coordinate System

Using (104)÷(108), we write the expressions for the kinetic and potential energies of the original system in the y_0,φ coordinates with a kinematic perturbation from the side of the reference surface:

$$T=\frac{1}{2}M\dot{y}_0^2+\frac{1}{2}J\dot{\phi}^2+\frac{1}{2}m_1[-i_1(\dot{y}_0-l_{10}\dot{\phi})+\dot{z}(1+i_1)]^2+\frac{1}{2}m_2[-i_2(\dot{y}_0+l_{20}\dot{\phi})+\dot{z}(1+i_2)]^2 \tag{131}$$

$$\Pi=\frac{1}{2}k_1(y_0-l_{10}\phi-z)^2+\frac{1}{2}k_2(y_0+l_{20}\phi-z)^2 \tag{132}$$

Let us carry out a number of similar calculations, as it is given in the previous sections and get the system of equations of motion in the y_0,φ coordinates. The coefficients of the equations in the operator form are given in Table 4.

Table 4. The coefficients of the equations of motion of the system in the y_0, φ coordinates in the operator form with the kinematic perturbation

a_{11}	a_{12}
$(M+m_1i_1^2+m_2i_2^2)p^2+k_1+k_2$	$(-m_1i_1^2l_{10}+m_2i_2^2l_{20})p^2-k_1l_{10}+k_2l_{20}$
a_{21}	a_{22}
$(-m_1i_1^2l_{10}+m_2i_2^2l_{20})p^2-k_1l_{10}+k_2l_{20}$	$(J+m_1i_1^2l_{10}^2+m_2i_2^2l_{20}^2)p^2+k_1l_{10}^2+k_2l_{20}^2$
Generalized forces	
Q_1''	$\bar{Q}_2''$
$\bar{z}\cdot[m_1i_1(1+i_1)+m_2i_2(1+i_2)]p^2+\bar{z}\cdot(k_1+k_2)$	$\bar{z}\cdot[-m_1i_1(1+i_1)l_{10}+m_2i_2(1+i_2)l_{20}]p^2-\bar{z}\cdot(k_1l_{10}-k_2l_{20})$

The structural diagram of the system has the form, as shown in Figure 12.

In the structural diagram (Figure 12), the following notation is agreed: a_{11}, a_{12}, a_{21}, a_{22} are given in Table 4; wherein:

$$R_1'=[m_1i_1(1+i_1)+m_2i_2(1+i_2)]p^2+k_1+k_2 \tag{133}$$

$$R_2'=[-m_1i_1(1+i_1)l_{10}+m_2i_2(1+i_2)l_{20}]p^2-k_1l_{10}+k_2l_{20} \tag{134}$$

With the kinematic perturbation z, the transfer functions of the system are:

Figure 12. Structural diagram of the original systems in Figure 7 in the coordinate system y0, φ under the kinematic action from the side of the supporting surface

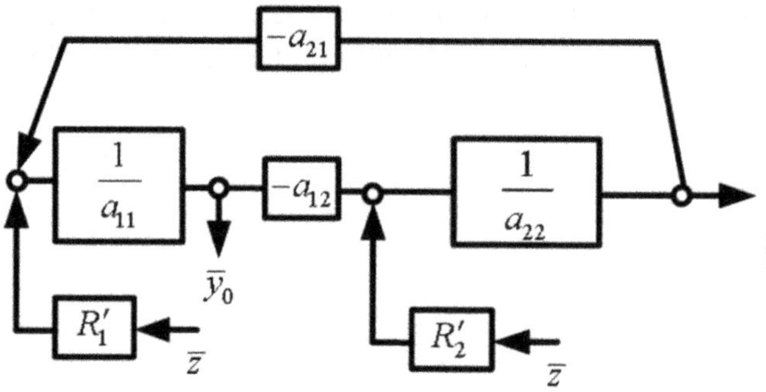

$$W_1''(p) = \frac{\bar{y}_0}{\bar{z}} = \frac{a_{22}R_1' + a_{12}R_2'}{A_0''(p)} \tag{135}$$

$$W_2''(p) = \frac{\bar{\phi}}{\bar{z}} = \frac{a_{11}R_2' + a_{21}R_1'}{A_0''(p)} \tag{136}$$

where $A_0''(p)$ is the characteristic equation, $A_0''(p) = a_{11}a_{22} - a_{12}^2$, in which a_{11}, a_{12}, a_{22} are determined from Table 4; R_1', R_2' are determined by expressions (133), (134).

Features of the dynamic properties of the system in the y_0,φ coordinates have their own characteristics, but the simultaneous coincidence of y*0* = 0 and φ = 0, as well as in the coordinate system y*1* and y2 is not possible with a given set of typical elements and attracted additional devices in the form of lever mechanisms.

Consideration of the transfer properties of lever linkages between the coordinates y_0 and φ is made on the basis of the transfer function of interpartial relations:

$$W_{12}''(p) = \frac{\bar{y}_0}{\bar{\phi}} = \frac{a_{22}R_1' + a_{12}R_2'}{a_{11}R_2' + a_{12}R_1'} \tag{137}$$

where a_{11}, a_{12}, a_{21}, a_{22} are given in Table 4.

In general, the dynamic properties of the system in coordinates y_0 and φ reflect the same features as in the system; the motion of which is sent to the coordinates y_1 and y_2 (in qualitative forms of representations). In this case, we note that the lever linkage in this case, in the physical signal is interpreted as "virtual", since it is implemented in the conventional screw mechanism. Detailed views on this issue are given in Eliseev et al. (2014b) and Lontsikh and Eliseev (2014).

Thus, the introduction of additional constraints in the form of lever mechanisms into the structure of mechanical oscillatory systems for various purposes significantly changes the dynamic properties of the systems, in particular, in problems of vibration protection.

The proposed method for constructing mathematical models based on the use of structural diagrams that are dynamically equivalent to automatic control systems has the ability to display the processes of forming dynamic constraints and to detail the manifestation of their properties with regard to specific structural and technical forms of mechanisms.

LEVERAGES IN THE DYNAMIC INTERACTIONS OF SYSTEM ELEMENTS

Ensuring the reliability and safety of operation of technological machines, equipment and equipment is an important scientific and technical direction in the development of modern engineering, although the necessary parameters of the dynamic quality of technical objects are formed at all stages of the life cycle. Of particular importance are preliminary exploratory research and development, within which the principles of constructing technical devices are developed, allowing to evaluate the possibilities of functioning, monitoring and controlling the dynamic state of engineering objects (Ivanov, 2007; Eliseev et al., 2015b; Khomenko & Eliseev, 2012). Modern technical means are quite complex systems, in which dynamic interactions of components, mechanisms and parts are characterized by a great variety, which is reflected in the variable forms of computational schemes in the form of mechanical oscillatory systems. Such systems often include mechanisms and devices with a large range of dynamic properties and forms of interaction of elements. Lever linkages are reflected in the dynamic properties of mechanical systems, in particular mechanical oscillatory systems, creating opportunities for implementing various dynamic modes, including dynamic vibration damping modes, redistribution of energy flows, organization of connections between partial structures of complex systems, etc. The results are presented research on the development of a generalized approach and a method for constructing mathematical models for determining and evaluating the dynamic properties of objects engineering industry as mechanical oscillatory systems with lever linkages and mechanisms with vibratory external influences. Lever linkages are diverse, they are implemented in various constructive and engineering forms depending on the features of machines and equipment and expand the functional properties of mechanical oscillatory systems, which creates opportunities for searching and developing new ways and means of controlling the dynamic state of engineering objects and the interaction of their elements.

Accounting for lever linkages and features of constructive and engineering forms of implementation of lever mechanisms requires the development of appropriate methodological positions in the formation and construction of mathematical models. In a generalized form, the possibilities of lever interactions are presented in the classification of articulation vibration isolators in Figure 13, which provides information on various forms of connection and coupling of lever mechanisms with an element of vibration protection systems (Eliseev et al., 2006; Eliseev & Artyunin, 2016; Khomenko et al., 2012).

When considering technical objects with several degrees of freedom, lever linkages and mechanisms can be considered as mechanical oscillatory systems with additional constraints (Belokobylsky et al., 2015), which is reflected in the development of structural methods of mathematical modeling. Within the framework of such a method, a mechanical oscillatory system is displayed with a structural diagram, which is dynamically equivalent to the automatic control system (Khomenko et al., 2014).

The development of such approaches allows us to consider the mechanisms in the structure of the mechanical oscillatory system and the emerging market relations in the interactions of elements as the introduction of additional constraints in relation to an object, the dynamic state of which is estimated (Eliseev et al., 2015a; Eliseev et al., 2013a).

Figure 13. The classification scheme of passive articulation vibration isolators

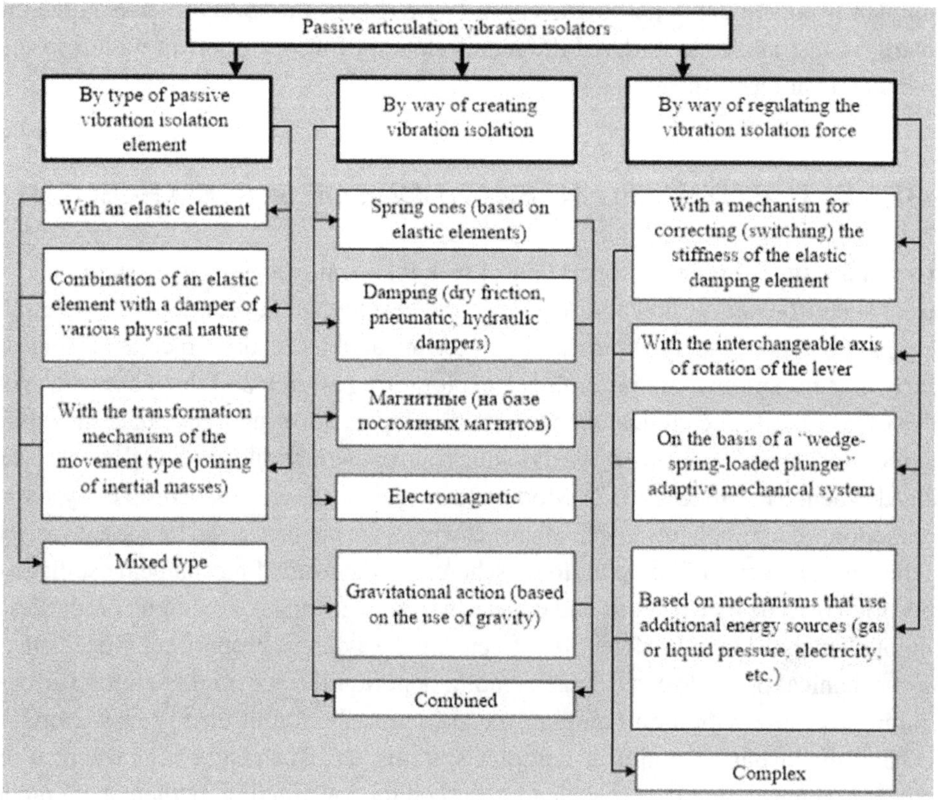

The dynamics of mechanical oscillatory systems with various mechanisms and devices can be interpreted as the task of introducing additional constraints into structural mathematical models of systems. The task of the research is to develop a scientific concept of searching and creating new ways and managing the dynamic state of various objects under vibration effects, using the methodological basis of structural mathematical modeling.

General Provisions and Features of the Structural Mathematical Modeling

As an example, we propose and detail a technology for the determination of interpartial constraints in a system with one or more degrees of freedom of the chain type. The interactions between partial systems are displayed as constraints in the structural diagrams shown in Figure 14. Details of the construction of structural models of mechanical oscillatory systems under force perturbation are given in Khomenko et al. (2014).

The structural diagram (Figure 14a) consists of two partial blocks (m_1, k_1, k_2 and m_2, k_2, k_3) connected in series, which reflects the properties of a mechanical oscillating system of a chain type. At the inputs of the partial systems are applied external harmonic perturbations $\tilde{Q}_1$ and $\tilde{Q}_2$. The structural diagrams in Figure 14 are constructed on the basis of previously obtained linear differential equations with constant coefficients, to which the Laplace transforms are applied under zero initial conditions and the subsequent transition using the rules of structural interpretations. In Figure 14b (when $\tilde{Q}_1 \neq 0$, $\tilde{Q}_2 = 0$)

Figure 14. Structural diagrams of a vibration protection system: a) detailed structural diagram; b) structural diagram with excluded coordinate $\tilde{y}_2\left(\tilde{Q}_1 \neq 0,\ \tilde{Q}_2 = 0\right)$*; c) scheme with excluded coordinate* $\tilde{y}_2\left(\tilde{Q}_1 \neq 0,\ \tilde{Q}_2 \neq 0\right)$

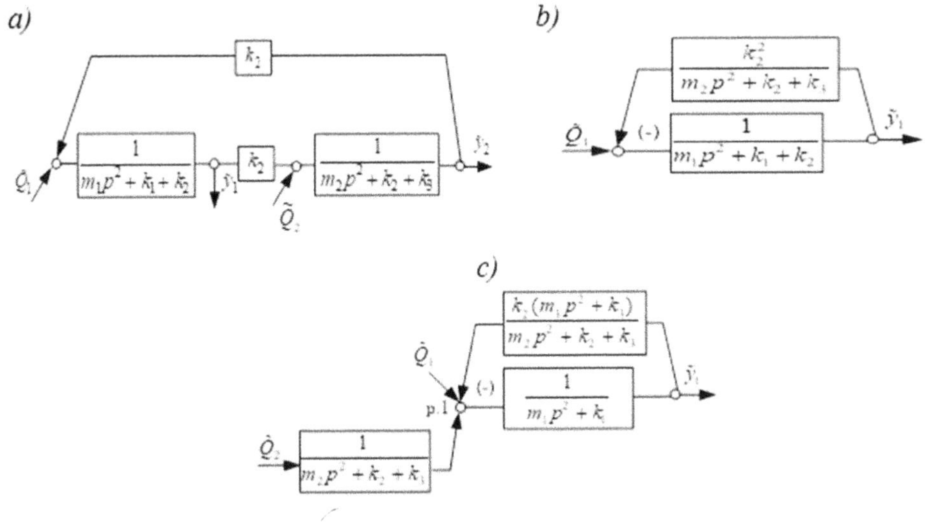

the $\tilde{y}_2$ coordinate is excluded, and the partial system is transformed into a negative feedback. Such a feedback is (in the physical sense) a quasi-spring, the dynamic stiffness of which depends on the frequency of the external force. In Figure 14c the original structural diagram is transformed with the selection of a block of the so-called basic system with a transfer function $\frac{1}{m_1 p^2 + k_1}$ reflecting the properties of the simplest system consisting of an object of mass m_1 and an elastic element k_1. All other connections can be viewed as complicated structures of a mechanical oscillatory system based on the parallel connection of additional springs of varying complexity. In Figure 14 it is shown that in the case of joint action of forces ($\tilde{Q}_1 \neq 0,\ \tilde{Q}_2 \neq 0$) external disturbance $\tilde{Q}_2$ can be brought to the input of the first partial system, but the action $\tilde{Q}_2$ is "corrected" by passing through the block with the transfer function $\frac{1}{m_2 p^2 + k_2 + k_3}$.

Note that preliminary obtaining a mathematical model in the form of differential equations of motion of a mechanical oscillatory system can be derived by any known method, for example, on the basis of the Lagrange equation of the second kind; The subsequent construction and transformation of structural mathematical models are carried out in accordance with the technology described in Khomenko et al. (2014), Eliseev et al. (2013a, 2015a). In Figure 14 the following notation is adopted: $p = j\omega$ is a complex variable; $j = \sqrt{-1}$; the sign "~" denotes the transform image of the Laplace variable.

Operator lever linkage transfer ratio:

$$i_{\text{gen}}(p) = \frac{\tilde{y}_2}{\tilde{y}_1} = \frac{k_2}{m_2 p^2 + k_2 + k_3} \tag{138}$$

can be called the transfer function of interpartial relations, which determines the properties of the virtual lever, provided that external forces $\tilde{Q}_1 \neq 0,\ \tilde{Q}_2 \neq 0$. The transfer ratio of the virtual lever is determined by the expression:

$$\underset{\tilde{Q}=\tilde{Q}_1+\tilde{Q}_2}{i_{\text{gen}}(p)} = \frac{\tilde{y}_2}{\tilde{y}_1} = \frac{m_1 p^2 + k_1 + 2k_2}{m_2 p^2 + 2k_2 + k_3} \quad (139)$$

When $p \to 0$, i.e. in a static state, (139) can be represented as:

$$\underset{p\to 0}{i'_{\text{gen}}(p)} = \frac{k_1 + 2k_2}{k_3 + 2k_2} \quad (140)$$

In turn, as $p \to \infty$, expression (139) is written:

$$\underset{p\to\infty}{i''_{\text{gen}}(p)} = \frac{m_1}{m_2} \quad (141)$$

The $\left|\frac{\tilde{y}_2}{\tilde{y}_1}\right| = \left|i_{\text{gen}}(p)\right|$ ratio module, defined by expression (6.138), depends on the frequency of the external influence, which is shown in Figure 15a-c.

Figure 15. The graphs of dependencies of the transfer ratio $i_{ob}(p)$ from the conditions of force disturbance with increasing frequency of external influence: a) the force perturbation variant $Q_1 \neq 0$, $Q_2 = 0$; b) the variant of $Q_1 = 0$, $Q_2 \neq 0$; c) the variant of $Q = Q_1 + Q_2$ – total action

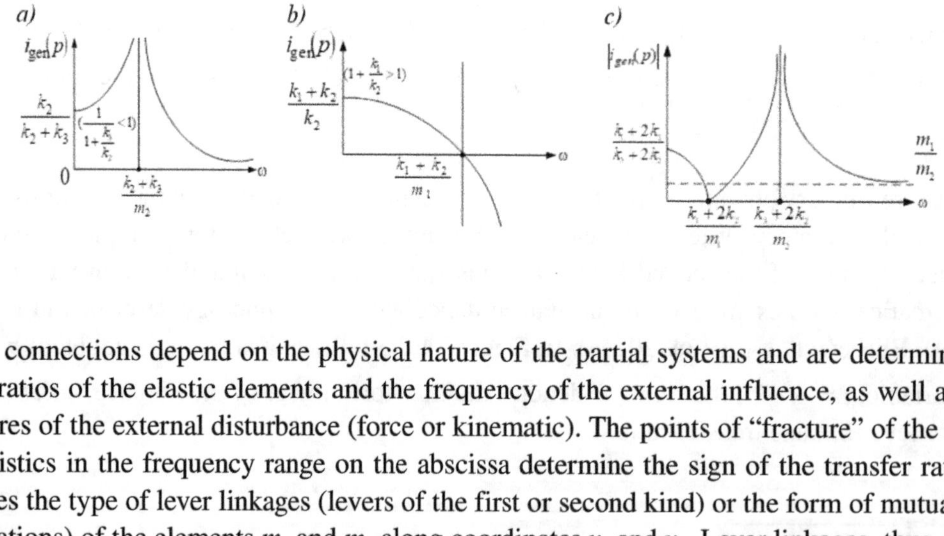

Lever connections depend on the physical nature of the partial systems and are determined by the stiffness ratios of the elastic elements and the frequency of the external influence, as well as the type and features of the external disturbance (force or kinematic). The points of "fracture" of the frequency characteristics in the frequency range on the abscissa determine the sign of the transfer ratio, which determines the type of lever linkages (levers of the first or second kind) or the form of mutual motions (or oscillations) of the elements m_1 and m_2 along coordinates y_1 and y_2. Lever linkages, thus, characterize also forms of own oscillations (Eliseev et al., 2015c). For a system with three degrees of freedom, lever linkages are characterized by more complex dependencies, which are based on the use of Cramer's formulas. In particular, the study of market relations is carried out separately by relationship $\frac{\tilde{y}_2}{\tilde{y}_1}$ and $\frac{\tilde{y}_3}{\tilde{y}_2}$, which is considered in more detail in Druzhinsky (1977) and Eliseev et al. (2013b).

An important circumstance for the evaluation of the dynamic properties of mechanical oscillatory systems is the behavior of the systems in the modes of dynamic oscillation damping. When passing through the frequency of dynamic damping, the sign of the lever linkage ratio of the lever linkage changes, i.e., there is a change in the form of vibrations (Eliseev et al., 2014a).

Dynamic interactions in mechanical oscillatory systems for various purposes, including vibration protection ones, can manifest themselves through ratios of oscillation amplitudes along the coordinates of the so-called coupled motions. These types of constraints are typical for static states. The relations between the displacements of coordinates depend not only on the configuration of the system, but also on the ratio of the stiffnesses.

The developed generalized approach to the assessment of lever linkages involves the consideration of the relationship between two coordinates in the forms characteristic of the interrelations of parameters in the levers of the first and second kinds. For systems of translational oscillatory motions, it is proposed to correlate lever linkages with the concept of a virtual lever. Transmission ratios of virtual levers depend on the parameters of the system, which can be both under the influence of static and dynamic disturbances of the force and kinematic types. In determining the transfer ratios of virtual levers or lever linkages, operator forms can be used. The implementation of lever linkages correlates with the ideas about the choice of certain frequency ranges in which a certain form of lever linkage dominates. Changing the frequency of the harmonic external disturbance affects the parameters of the transfer ratio. Cumulative interactions of external factors have a significant impact on the parameters of lever linkages, which implies the possibility of evaluating changes in dynamic properties by highlighting the coefficients of links of external influences.

The concepts of lever linkages form the basis for determining the forms of joint interconnected motions (or modes of oscillations) of elements in systems with several degrees of freedom.

Features of the manifestation of lever linkages are predetermined by the structure of the initial system and the composition of the partial blocks. The partial system of angular oscillations is usually associated with "spatial" forms of manifestation of motions, and the parameters of the motion depend on the location of the typical elements of the system, the type of their connection and the places where external disturbances are applied. The axis of rotation of the mass-and-inertia element can be variously oriented with respect to the direction of motion of the object.

There are also possible generalized ideas about the forms of partial systems, if we take into account that a generalized coordinate can characterize screw motions. By way of example, Figure 16a-z gives possible types and features of the combination of partial systems in the computational schemes of machine engineering objects.

The manifestation of lever linkages is characteristic not only for systems with two degrees of freedom, but also for mechanical oscillatory systems with one degree of freedom, which, in particular, was reflected in (Eliseev et al., 2013a).

Figure 16. Variants of the representation of partial systems and types of connections between partial systems: a) translational; b) rotational; c) consecutive elastic connection of two translational partial systems; d) connection of the translational and rotational systems; e) rotational system of lever type; f) connection of two rotational systems; g) connection of the rotational system of lever type and systems of translational type with a solid body (and not a material point); h) connection of the rotational system of the lever type with the point of rotation on a solid body making a translational motion

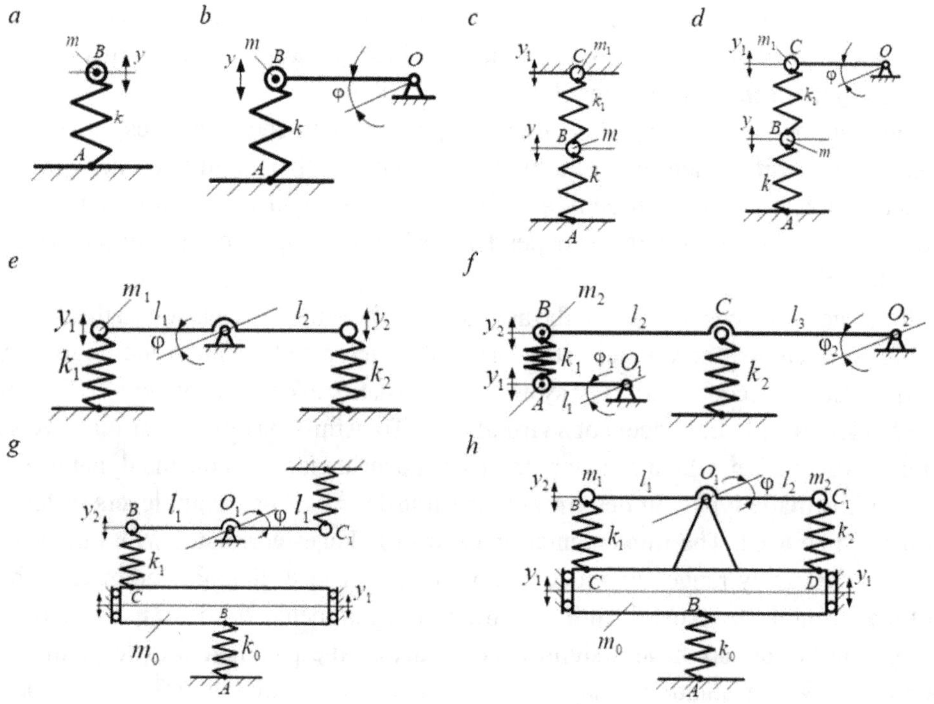

With force disturbances, the ratio of static and dynamic reactions at the attachment points of the elastic elements with the supporting surfaces can be chosen as parameters, whose ratio a lever linkage reflects. In systems with several degrees of freedom, the plots of dependencies of lever linkages no longer coincide with the amplitude-frequency characteristics of the system and have significant differences.

The tasks of forming lever linkages in a system with two degrees of freedom, where a solid body on elastic supports k_1 and k_2 has coordinates y_0 and φ (y_0 characterizes the motion of the center of gravity, φ is the angle of rotation relative to the center of gravity) have their own characteristics. The dimension of the transfer ratio of the lever is determined by the ratio of the angle of rotation and displacement along the vertical, which is typical for a screw pair.

The above considerations about the forms and properties of lever linkages that manifest themselves in mechanical oscillatory systems can be viewed as some scientific concept of building and developing principles for controlling the dynamic state of systems, in which the settings can be determined not only by the parameters of the system, but also have adaptive properties.

Transformation of Structural Mathematical Models for Systems With Partial Blocks of Various Forms

Dynamic interactions of elements have their own features in different coordinate systems. The computational scheme with the object of protection with the mass M_0 and the intermediate solid on elastic supports (dynamic vibration damper) is shown in Figure 17, and its structural diagram – in Figure 18.

Figure 17. The computational scheme of the system with a dynamic damper in the form of a solid on elastic supports (k_2, k_3)

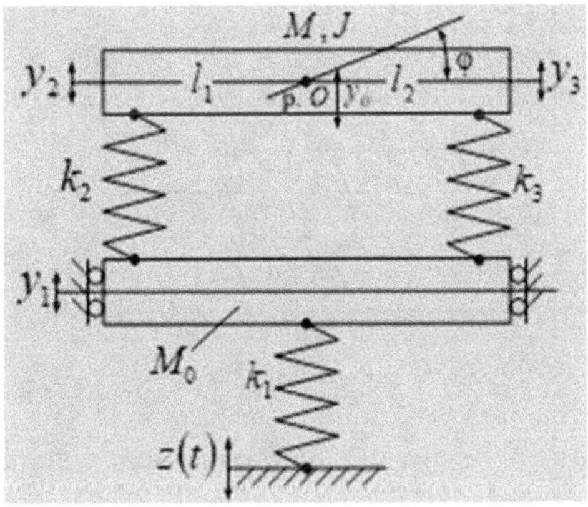

In Eliseev et al. (2014a), the methodology of structural transformations and determination of the main characteristics of a mechanical oscillatory system is considered. Analytical relations for dynamic stiffnesses (i.e., parameters of generalized springs) are obtained.

Figure 18. The structural diagram, analogous of the equation of motion of the mechanical system shown in Figure 17

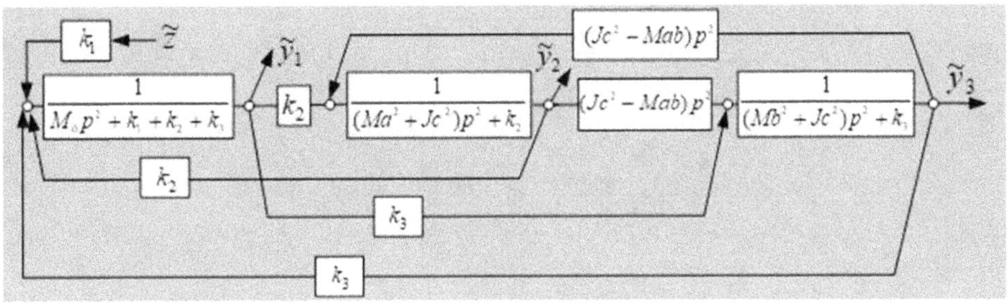

It is shown that the generalized springs in structural transformations behave like ordinary elastic elements, which allows us to consider them as quasi-springs (Khomenko et al., 2015). The transformation of computational schemes with other variants of the distribution of forms of motion of partial blocks gives the possibility of reduction to equivalent structures in an operator form characteristic of systems of a valuable type. In Lenk (1978), a number of techniques for distinguishing lever linkages in systems with solid bodies with angular motions are presented. The necessary simplifications are achieved when the conditions of smallness (or zeroing) of the moment of inertia of a solid are satisfied. Figure 19 shows one of the computational schemes in which elements of the system execute angular motions.

Figure 19. The computational scheme of the system with partial blocks of rotary type (levers of the second kind)

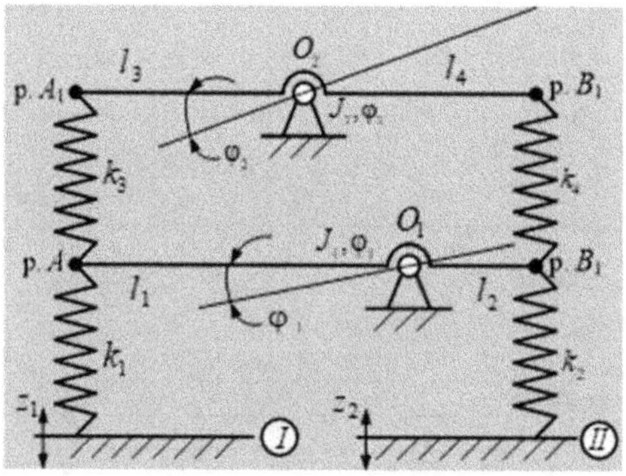

Figure 20 presents the dynamically equivalent computational scheme of a system with quasi-translational type motions.

Figure 20. The reduced computational scheme, equivalent to the original system in Figure 19

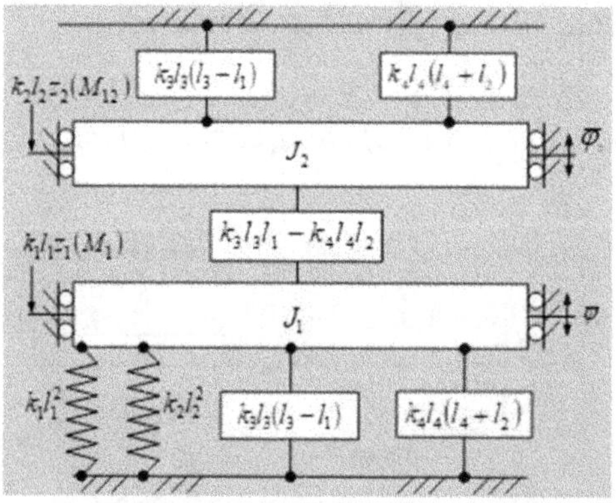

In this case, the problems of dynamic synthesis can be solved in a comparative comparison of the forms of implementation of certain types of motion. In relation to the computational scheme in Figure 20 the transfer ratio at the input action $\tilde{z}_1$ and output action $\tilde{y}_1$ is:

$$W_1(p)=\frac{\tilde{y}_1}{\tilde{z}_1}=\frac{k_1\cdot\left(m_2p^2+k_3+k_4i_2^2\right)}{\left(\begin{array}{l}\left[m_1p^2+k_1+k_2i_1^2+i_{13}^2\cdot\left(k_3+k_4i_2^2\right)\right]\times\\ \times\left(m_2p^2+k_3+k_4i_2^2\right)-\left(k_3-k_4i_1i_2\right)^2\end{array}\right)} \quad (142)$$

where $i_1=\frac{l_2}{l_1}$, $i_2=\frac{l_4}{l_3}$ are the transfer ratios of lever linkages.

Thus, a generalized approach has been developed in problems of dynamics and a method for constructing and evaluating the properties of mathematical models of various mechanical oscillatory systems, reflecting the specificity of mechanical systems, taking into account the geometric features of the arrangement of elements and the various forms of their interaction, taking into account lever linkages. A number of proposals made at the level of inventions. The research results are of interest for solving a wide range of dynamic tasks for mechanisms and machines for various purposes, which work in conditions of vibration and dynamic forces arising at the object of protection.

REFERENCES

Baranovsky, A. M. (2000). *Theoretical foundations of effective vibration isolation on ships* [PhD thesis]. Novosibirsk.

Belokobylsky, S. V., Eliseev, S. V., Kashuba, V. B., Nguyen, D. H., & Tsygan, V. V. (2015). On relations between motion coordinates in mechanical oscillatory systems with lever devices. Systems. Methods. *Technology*, 2(26), 7–13.

Belokobylsky, S. V., Eliseev, S. V., & Sitov, I. S. (2013). *Dynamics of mechanical systems: Lever and inertial-elastic constraints*. Polytekhnika Publ.

De Silva, C. W. (2000). *Vibration: Fundamentals and Practice*. CRC Press.

Druzhinsky, I. A. (1977). *Mechanical chains*. Mashinostroenie Publ.

Eliseev, S. V., & Artyunin, A. I. (2016). *Applied theory of oscillations in problems of the dynamics of linear mechanical systems*. Nauka Publ.

Eliseev, S. V., Artyunin, A. I., & Kaimov, E. V. (2013a). Equivalent transformations in the structures of mechanical oscillatory systems containing material points. In *Information and mathematical technologies in science and management: Proceedings of XVIII Baikal All-Russian conf. (Irkutsk, July 01–10, 2013) in 3 vol.* (*Vol. 1*, pp. 173-182). Irkutsk: ISEM of the SB RAS.

Eliseev, S. V., Artyunin, A. I., Kaimov, E. V., & Eliseev, A. V. (2013b). Pat. 136112 Russian Federation, IPC F16F 15/04. A damping device.

Eliseev, S. V., Artyunin, A. I., Kaimov, E. V., & Eliseev, A. V. (2014a). Pat. 142137 Russian Federation, IPC F16F 15/02. Device for regulating the elastic-dissipative properties of a vibration protection system.

Eliseev, S. V., Belokobylsky, S. V. & Kashuba, V. B. (2011a). Mathematical modeling in mechanical oscillatory systems: Mechatronic approaches. *Problems of Mechanical Engineering and Automation, 3*, 70-78.

Eliseev, S. V., Bolshakov, R. S., & Kinash, N. Zh. (2015a). Leverage: Virtual leverage mechanisms, features of oscillatory processes. *News of South-West State University, 2*(59), 8–15.

Eliseev, S. V., & Khomenko, A. P. (2014). *Dynamic oscillation suppression: Feedback concept and structural methods of mathematical modeling*. Nauka Publ.

Eliseev, S. V., Kinash, N. Zh., & Kaimov, E. V. (2015b). Lever linkages of mechanical oscillatory systems. *Bulletin of VELNII, 1*(69), 112–126.

Eliseev, S. V., Kovyrshin, S. V. & Bolshakov, R. S. (2012a). Features of the construction of compact elastic elements in mechanical oscillatory systems: Interactions with elements of the system and form of the compound. *Modern Technologies, System Analysis, Modeling, 4*(36), 61-70.

Eliseev, S. V., Kuznetsov, N. K., & Kaimov, E. V. (2015c). On the question of the theory of lever constraints in the dynamics of mechanical oscillatory systems. *Bulletin of the Irkutsk State Technical University, 12*(107), 30–40.

Eliseev, S. V., Lukyanov, A. V., Reznik, Y. N., & Khomenko, A. P. (2006). *Dynamics of mechanical systems with additional ties*. Irkutsk State University Publ.

Eliseev, S. V., Moscovskikh, A. O., Bolshakov, R. S., & Savchenko, A. A. (2012b). Possibilities of integration of methods of the theory of chains and the theory of automatic control in problems of the dynamics of machines. *Science and Education, 6*, 277–296.

Eliseev, S. V., Parshuta, E. A., Kaimov, E. V., & Kinash, N. Zh. (2014b). Mechanisms in the structure of vibration protection systems: Mathematical models, evaluation of dynamic properties. *Bulletin of VSGUT, 6*(51), 37–44.

Eliseev, S. V., Reznik, Y. N., & Khomenko, A. P. (2011b). *Mechatronic approaches in the dynamics of mechanical oscillatory systems*. Nauka Publ.

Eliseev, S. V., Reznik, Y. N., Khomenko, A. P., & Zasyadko, A. A. (2008). *Dynamic synthesis in generalized tasks of vibration protection and vibration isolation of technical objects*. Publishing House of Irk. State University.

Fomichev, P. A. (2010). *Vibration isolation of ship power plants with electropneumo-hydraulic supports* [PhD thesis]. Novosibirsk.

Harris, S. M., & Piersol, A. G. (2002). *Shock and Vibration Handbook*. McGraw Hill Book Co.

Ivanov, B. G. (2007). *Development of methods for calculating the dynamics and strength of transport equipment aggregates with articulated kinematic relations* [PhD thesis]. Samara.

Karnovsky, I. A., & Lebed', E. (2016). *Theory of Vibration Protection*. Springer International Publishing. doi:10.1007/978-3-319-28020-2

Khomenko, A. P. & Eliseev, S. V. (2010). Vibration protection systems with joints: Technology of construction of mathematical models. *Modern Technologies, System Analysis, Modeling, 4*(28), 8-18.

Khomenko, A. P., & Eliseev, S. V. (2012). Non-traditional approaches to the construction of mathematical models of mechanical oscillatory systems with lever linkages. *Izvestia Transsiba, 4*(12), 75–86.

Khomenko, A. P. & Eliseev, S. V. (2013). Quasi-elements in mechanical oscillatory systems: Features of systems with the exclusion of variables of the dynamic state. *Modern Technologies, System Analysis, Modeling, 2*(38), 8-17.

Khomenko, A. P., Eliseev, S. V., Artyunin, A. I., Kaimov, E. V., & Eliseev, A. V. (2015). A damping device. Pat. 150331 Russian Federation, IPC F16F 15/04.

Khomenko, A. P., Eliseev, S. V. & Ermoshenko, Yu. V. (2012). *System analysis and mathematical modeling in the mechatronics of vibration protection oscillatory systems.* Irkutsk: IrGUPS Publ.

Khomenko, A. P., Eliseev, S. V., & Kaimov, E. V. (2014). Virtual lever mechanism: Dynamic oscillation of oscillations as a form of manifestation of lever linkages. *News of Transsib, 4*(20), 61–71.

Kolovsky, M. Z. (1976). *Automatic control of vibration protection systems.* Nauka Publ.

Kuznetsov, N. K. (2009). *Dynamics of controlled machines with additional constraints.* ISTU Publ.

Lenk, A. (1978). *Electromechanical systems. Systems with lumped parameters.* Mir Publ.

Lontsikh, P. A., & Eliseev, S. V. (2014). *Dynamic quality of machinery and equipment as a tool to ensure the reliability of production and competitiveness of processes.* ISTU Publ.

Sun, X., & Jing, X. (2015). Multi-direction vibration isolation with quasi-zero stiffness by employing geometrical nonlinearity. *Mechanical Systems and Signal Processing, 62-63*, 149–163. doi:10.1016/j.ymssp.2015.01.026

ADDITIONAL READING

Eliseev, A.V. & Kopylov, Y.R. (2015). Modeling contact interactions in vibration technological processes and machines with account for non-sustaining connections. *Modern High Technologies, 1*(1), 19-22.

Eliseev, A.V. & Kuznetsov, N.K. (2022). Technology of structural mathematical modeling of technical objects under vibrational loads: forms of interactions and dynamic invariants. *iPolytech Journal, 26*(3), 368-385.

Eliseev, A. V., & Orlenko, A. I. (2018). On mathematical modeling of continuous brush throwing modes in contact interactions with a collector. *Bulletin of Irkutsk State Technical University, 22*(1), 10–22. doi:10.21285/1814-3520-2018-1-10-22

Eliseev, S. V., & Eliseev, A. V. (2012). Generalized representation of dynamic damping of vibrations as an additional feedback in mechanical systems with non-sustaining connections. Modern Technologies. System Analysis. *Modeling, 2*, 36.

Eliseev, S. V., & Eliseev, A. V. (2012). Generalized approaches to determining contact reactions in solids under static loads with account for non-sustaining connections. Modern Technologies. System Analysis. *Modeling*, *4*, 51.

Eliseev, S. V., Trofimov, A. N., & Eliseev, A. V. (2012). Generalized representation of dynamic damping of vibrations as an additional feedback. *Bulletin of Vladivostok State Technical University*, *3*(38), 30.

Chapter 7
Some Directions of Development of Methods of Structural Mathematical Modeling:
Structural Representations and Motion Transformation Devices

ABSTRACT

The chapter discusses the dynamic properties of mechanical systems and the possibilities of their evaluation. The features of using the characteristic frequency equation, which allows determining the natural frequencies of the system, are shown. Methods of accounting for additional inertial type connections that can be used to improve the dynamic properties of technical systems are presented. The ideas about the possibility of including motion transformations devices in the system in order to change dynamic properties are presented.

INTRODUCTION

Mathematical modeling is widely used in solving various problems related to the assessment of the dynamic state of technical objects. The methods of structural mathematical modeling are a rather well-developed direction determined by an interdisciplinary space defined by the methods of the theory of systems, the theory of automatic control, the theory of circuits, and the theory of graphs in their various forms of applications. The dynamics of mechanical oscillatory systems with lumped parameters is reflected in many works on the theory of linear and nonlinear oscillations, which, in a certain sense, predetermined interest in the formation and features of the constraints arising in dynamic interactions between elements of the system during periodic influences. There are a number of well-developed applications related to

DOI: 10.4018/978-1-6684-7237-8.ch007

solving the problems of the theory of vibration engineering, vibration technologies, problems of transport dynamics, vibration protection and vibration isolation of technical objects (Samara & Mikhailov, 2005; Lontsikh & Eliseev, 2014; Eliseev et al., 2011; Eliseev & Khomenko, 2014; Eliseev et al., 2008).

CHARACTERISTIC FREQUENCY EQUATION: STRUCTURE, DYNAMIC STIFFNESS, FEATURES OF THE INTERACTION OF SYSTEM ELEMENTS

Interpretation of dynamic interactions of elements of mechanical systems, displayed by structural diagrams of dynamically equivalent automatic control systems, became the basis for the formation of non-traditional approaches, which initiated the introduction of a number of new issues related to the ideas of expanding ideas about a set of typical elements, ways of connecting them, and possibilities of structural transformations and identifying dynamic features (Khomenko et al., 2012; Belokobylsky et al., 2013; Khomenko et al., 2016). Structural mathematical modeling considers the connections and regulatory influences between elements, using tools from automatic control and mechanical circuits. This approach can expand the set of elements, introduce constraints, and transform models. One direction is to use dynamically equivalent automatic control systems as mathematical models, based on interactions between elements of mechanical and control systems. This approach is generalizable to systems with distributed parameters.

Figure 1a-f gives various interpretations in the representations of the initial physical model in the form of a mechanical oscillatory system with one degree of freedom. Such models are widely used in the tasks of machine dynamics and the protection of equipment and apparatus from vibrations, impacts, etc. The mechanical system in Figure 1a consists of the so-called typical mechanical elements in the form of a spring k, a damper bp and a mass-and-inertia unit m. A mass-and-inertia element can be considered as an object of control (in vibration protection tasks, an object of protection), which predetermines the possibility of using the analogy between automatic control systems.

The mathematical model of the system (Figure 1a) with a force perturbation in the form of a harmonic force Q has the form

$$m\ddot{y} + b\dot{y} + ky = Q(t). \tag{1}$$

After the Laplace transform with respect to (1) with zero initial conditions, the differential equation (1) transforms to

$$mp^2\bar{y} + bp\bar{y} + k\bar{y} = \bar{Q}, \tag{2}$$

where $p = j\omega$ is a complex variable; the <–> symbol means a Laplace transform image (Eliseev et al., 2011).

By analogy with the basics of the theory of automatic control, the structure of the system in Figure 1a is transformed into a structural diagram, shown in Figure 1b. Typical elements m, b and k on the structural diagram (Figure 1b) correspond to typical units of the automatic control system, respectively, in the form of units with the transfer function of the integrating unit of the second kind (protection object) – mp^2, the differentiating unit of the first kind (damper) – bp and the amplifying unit (spring) – k.

The structural diagram in Figure 1b displays not only the composition of the systems (the presence of typical units), but also couplings (direct, feedback, etc.). Typical units in the structural diagram can be interconnected and form the appropriate couplings, as shown in Figure 1c and d.

Figure 1. The computational scheme and structural diagrams of a mechanical oscillatory system with one degree of freedom in case of force external disturbance: a) computational scheme, b) structural diagram with the display of the action of two elements; c) structural diagram with the generalized element (k + bp), d) structural diagram with elastic and damping elements as negative feedbacks, e) structural diagram of the system in the generalized form, f) diagram reflecting features of the transfer element from the direct coupling to the feedback one

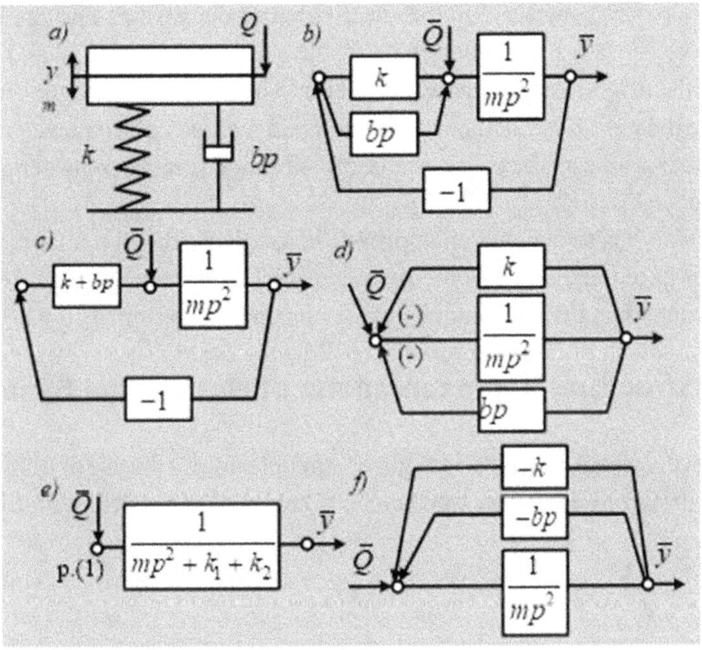

The structural diagram of the initial mechanical oscillatory system in Figure 1a can be presented in a generalized form, as shown in Figure 1e. Note that the elements *bp* and *k* (Figure 1e) can obtain the "status" of negative feedbacks. Physically, negative feedback (Fig. 7.1, *f*) is interpreted as the stiffness of the elastic element (*k*). Since at the input to the adder (Figure 1f) a force action $k\overline{y}$ is formed, the inverse negative relationship (–*k*) reflects the action of the elastic element upon the return of the object (*m*) to the initial position in the presence of deviation. The damper, as a unit with the transfer function *bp*, in the physical sense reflects the action of the force of viscous resistance. In this case, the negative feedback with the transfer function of the first-order differentiating unit *bp*, as well as the spring *k*), forms a force action.

Thus, at p. (1) in Figure 1f you can make the equation of kinetostatics:

$$\overline{y}mp^2 + \overline{y}bp + \overline{y}k = \overline{Q} \tag{3}$$

Expression (3) is a form of representation of the d'Alembert principle.

For further research, it is important that from (3) it follows that elements (bp) and (k) can be represented as units, for which the input signal is the displacement $\overline{y}$, and the output is a certain force factor. The latter can be interpreted as the possibility of considering the named elements in the "status" of the spring with the corresponding stiffness. If for the spring k the stiffness of the element does not depend on ω (that is, on the perturbation frequency), then for the element bp, the "stiffness" of the element depends on the frequency ω. In this case, the stiffness can be called dynamic. In systems with several degrees of freedom in vibration protection tasks, one of the mass-and-inertia elements is selected as the object of protection, which is displayed by the type of transfer function of the integrating unit of the second order. As for the rest of the mass-and-inertia elements (they can be called intermediate), they take the form of second-order differentiating units with properties to enter into connections between themselves and the supporting surfaces by the same rules as the springs and the damper. Such issues were considered in more detail in Eliseev et al. (2008), Khomenko et al. (2012), and Eliseev et al. (2006).

The translation of the original mathematical model into a structural analogue and the corresponding transformations create the possibility of using the developed methods of frequency analysis and dynamic synthesis of systems for various applications, including vibration protection systems (Druzhinsky, 1977; Eliseev, 2012; Belokobylsky, 2015).

The task of the research is to develop an approach to assessing the dynamic properties of mechanical oscillatory systems based on the use of the frequency characteristic equation and concepts about the dynamic stiffness of elements, their structural formations and the system as a whole.

Development of Structural Representations of Mechanical Systems

Transition to structural diagrams, as shown in Figure 1, predetermines the use of transfer functions (Figure 1e). In its simplest form (Figure 1a and b), the transfer function of the system with a force perturbation is:

$$W(p) = \frac{\overline{y}}{\overline{Q}} = \frac{1}{mp^2 + bp + k} \tag{4}$$

The denominator of the transfer function (4) is used as the frequency characteristic equation:

$$mp^2 + bp + k = 0 \tag{5}$$

By substituting $p = j\omega$, the equation can be solved by determining the natural frequency. In the simplest case ($b = 0$), the natural frequency is determined by:

$$\omega^2 = \frac{k}{m} \tag{6}$$

that follows from the relation:

$$-m\omega 2^{+}\mathrm{k} = 0 \tag{7}$$

If we make an inversion (4), then taking into account (7), it can be noted that it is at the natural frequency that the dynamic stiffness of the system as a whole takes on a zero value. In this postulate, it is assumed that when considering the force loading of the system, the mass-and-inertia element, in which the dynamic stiffness, defined as the ratio of the displacement of the protected object to the applied force, can take values equal to zero (for $b = 0$ or $b \to 0$).

Thus, from the frequency equation (5), under appropriate conditions, the general property of the system can be determined, which is formed with a dynamic stiffness equal to zero and the presence of a harmonic perturbing force. The same dynamic effect is also considered as resonant oscillations. In the framework of the proposed approach, dynamic stiffness is a property that extends not only to individual elements of the system, but also to the system as a whole. The latter is associated with the concept of the transfer function, in which the input action (or signal) and the output signal (in this case, the displacement y) are distinguished. The above concept of the transfer function can be extended by varying the concepts of input and output signals, as shown in (Belokobylsky, 2015). In the physical sense, the frequency characteristic equation (5) in the operator form is the sum of the dynamic stiffness correlated with the parameters of the mass-and-inertia element m. If this sum is zero at a certain frequency of an external force, then this means that the dynamic stiffness of the system as a whole is zero (in this case, the system is represented by a material point with mass m and an elastic element k for $b = 0$). If there is a wider set of typical elements, local (or particular) dynamic stiffnesses may not be equal to zero. An important circumstance is the existence of a connection between such dynamic effects as a decrease in the stiffness of the systems as a whole to zero (or to some value) at certain frequencies, which is associated with ideas about resonant processes. Such approaches, in particular, form the basis of methods for determining the natural frequencies of technical objects.

Systems With Two Degrees of Freedom

Mechanical systems with two degrees of freedom are quite diverse in their physical nature and constructive and engineering forms, which is reflected in the variety of mathematical models and coordinate systems used. Figure 2a-d presents mechanical oscillatory systems of a chain type (Figure 2a), a solid body on elastic supports (Figure 2b), as well as their structural diagrams in Figure 2c-e. To obtain structural mathematical models, the methods described in Eliseev et al. (2011) and Eliseev and Khomenko (2014) are used. When considering the motions of a solid body on elastic supports (Figure 2b), two coordinate systems y_1, y_2 and y_0, φ, are used, which is due to the selection of the center of mass, taking into account the moment of inertia of the solid and angular oscillations. Systems with two degrees of freedom are composed of partial blocks (systems) and corresponding to the chosen coordinate system of forms of interpartial constraints. In the general case, typical elements of mechanical oscillatory systems have the same form as typical elements of systems with one degree of freedom; however, the connections between elements, including partial systems, are very diverse. Note also that the set of typical elementary units of mechanical systems can be expanded by introducing units with transfer functions in the form of second-order differentiating units, as well as at the expense of other elements, the transfer functions of which correspond to the elements of transfer functions of similar elements. In this case, a typical element with a certain transfer function is a device that, when described, does not break down into separate components, that is, in a sense, a typical element is considered as "indivisible". When using rules for transforming structural models, typical elementary units can be used to build more complex structures,

Figure 2. Systems with two degrees of freedom: a) computational scheme of a chain type; b) computational scheme of a solid system body; c) structural diagram of the chain type in the coordinates y_1, y_2; d) system with a solid body in coordinates y_0, φ; e) structural diagram of the system in coordinates y_1 and y_2

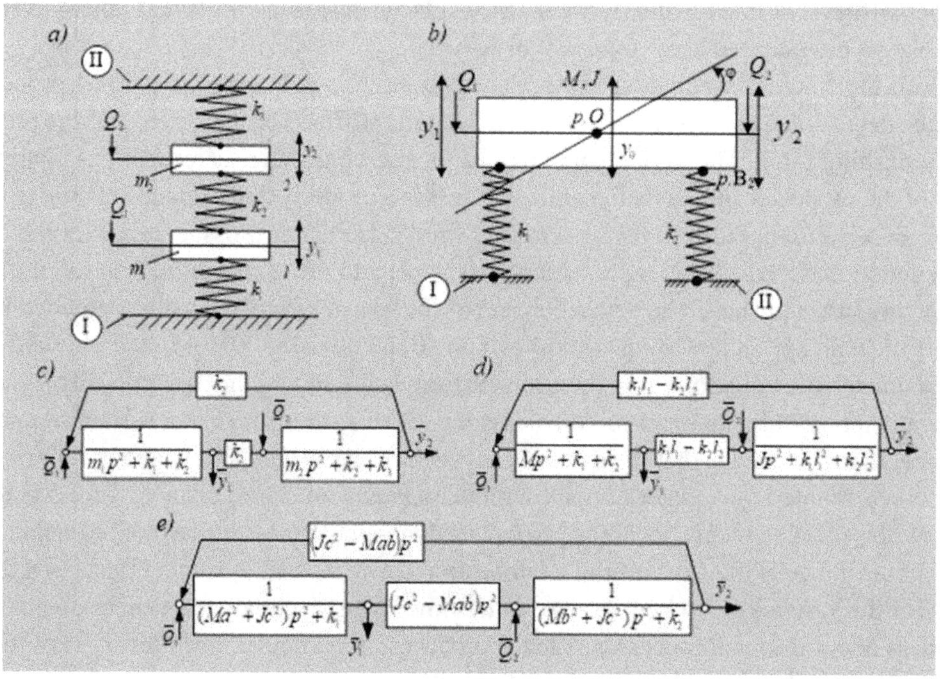

which can consist of several related elements and have dynamic stiffness. In structural transformations such composite structures are called quasi-springs.

Quasi-springs in structural transformations behave the same as ordinary typical elements, but their dynamic stiffness generally depends on the frequency of the external force (Eliseev, 2013).

In systems with solids (Figure 2b), the coordinates y_1, y_2 and y_0, φ are related by the relations:

$$y_0 = ay_1 + by_2, \phi = c(y_2 - y_1), y_1 = y_0 - l_1\phi,$$
$$y_2 = y_0 + l_2\phi, a = \frac{l_2}{l_1 + l_2}, b = \frac{l_1}{l_1 + l_2}, c = \frac{1}{l_1 + l_2}. \quad (8)$$

The parameters of the structural mathematical models are presented in Table 1.

As generalized forces in the systems in Figure 2a and b one can use both force and kinematic perturbations from the side of supporting surfaces *I* and *II*.

1. For the structural diagram in Fig. 7.2, *c* in case of force disturbance $Q_1 \neq 0$ transfer functions are defined by the expressions:

Table 1. The coefficients of the structural mathematical models of the systems in Figure 2a and b

The chain type system (Figure 2a)	
a_{11}	a_{12}
$m_1p^2 + k_1 + k_2$	$-k_2$
a_{21}	a_{22}
$-k_2$	$m_2p^2 + k_2 + k_3$
Generalized forces	
Q_1	Q_2
The system with a solid body (coordinates y_1, y_2) (Figure 2b)	
a_{11}	a_{12}
$(Ma^2 + Jc^2)p^2 + k_1$	$(Mab - Jc^2)p^2$
a_{21}	a_{22}
$(Mab - Jc^2)p^2$	$(Mb^2 + Jc^2)p^2 + k_2$
Generalized forces	
Q_1	Q_2
The system with a solid body (coordinates y_0, φ) (Figure 2b)	
a_{11}	a_{12}
$Mp^2 + k_1 + k_2$	$k_2l_2 - k_1l_1$
a_{21}	a_{22}
$k_2l_2 - k_1l_1$	$Jp^2p^2 + k_1l_1^2 + k_2l_2^2$
Generalized forces	
Q_1	Q_2

$$W_1(p) = \frac{\bar{y}_1}{\bar{Q}_1} = \frac{m_2p^2 + k_2 + k_3}{A(p)} \tag{9}$$

$$W_2(p) = \frac{\bar{y}_2}{\bar{Q}_1} = \frac{k_2}{A(p)} \tag{10}$$

where

$$A(p) = (m_1p^2 + k_1 + k_2)(m_2p^2 + k_2 + k_3) - k_2^2 \tag{11}$$

Expression (11) determines the characteristic frequency equation for the mechanical oscillatory system in Figure 2a Comparing (9) and (10) shows that the transfer functions may be different depending on the choice of points of application of external disturbances and the choice of output signals, giving an idea of the dynamic state of the system. When solving specific problems of dynamics, the computational scheme of the system takes on a certain form, for example, in vibration protection tasks, which is accompanied by the choice of the object of protection. Assuming that the object of protection from external influence Q_1 is related to the mass-and-inertia element m_1, we will reduce the frequency equation (7.11) to the form:

$$(m_1 p^2 + k_1 + k_2) - \frac{k_2^2}{(m_2 p^2 + k_2 + k_3)} = 0 \tag{12}$$

Convert (12):

$$m_1 p^2 + k_1 + \frac{k_2 (m_2 p^2 + k_3)}{m_2 p^2 + k_2 + k_3} = 0 \tag{13}$$

Structural mathematical model of the mechanical system shown in Figure 2a can be obtained by transforming the structural diagram in Figure 2, as shown in Figure 3a and b.

Figure 3. Variants of the structural diagrams of the mechanical oscillatory system of chain type (Figure 2) with the exclusion of the coordinate y_1: a) structural diagram with the exclusion of the coordinate y_2 with the base model m_1, k_1; b) structural diagram with the distinguishing of the object of protection; c) structural diagram with the generalized feedback with respect to the object of protection

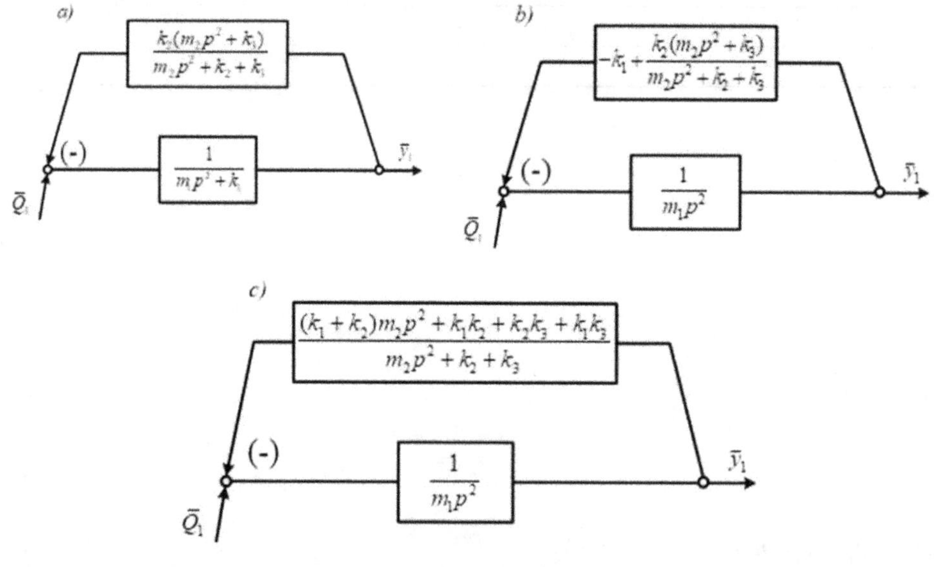

Comparison of structural diagrams in Figure 3 allows you to track the characteristics of negative feedback in relation to the object of protection (m). In Figure 3, and the negative feedback, in the physical sense, is the dynamic stiffness of the quasi-spring, which reflects the dynamic properties of a certain part of the mechanical system associated with the intermediate mass m_2.

2. Dynamic stiffness of quasi-springs

$$\bar{k}_{red}(p)=\frac{k_2(m_2p^2+k_3)}{m_2p^2+k_2+k_3} \qquad (14)$$

may be zero at the frequency:

$$\omega_1^2=\frac{k_3}{m_2} \qquad (15)$$

In this case, the system (Figure 2a) operates under the action of an external force $Q_1 \neq 0$ in this way, if the initial system was transformed into a system with one degree of freedom with the elements m_1 и k_1. Feedback (15), in a certain sense, neutralizes the dynamic action of a part of the system. On the other hand, the dynamic stiffness of a quasi-spring at a frequency:

$$\omega_2^2=\frac{k_2+k_3}{m_2} \qquad (16)$$

takes an infinitely large value, which corresponds to the dynamic damping mode. In this case, motion along the y_1 coordinate stops ($y_1 = 0$).

In the transition to the structural diagram in Figure 3b, the object of protection is considered in the scheme of supporting two parallel springs with stiffnesses k_1 and k_2. The total dynamic stiffness of such a quasi-spring will be:

$$\bar{k}'_{red}(p)=\frac{(k_1+k_2)m_2p^2+k_1k_2+k_2k_3+k_1k_3}{m_2p^2+k_2+k_3} \qquad (17)$$

The dynamic stiffness of the system in such conditions is zero at the frequency:

$$\omega_3^2=\frac{k_1k_2+k_2k_3+k_1k_3}{(k_1+k_2)m_2} \qquad (18)$$

At the frequency ω_2 from (16), the dynamic damping mode is preserved.

Thus, when the dynamic stiffness of the quasi-spring is "nullified", if we consider in terms of the successive release of the object of protection m_1 from additional constraints, various characteristic forms of self-organization of the system elements motion will be observed.

Figure 4. The schematic diagram of the graphical solution of the condition (19)

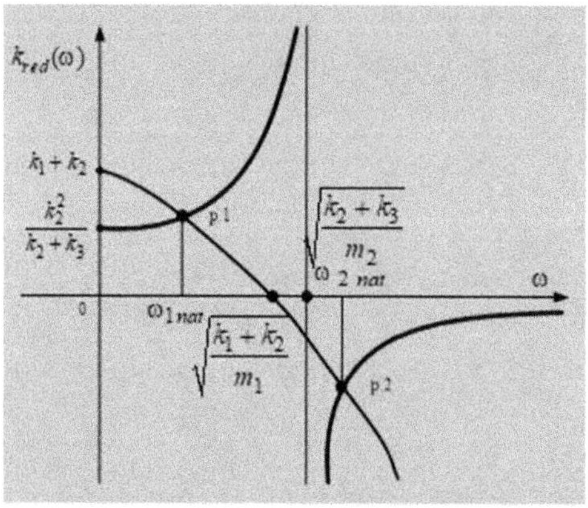

3. On the basis of (12) the condition can be made:

$$m_1 p^2 + k_1 + k_2 = \frac{k_2^2}{m_2 p^2 + k_2 + k_3} \tag{19}$$

which allows determining, on the basis of the graphic-analytical approach with the substitution of $p^2 = -\omega^2$, the values of natural frequencies as shown in Figure 4.

From the intersection of the dependency plots that make up condition (19) and denoted by p. *1* and 2, which determine the frequency at which the dynamic stiffness of the system, in general, will be zero. Such intersection points determine the natural frequency. Figure 4 also shows partial frequencies $n_1^2 = \frac{k_1 + k_2}{m_1}$ and $n_2^2 = \frac{k_2 + k_3}{m_2}$, whose location corresponds to the known provisions of the theory of oscillations (Babakov, 1968).

4. The physical meaning of the frequency characteristic equation in the form (12) corresponds to the concepts of the summation of all forces acting on the object of protection. Moreover, each of the elements (12) is essentially a product of the dynamic stiffness reduced to m_1 by $\bar{y}_1$, since $\bar{y}_2$ is eliminated by structural transformations. Technically, this also coincides with the known provisions of the d'Alembert principle.

The peculiarity of the considered chain systems is the homogeneity of the physical nature of the partial system. In this case (Figure 2a), the partial systems execute rectilinear small oscillations, and the connections between the partial systems are inertial.

When considering systems with a large number of degrees of freedom, the method of determining dynamic properties and their connection with frequencies, types of joint forms of motion of elements, remain the same, taking into account the complications introduced by increasing the order of the frequency characteristic equation.

Systems With Two Degrees of Freedom Containing Solids

When considering the mechanical system in Figure 2b, it can be noted that the shape of the partial systems depends on the choice of the coordinate system.

If y_1 and y_2 are selected as the motion coordinates, the structural diagram in Figure 2d has partial systems with motions of the same type. The structure of the partial system 2 includes the reduced mass $Ma^2 + Jc^2$, attributable to the coordinate y_1. For the second partial system, we have the reduced mass $Mb^2 + Jc^2$, attributable to the coordinate y_2. Constraints between partial systems are elastic.

Note that the partial systems with coordinates y_1 and y_2, actually reflect small angular oscillations, designated y_1 and y_2, with respectively "stopped" motions along y_2 and y_1, which is quite acceptable at small angles of rotation relative to the center of masses.

Assuming that the external influence in the system is carried out by the harmonic force Q_1 applied at the point with the coordinates y_1 ($Q_2 = 0$), we write the transfer functions:

$$W_3(p) = \frac{\bar{y}_1}{\bar{Q}_1} = \frac{(Mb^2 + Jc^2)p^2 + k_2}{A_1(p)} \tag{20}$$

$$W_4(p) = \frac{\bar{y}_2}{\bar{Q}_1} = \frac{(Jc^2 - Mab)p^2}{A_1(p)} \tag{21}$$

where

$$A_1(p) = [(Ma^2 + Jc^2)p^2 + k_1][(Mb^2 + Jc^2)p^2 + k_2] - [(Jc^2 - Mab)p^2]^2. \tag{22}$$

From (22) we obtain the frequency equation in the coordinates y_1, y_2:

$$[(Ma^2 + Jc^2)p^2 + k_1][(Mb^2 + Jc^2)p^2 + k_2] - [(Jc^2 - Mab)p^2]^2 = 0. \tag{23}$$

Figure 5a-c shows options of transformations of the structural diagram in Figure 2d with the exclusion of the coordinate φ.

From the analysis of the structural diagram in Figure 5c can be done by analogy with the previous calculations that a selected object, whose dynamic state is estimated, is determined by the transfer function of the integrating unit of the second order. The object is enveloped by the feedback, which has a fractional rational transfer function. At the frequency:

$$\omega_3 = \sqrt{\frac{k_2}{Mb^2 + Jc^2}} \tag{24}$$

a dynamic damping mode is possible in the system. In this case, the condition $y_1 = 0$ under the action of a force Q_1. In turn, the numerator of the feedback transfer function is a biquadratic equation for p^2, which predetermines the possibility of obtaining the zero value of the numerator of the feedback transfer function. With such a frequency of external influence $\bar{Q}_1$, the system has the parameters of the motion of a system with one degree of freedom, which is manifested in certain forms of joint motions (Eliseev et al., 2009).

Figure 5. Options for the formation of feedbacks in structural diagrams that use the coordinates y_1 and y_2: a) structural diagram of the system with the exclusion of the y_2 coordinate; b) structural diagram with the distinguishing of feedback in relation to an object with a reduced mass $(Ma^2 + Jc^2)p^2$; c) structural diagram with a generalized feedback

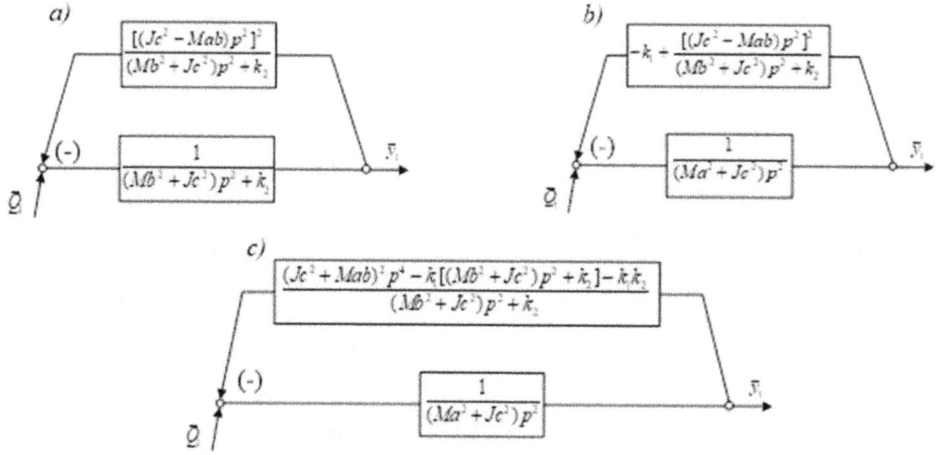

Note that the feedback (Figure 5c) determines the value of the dynamic stiffness not of the system as a whole, but only of its part, which takes the form of a quasi-spring.

To determine the natural frequencies, we write the frequency characteristic equation of the system (23) in the form:

$$(Ma^2 + Jc^2)p^2 + k_1 - \frac{\left[(Jc^2 - Mab)p^2\right]^2}{(Mb^2 + Jc^2)p^2 + k_2} = 0 \tag{25}$$

Since $p = j\omega$, then (25) transforms:

$$k_1 - (Ma^2 + Jc^2)\omega^2 = \frac{\left[(Jc^2 - Mab)\right]^2 \omega^4}{k_2 - (Mb^2 + Jc^2)\omega^2} \tag{26}$$

With a graphic-analytical solution, the intersection of the dependency graphs for the left and right sides of condition (26) determines the natural vibration frequencies (Figure 6).

On the ordinate axis in Figure 6, the values of dynamic stiffness of corresponding fragments of system are plotted. The intersection of the graphs determines the total value of the dynamic stiffness of the entire system, equal to zero.

Figure 6. The schematic diagram for determining the natural oscillation frequencies in a system with two degrees of freedom with a solid body in the coordinates y_1, y_2

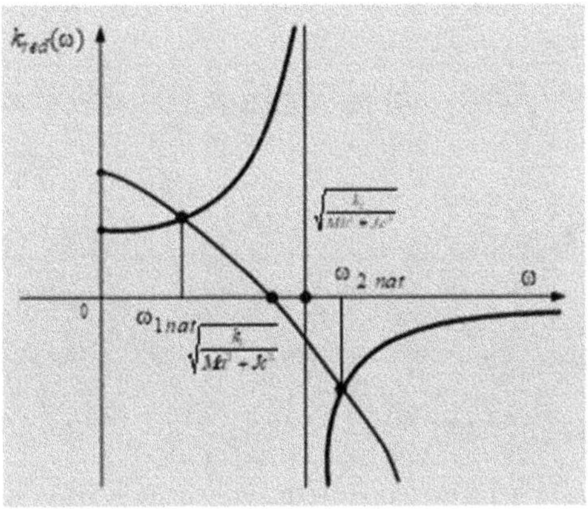

Note that the frequency characteristic equation (23) in this case is the sum of the reduced forces acting on the system as a whole when choosing a material point with a reduced mass $Ma^2 + Jc^2$ as a point of reduction with the coordinate y_1. Similar results can be obtained when choosing a point of reduction with the coordinate y_2 and the reduced mass $Mb^2 + Jc^2$.

When choosing the coordinate system y_0 and φ, the characteristic frequency equation takes the form

$$(Mp^2 + k_1 + k_2)(Jp^2 + k_1l_1^2 + k_2l_2^2) - (k_1l_1^2 - k_2l_2^2)^2 = 0. \tag{27}$$

Let us transform (27):

$$Mp^2 + k_1 + k_2 - \frac{(k_1l_1^2 - k_2l_2^2)^2}{Jp^2 + k_1l_1^2 + k_2l_2^2} = 0 \tag{28}$$

The schematic diagram of the graphic-analytical solution is presented in Figure 7 and determined by the condition:

$$k_1 + k_2 - M\omega^2 = \frac{(k_1l_1^2 - k_2l_2^2)^2}{k_1l_1^2 + k_2l_2^2 - J\omega^2} \tag{28}$$

From analysis (26) it follows that the original system (Figure 2b) in the coordinates y_0 and φ has partial frequencies:

$$n_3^2 = \frac{k_1 + k_2}{M} \tag{29}$$

$$n_4^2 = \frac{k_1 l_1^2 + k_2 l_2^2}{J} \tag{30}$$

Figure 7. The schematic diagram of determining the frequency of its natural oscillations of the system in coordinates y_0, φ

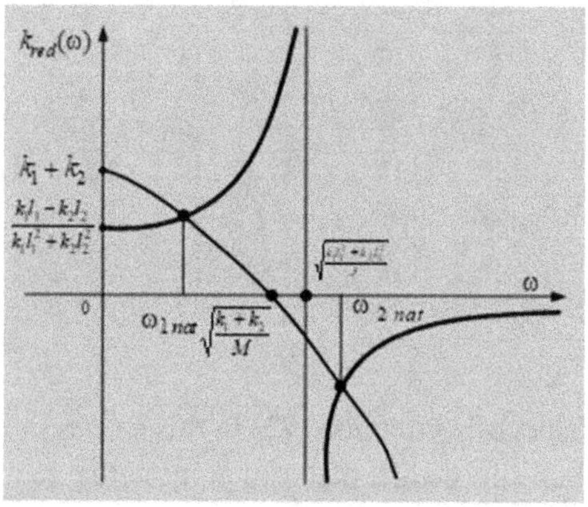

At a frequency n_4^2 of (30), the mode of dynamic oscillation damping is possible, when, under the action of Q_1, motion along the coordinate y_2 does not occur. At the same time, there is the value of the frequency of the external force, at which the feedback with respect to an object with mass M takes on a zero value. In this case:

$$\omega_5^2 = \frac{(k_1 + k_2)(k_1 l_1^2 + k_2 l_2^2) - (k_1 l_1 - k_2 l_2)^2}{J(k_1 + k_2)} = \frac{k_1 k_2 (l_1 + l_2)^2}{(k_1 + k_2) J}. \tag{31}$$

At this frequency, the system moves under the action of an external disturbance as a system with one degree of freedom. In this case, the dynamic stiffness of the part of the system that forms the quasi-spring will be zero, which, in the physical interpretation, manifests itself as some form of joint motion (a form of self-organization of joint motion of the system elements). In this case, the dynamic stiffness of the system, in general, is not zero (at this frequency).

The physical meaning of the frequency characteristic equation (27) when multiplied by the coordinate $\overline{y}_0$ is that it is the sum of the reduced forces that arise when the system moves as a material point, having a mass M, moving along the coordinate y_0.

Similar results can be obtained using the angular coordinates. In this case, the characteristic equation represents the sum of the moments of forces arising from the motion relative to the center of reduction (i.e., relative to the center of mass – p. O).

In this case, the material solid, having the moment of inertia J and mass M, has a fixed point of rotation (p. O), where the mass is concentrated. The use of the moment of inertia J in the calculations assumes an understanding of the organization of joint motions as actions occurring in a certain space, that is, taking into account the spatial metrics of the location of the acting forces.

Methodological positions in the specification of ideas about interactions of system elements with regard to the spatial metric are reflected in Kaimov et al. (2015), Eliseev et al. (2014, 2015), Khomenko et al. (2014).

In the cases considered, systems with two degrees of freedom with a solid body in different coordinate systems had different frequency characteristic equations, since the choice of the coordinate system predetermines the type of partial systems, features of interpartial constraints, which is quite clearly reflected in the structural mathematical models shown in Figure 2d and e.

The characteristic frequency equation derived from the structural diagram in Figure 2d has the form:

$$\left[(Ma^2 + Jc^2)p^2 + k_1\right]\left[(Mb^2 + Jc^2)p^2 + k_2\right] - \left[(Jc^2 - Mab)p^2\right]^2 = 0. \tag{32}$$

The expression (32) can be reduced to a form suitable for transformations:

$$\left[(Ma^2 + Jc^2 - Mab)p^2 + k_1\right]\left[\left(Mb^2 + Jc^2 - Mab\right)p^2 + k_2\right] - \left[(Jc^2 - Mab)p^2\right]^2 = 0. \tag{33}$$

After substituting the values of a, b, and c from expression (8), the characteristic equation in the coordinate systems y_1, y_2, and y_0, φ takes on the identical form.

Systems With Three Degrees of Freedom

Developing the described approach, we take as an initial mechanical oscillatory system with three degrees of freedom (Figure 8), the mathematical model of which is presented in Table 2.

Figure 8. The computational scheme of the system with three degrees of freedom

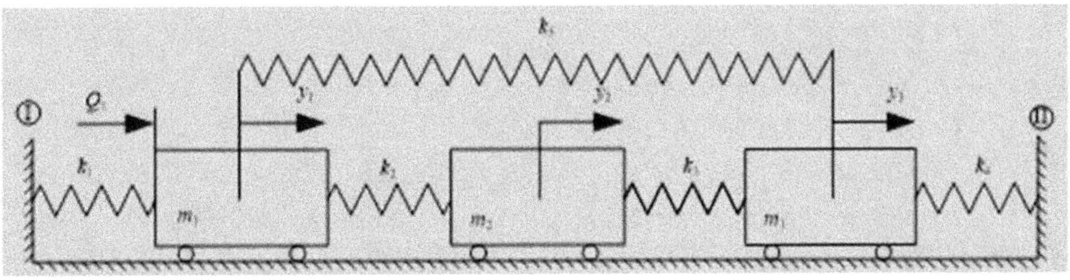

The system under consideration is closed in the sense that the connected elements of the system m_1, m_2, m_3 form a closed contour. The system has linear properties and makes small oscillations in the absence of friction forces relative to the position of stable equilibrium under the action of external harmonic force (Table 2).

Table 2 of the coefficients can be used to build a structural diagram, for example, for the case of a force disturbance Q_1. The structural diagram of the system (or its structural mathematical model) is shown in Figure 9.

The transfer functions of the system and the frequency characteristic equation can be determined directly from the structural diagram. In particular, the characteristic equation in this case is:

$$a_{11}a_{22}a_{33} - a_{11}a_{23}^2 - a_{22}a_{13}^2 - a_{33}a_{12}^2 + 2a_{12}a_{23}a_{31} = 0. \tag{34}$$

Expression (34) gives an idea that the system can be built from partial systems a_{11}, a_{22}, a_{33}, which enter into interactions with each other through certain connections (Figure 9).

Table 2. The coefficients of the equations of motion of the system in coordinates y_1, y_2, y_3

a_{11}	a_{12}	a_{13}
$m_1p^2 + k_1 + k_2 + k_5$	$-k_2$	$-k_5$
a_{21}	a_{22}	a_{23}
$-k_2$	$m_2p^2 + k_2 + k_3$	$-k_3$
a_{31}	a_{32}	a_{33}
$-k_5$	$-k_3$	$m_3p^2 + k_3 + k_4 + k_5$
	Generalized forces	
Q_1	0	0

If you select an object to control the dynamic state (for example, a_{11}, for vibration protection), then equation (34) can be converted to:

$$a_{11} - \frac{a_{22}a_{13}^2}{[m_2m_3]} - \frac{a_{33}a_{21}^2}{[m_2m_3]} + \frac{2a_{12}a_{23}a_{31}}{[m_2m_3]} = 0 \tag{35}$$

where

$$[m_2m_3] = a_{22}a_{33} - a_{23}^2 \tag{36}$$

As in previous considerations, the characteristic frequency equation can be interpreted as the sum of the dynamic stiffnesses of its individual components (or quasi-springs). Such approaches have been developed in a number of papers, for example (Eliseev et al., 2014, 2015; Khomenko et al., 2014; Belokobylsky et al., 2015). We note here that expression (36) is a characteristic equation of a system with two degrees of freedom (m_2, m_3, k_2, k_3, k_4).

Figure 9. The structural mathematical model (structural diagram) for a mechanical oscillatory system

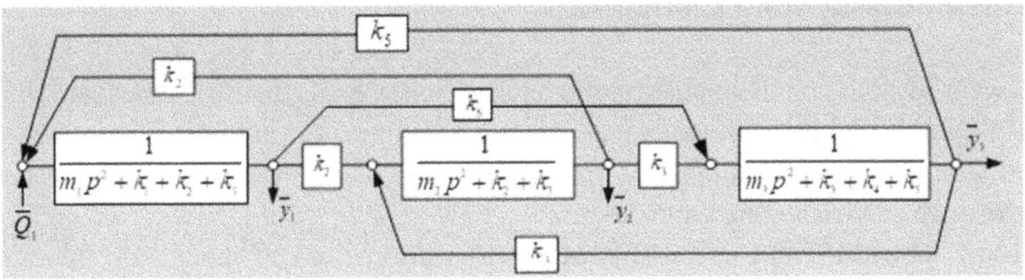

Structural mathematical modeling uses tools from the theory of automatic control and mechanical circuits, giving it advantages over other methods. The model allows for consideration of connections between elements and regulatory influences. This approach can lead to new results by expanding the set of typical elements, introducing constraints, and transforming models.

Dynamic stiffness of an elastic element or some structural formation of typical elements is a generalization of the properties of not only an elastic element in the form of an idealized spring, but includes ideas about other forms of interaction under the condition that the input action in the form of displacement initiates the output signal in the form of force. Actually, such representations form a set of typical elements in linear systems, when mass-and-inertia, dissipative, elastic and other elements are distinguished.

ON THE POSSIBILITIES OF USING ADDITIONAL CONSTRAINTS OF INERTIAL TYPE IN THE PROBLEMS OF THE DYNAMICS OF TECHNICAL SYSTEMS

Attention to the issues of ensuring the reliability of machines, improving their operational safety and dynamic quality stimulates the search and development of technical means of monitoring and controlling the dynamic state of technical systems (Lontsikh & Eliseev, 2014; Eliseev & Khomenko, 2014; Kuznetsov, 2009). A number of studies of theoretical and practical orientation are reflected in the works (Belokobylsky et al., 2012; Eliseev et al., 2013b), related to the assessment of the possibilities of using elements of an unconventional type in the form of motion transformation devices. Such elements can be implemented, in particular, on the basis of various mechanisms (screw, non-locking, toothed, articulation, etc.). A detailed description of the features of the dynamics of mechanical oscillatory systems with such elements is presented in recent years (Ivanov, 2007; Lavrus', 2006; Eliseev et al., 2013a). Considerable attention to the assessment of the dynamic properties of motion transformation devices has been given in studies that develop the ideas of structural mathematical modeling, in which motion transformation devices are considered at the level of typical second-order differentiating units. The features of the mentioned structural approaches include the use of dynamic analogies in the interactions of electrical, mechanical systems, as well as automatic control systems. To date, a sufficient variety of constructive and engineering solutions has been developed to predict the development of research areas. In this regard, of particular interest are the possibilities of using such effects from using the motion transformation device, such as the introduction of additional constraints to accelerate, changing the parameters of the reduced masses and stiffnesses, creating additional effects of dynamic oscillation damping, "locking" the interaction of elements at certain frequencies, the formation of non-traditional types of interpartial constraints, etc.

General Provisions: The Features of Dynamic Interactions of the System Elements

In the framework of structural mathematical modeling (Eliseev et al., 2012b), a mechanical oscillatory system is displayed by a dynamically equivalent structural diagram of an automatic control system, which is associated with performing a number of operations using Laplace transforms. The methodological foundations of such approaches are quite detailed.

Figure 10b-f shows different forms of structural transformations of models on the example of a vibration protection system (Fig. 7.10 *a*) with one degree of freedom. In Figure 10a-f, the following designations are adopted: *m* is the object of protection with mass *m*, *k* is the spring stiffness coefficient, *b* is the viscous friction coefficient, *L* is the reduced mass of the motion transformation device, *I* and *II* are support surfaces. The system considers force *Q* and kinematic harmonic perturbations $z_1(t)$ and $z_2(t)$. The system has linear properties and performs small oscillations relative to the position of static equilibrium. In Fig. 7.10, it is assumed that $p = j\omega$ is a complex variable (the Laplace transform is performed for zero initial conditions) (Eliseev et al., 2011).

The construction of the structural diagram in Figure 10b requires the prior construction of a mathematical model in the form of a differential equation. The corresponding expressions for the kinetic and potential energies can be written:

$$T = \frac{1}{2}m\dot{y}^2 + \frac{1}{2}L_1(\dot{y} - \dot{z}_1)^2 + \frac{1}{2}L_2(\dot{y} - \dot{z}_2)^2 \quad (37)$$

$$\Pi = \frac{1}{2}k_1(y - z_1)^2 + \frac{1}{2}k_2(y - z_2)^2 \quad (38)$$

Figure 10. The calculation and structural diagram of the system: a) scheme with a full set of elementary type units; b) structural diagram with a full set of standard units; c) structural diagram with the motion transformation device in the circuit of negative feedback; d) structural diagram in a generalized form; e) structural diagram with an object with a reduced mass; f) computational scheme with generalized elastic elements with stiffnesses k_{red1} and k_{red2}

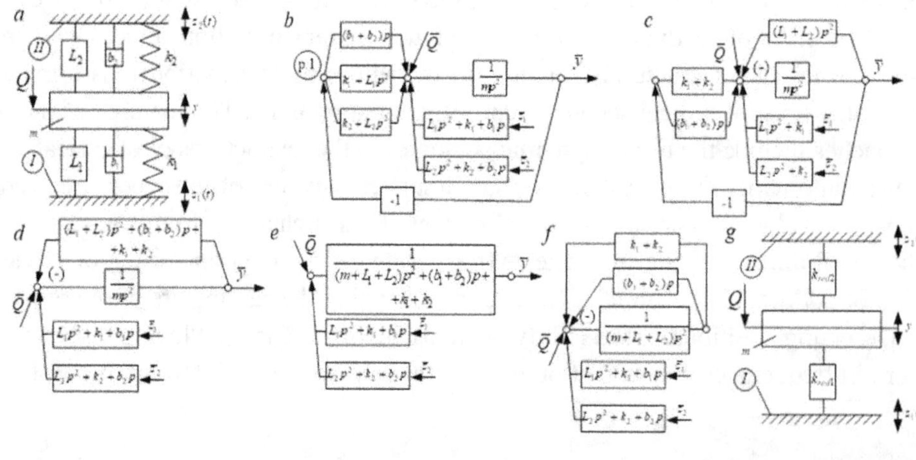

The expression for oscillation energy is taken in the form:

$$\Phi = \frac{1}{2}b_1(\dot{y} - \dot{z}_1)^2 + \frac{1}{2}b_2(\dot{y} - \dot{z}_2)^2 \tag{39}$$

The equation of motion will be obtained on the basis of the use of the Lagrange equation of the second kind:

$$\begin{aligned} &\bar{y}(m + L_1 + L_2)p^2 + \bar{y}(b_1 + b_2)p + \bar{y}(k_1 + k_2) = \\ &= \bar{Q} + \bar{z}_1(L_1p^2 + b_1p + k_1) + \bar{z}_2(L_2p^2 + b_2p + k_2) \end{aligned} \tag{40}$$

Equation (40) is the basis for constructing a structural mathematical model. Figure 10a shows the computational scheme, which as a result of formal transformations is transformed into a structural diagram in Figure 10b, which is a mathematical model of another kind, but this model, as well as equation (40); Laplace transforms are performed under zero initial conditions. The <–> sign means a Laplace transform image. Note that typical elements within the framework of the structural model are interpreted as the corresponding feedback chains. Since the feedback chain has the $\bar{y}$ coordinate at the input, and the transfer function of each feedback branch, in the physical sense, corresponds to the "stiffness coefficient" of the element, then at the output of the branch at p. *1* (see Figure 10b), the feedback branch will already correspond to the constraint reaction, the value of which is equal to the product of the reduced stiffness of the unit by displacement.

In this case, it is obvious that typical elements with transfer functions L_1p^2, L_2p^2, b_1p, b_2p, k_1, and k_2 are distinguished by a certain feature: there is an displacement at the input of any typical unit, and a force is formed at the output. In such conditions, the rules of elementary units switching are predetermined as "like the rules of parallel and consecutive connection of springs" (Eliseev et al., 2012b). The transfer functions for the system in Figure 10a are determined by any of the structural diagrams in Figure 10c-f.

The transfer functions of the system are determined with one input signal, assuming that other input influences will be zero. In this way:

$$W_1(p)_{\bar{Q}\neq 0, \bar{z}_1 = 0, \bar{z}_2 = 0} = \frac{\bar{y}}{\bar{Q}} = \frac{1}{A(p)} \tag{41}$$

$$W_2(p)_{\bar{z}_1\neq 0, \bar{Q} = 0, \bar{z}_2 = 0} = \frac{\bar{y}}{\bar{z}_1} = \frac{L_1p^2 + b_1p + k_1}{A(p)} \tag{42}$$

$$W_3(p)_{\bar{z}_2\neq 0, \bar{Q} = 0, \bar{z}_1 = 0} = \frac{\bar{y}}{\bar{z}_2} = \frac{L_2p^2 + b_2p + k_2}{A(p)} \tag{43}$$

$$\underset{\bar{Q}=0,\bar{z}_1=\bar{z}_2=z}{W_3(p)} = \frac{\bar{y}}{\bar{z}} = \frac{(L_1+L_2)p^2+(b_1+b_2)p+k_1+k_2}{A(p)} \tag{44}$$

where

$$A(p)=(m+L_1+L_2)p^2+(b_1+b_2)p+k_1+k_2 \tag{45}$$

is a characteristic frequency equation.

Assessment of the dynamic properties of the vibration protection system in Figure 10a is carried out on the basis of the frequency characteristics obtained by substituting in the expression (7.41)–(7.44) $p = j\omega$ ($j = \sqrt{-1}$), as is done in the frequency methods of analyzing systems (Kim, 2003; Eliseev et al., 2012a).

Features of the Switching of Typical Elements of Mechanical Oscillatory Systems

From the above it follows that each of the typical elements has its own transfer function of an elastic, dissipative, or mass-and-inertia (motion transformation device) unit. The object of protection (Figure 10b-f) has the transfer function of an integrating unit of the second order, and the transfer functions of the system (41)–(44) are formed using the convolution rules of elements with feedback (Kim, 2003).

Consider a number of cases of combining motion transformation device with typical elements.

The elastic element k_1 and Lp^2 are consecutive and parallel connections:

$$k_{red1} = k+Lp^2 \tag{46}$$

$$k_{red2} = \frac{k \cdot Lp^2}{k+Lp^2} \tag{47}$$

The dissipative element and the motion transformation device in parallel and consecutive connections are:

$$k'_{red} = bp+Lp^2 \tag{48}$$

$$k''_{red} = \frac{bLp^3}{b+Lp} \tag{49}$$

Compounds L_1p^2 and L_2p^2 respectively have forms:

$$k'''_{red} = (L_1+L_2)p^2 \tag{50}$$

$$k_{red}^{IV} = \frac{L_1 L_2 p^2}{L_1 + L_2}. \tag{51}$$

Similarly, other forms of a bundle of typical units in more complex configurations can be built. Such formations of elements form a generalized spring, and the vibration protection system can be reduced to the basic form, when the initial system, for example, in Figure 10a is reduced to a simple form. In this case, the system has a protection object with mass m and one generalized spring with dynamic stiffness:

$$k_{red\ o} = (L_1 + L_2)p^2 + (b_1 + b_2)p + k_1 + k_2 \tag{52}$$

The dynamic stiffness $k_{red\ o}$ depends on the frequency of the external disturbance and can take a zero value at a frequency that can be found from the solution of equation (52).

In this case, $k_{red\ o}$ is represented as some complex unit reflecting the properties of quasi-springs. The properties of quasi-springs are similar to ordinary elements in switching operations.

Under the action of an external force perturbation Q in a mechanical system (Figure 10a), a mode of forced oscillations arises, characterized by a transfer function:

$$W_Q(p) = \frac{\bar{y}}{\bar{Q}} = \frac{1}{(m+L)p^2 + k_1 + k_2} \tag{53}$$

The frequency of resonant oscillations is determined from the expression:

$$\omega_{res}^2 = \frac{k_1 + k_2}{m + L} \tag{54}$$

The amplitude-frequency characteristics of the system with a force disturbance has the usual form. With kinematic perturbations, for example, with $z_1 \neq 0$ ($Q = 0$, $z_2 = 0$), the transfer function of the system will be determined:

$$W_{z_1}(p) = \frac{\bar{y}}{\bar{z}_1} = \frac{Lp^2 + k_1}{(m+L)p^2 + k_1 + k_2} \tag{55}$$

Under the condition $Lp^2 + k$, the dynamic damping mode is implemented in the system, but the dynamic damping frequency can be correlated with the following conditions:

$$\omega_{dyn}^2 < \omega_{res}^2, \omega_{dyn}^2 > \omega_{res}^2, \omega_{dyn}^2 = \omega_{res}^2.$$

Figure 11 shows the corresponding amplitude-frequency characteristics. From the expression (55) it follows that $\underset{p\to\infty}{W(p)} = \frac{1}{m+L}$, and $\underset{p\to 0}{W(p)} = \frac{k_1}{k_1+k_2}$ as $p \to 0$. If the numerator and denominator (55) both become equal to zero, $\frac{k_1}{L} = \frac{k_1+k_2}{m+L}$, then it follows that between the values of L and m there is a relation:

$$L = \frac{k_1 m}{k_2} \tag{56}$$

When there is a disturbance from the side of the support surface *II* (Figure 10a):

$$\underset{z_2\neq 0}{W(p)} = \frac{\bar{y}}{\bar{z}_2} = \frac{Lp^2 + k_2}{(m+L)p^2 + k_1 + k_2} \tag{57}$$

From (57) it follows that cases $\omega_{2\,res}^2 < \frac{k_1+k_2}{L+m}$, $\omega_{1\,res}^2 < \frac{k_1+k_2}{L+m}$, $\omega_{2res}^2 = \frac{k_1+k_2}{L+m}$ are possible. The latter gives relations between L and m, with a perturbation of $z_2 \neq 0$, the relation:

$$m = \frac{k_1}{k_2} L \tag{58}$$

whereas with the perturbation $z_1 \neq 0$ there is:

$$L = \frac{k_2}{k_1} m \tag{59}$$

Figure 11. Amplitude-frequency characteristics of the system: a) with disturbance from the side of the supporting surface $\bar{z}_1 \neq 0$ (dynamic damping before the resonance – 1) dynamic damping after the resonance, 2) xxx – frequencies of the dynamic damping and resonance coincide); b) similar frequency characteristics with the kinematic perturbation $\bar{z}_1 \neq 0$ ($\bar{z}_2 = 0$, $\bar{Q} = 0$)

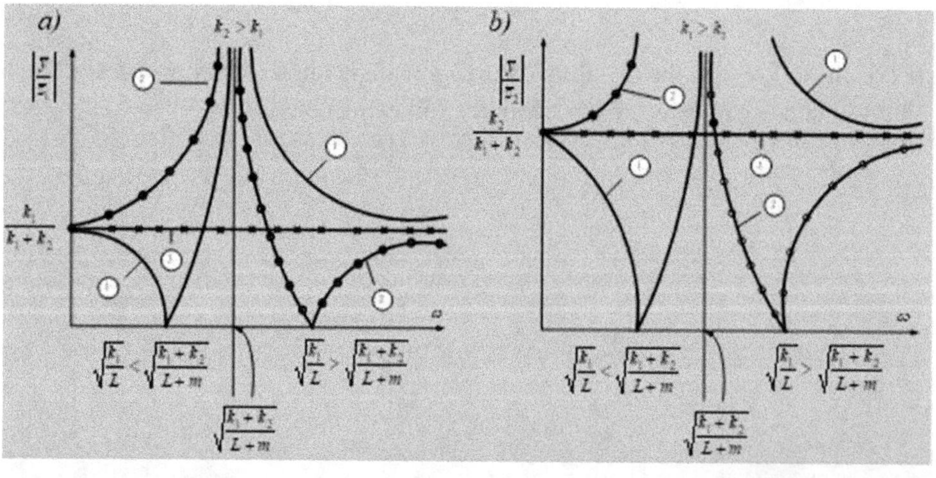

Figure 11a and b shows the amplitude-frequency characteristics of the system with kinematic perturbations $z_1(t)$ and $z_2(t)$, respectively.

Note that with a certain ratio of parameters, when $m = \frac{k_2}{k_1} L$ $(z_1 \neq 0)$ and with $m = \frac{k_1}{k_2} L$ $(z_2 \neq 0)$, the "locking" mode may occur at all frequencies. The ratios between the oscillation amplitudes of the protected object and the supporting surface are determined as $\frac{k_1}{k_1 + k_2}$ when $z_1 \neq 0$, and $\frac{k_2}{k_1 + k_2}$ when $z_2 \neq 0$. These conditions are met at all frequencies of the range (the friction forces are not taken into account in the calculations; otherwise, the problem requires different solutions to be solved) .

If the critical relations between the inertial parameters m and L (expressions (58), (59)) do not hold, then the blocking effect is implemented at high frequencies ($p \to \infty$). The limits of the ratio of the amplitudes as $p \to \infty$ are determined by the same values as in the critical case. The systems are implemented as shown in Fig. 7.11, and dynamic oscillation damping modes that can be implemented in the resonance and pre-resonance frequency ranges of external influences. A new effect can be considered the circumstance when the difference between the oscillation amplitudes of the object of protection is a certain amount. For example, if $z_1 = z_0 \sin\omega t$, then the amount of compression of the spring k*1* is $z_{10} - y\frac{k_1}{k_1 + k_2}$, which allows you to determine a certain amount of potential energy that the system has in the "locking" mode. Similarly, the spring k_2 will be subjected to deformation, for which the potential energy is determined by the value of the oscillation amplitude $\frac{y_2 k_1}{k_1 + k_2}$. This "stock" of the energy of elastic elements is determined by the relations between the values of m and L at the frequency of critical interactions ("locking"). This condition is observed at different frequencies. When excited from the side of the supporting surface *II*, a similar interaction pattern will be observed.

In a certain sense, the "locking" mode can be viewed as a mode of "capturing" energy and storing it for the period of forced oscillations when performing the necessary relationships between the parameters of the system. This effect can be achieved, for example, by adjusting the magnitude of the reduced mass L, which is possible if there are controlled devices that create certain torques on the nut-flywheel if the non-locking screw mechanism is used as the motion transformation device (Khomenko, 2012; Eliseev et al., 1990).

The Motion Transformation Device in a System With Two Degrees of Freedom

Consider a chain mechanical oscillatory system with two degrees of freedom, which contains elastic elements k_1, k_2, k_3 and mass-and-inertia elements m_1 and m_2, as shown in Figure 12.

Let us find the expressions for the kinetic and potential energies under the assumption that the system performs free oscillations and there is a relationship between the coordinates y_2 and y_1:

$$y_2/y_1 = a. \tag{60}$$

Then

Figure 12. The computational scheme of vibration protection system with two degrees of freedom

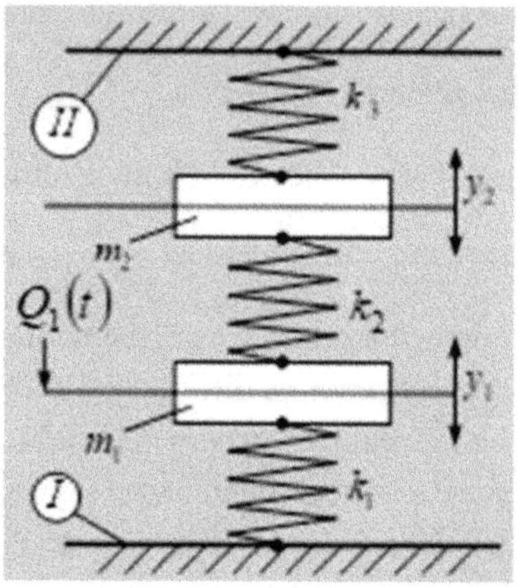

$$\mathrm{T} = \frac{1}{2} m_1 \dot{y}_1^2 + \frac{1}{2} m_2 \dot{y}_2^2 = \frac{1}{2} \dot{y}_1^2 (m_1 + a^2 m_2) \tag{61}$$

$$\Pi = \frac{1}{2} k_1 y_1^2 + \frac{1}{2} k_1 (y_2 - y_1)^2 + \frac{1}{2} k_3 y_2^2 = \frac{1}{2} y_1^2 (k_1 + k_2 + (k_2 + k_3) a^2 - 2 k_2 a) \tag{62}$$

Let us introduce into consideration the frequency energy function (by analogy with the approximate energy method of determining the lower natural frequencies), which has the form:

$$\omega^2(a) = \frac{\Pi(a)}{\mathrm{T}(a)} = \frac{(k_1 + k_2 + (k_2 + k_3) a^2 - 2 k_2 a)}{m_1 + a^2 m_2} \tag{63}$$

Figure 13 shows a graph of the dependence $\omega^2(a)$ for the given parameters of the computational scheme ($k_1 = k$, $k_2 = 2k$, $k_3 = 3k$, $m_1 = m_2 = m$). The graph in Figure 13 is constructed under the assumption that $a > 0$, that is, the form of relations between the coordinates of motion y_1 and y_2, which corresponds to the first form of the main oscillations, is taken into consideration. The graph has a characteristic shape and, with a maximum value, has two parameters ω^2, which corresponds to the fundamental frequency of natural oscillations (ω^2_{1nat}), and also determines the value of the coefficient of distribution of the amplitudes of the oscillations.

For the subsequent analysis, we construct a structural mathematical model of the system (Figure 14). The structural diagram of the system has the form as in Figure 14. Using the presented scheme, we write the expressions for the transfer functions.

In accordance with the scheme of the system shown in Figure 14, we write that:

$$W_1'(p) = \frac{\bar{y}_1}{\bar{Q}_1} = \frac{m_2 p^2 + k_2 + k_3}{A_0} \tag{64}$$

Figure 13. The graph of dependency of $\omega^2(a)$ on the values of the parameter a

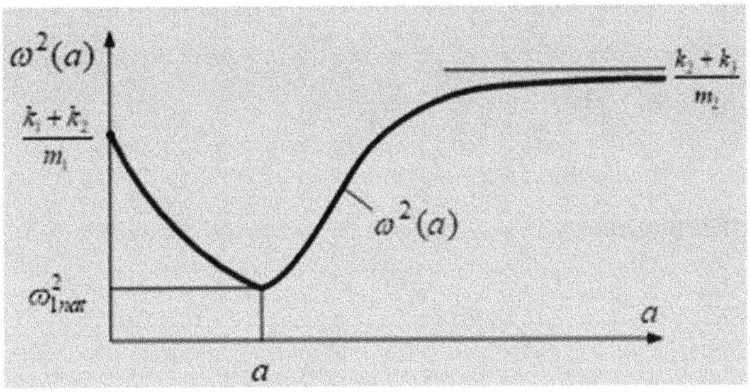

Figure 14. The structural mathematical model for the calculation scheme of the system shown in Figure 10

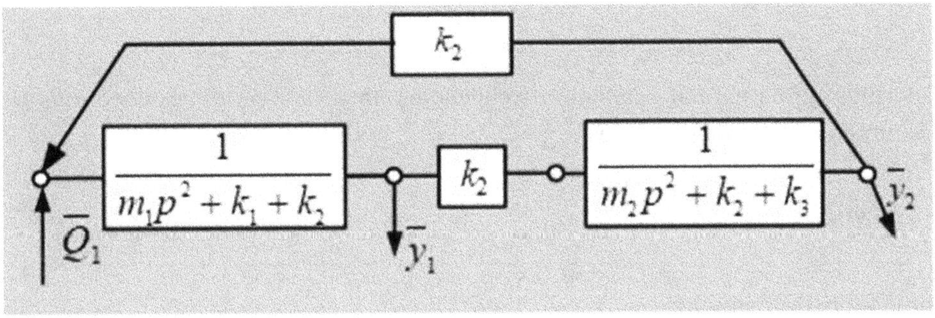

$$W_2'(p)=\frac{\bar{y}_2}{\bar{Q}_1}=\frac{k_2}{A_0} \tag{65}$$

$$W_{12}(p)=\frac{\bar{y}_2}{\bar{y}_1}=\frac{k_2}{m_2p^2+k_2+k_3} \tag{66}$$

where

$$A_0=(m_1p^2+k_1+k_2)(m_2p^2+k_2+k_3)-k_2^2 \tag{67}$$

Expression (67) defines the characteristic frequency equation. Using (67), one can find the natural oscillation frequencies analytically, in this case:

$$\omega_{1,2\ nat} = \frac{n_1^2}{2} + \frac{n_2^2}{2} \pm \sqrt{\frac{[m_1(k_2+k_3) - m_2(k_1+k_2)]^2 + 4m_1m_2k_2^2}{(m_1m_2)^2}} = \\ = \frac{n_1^2}{2} + \frac{n_2^2}{2} \pm \sqrt{(n_1^2 - n_2^2)^2\left[1 + \frac{4k_2^2}{m_1m_2(n_1^2 - n_2^2)^2}\right]}. \tag{68}$$

Here n_1, n_2 are partial frequencies:

$$n_1^2 = \frac{k_1 + k_2}{m_1} \tag{69}$$

$$n_2^2 = \frac{k_2 + k_3}{m_2} \tag{70}$$

When determining the natural oscillation frequencies, there are certain connections between the partial and natural oscillation frequencies.

$$\omega_{1\,nat}^2 < n_1 < n_2 < \omega_{2\,nat}^2 \tag{71}$$

At the same time, the condition:

$$\omega_{1\,nat}^2 + \omega_{2\,nat}^2 = n_1^2 + n_2^2 \tag{72}$$

We note that conditions (71) and (72) are fulfilled in certain coordinate systems, which requires attention to assessing the type of interpartial constraints that can be elastic and inertial.

Since the initial system has two degrees of freedom, then, taking into account the peculiarities of free vibrations in the chain type structure, the distribution coefficient of the oscillation amplitudes a will be negative, thus it can be assumed that $a < 0$. In this case, the $\omega^2(a)$ plot (Figure 15) takes a different form, but the presence of extremes of the frequency energy function $\omega^2(a)$ is preserved.

It should be noted that on the graph (Figure 15), in the region of positive values of a, an assymptotic motion of $\omega^2(a)$ is observed as $a \to \infty$ towards the limit $\omega^2(a) \to \frac{k_2 + k_3}{m_2} = n_2^2$, that is, towards the partial system. When $a = 0$, $\omega^2(a) = \frac{k_1 + k_2}{m_1}$, which also corresponds to the value of the partial frequency. In turn, with the decrease of a, when the negative value of $a \to -\infty$, the function is $\omega^2(a) \to \frac{k_2 + k_3}{m_2} = n_2^2$, while for $a = 0$, $\omega^2(a)$ has a meaning of $\frac{k_1 + k_2}{m_1}$. The above-noted allows you to enter into consideration a generalized representation of the frequency energy function in the form of a graph in Figure 16.

Figure 15. The graph of dependency of $\omega^2(a)$ in the region of negative values of the parameter a

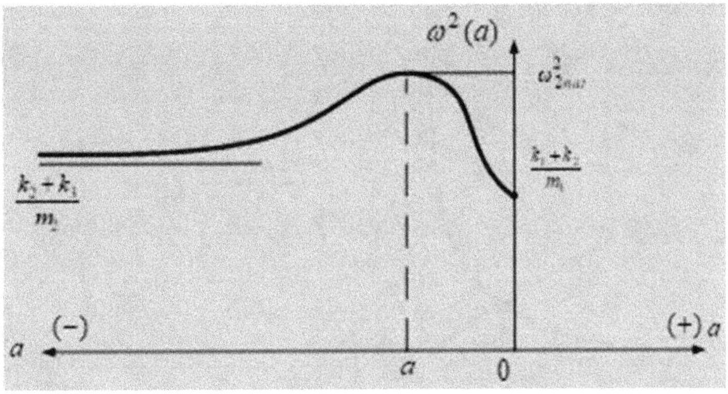

Figure 16. The graph of the generalized form of the dependence $\omega^2(a)$

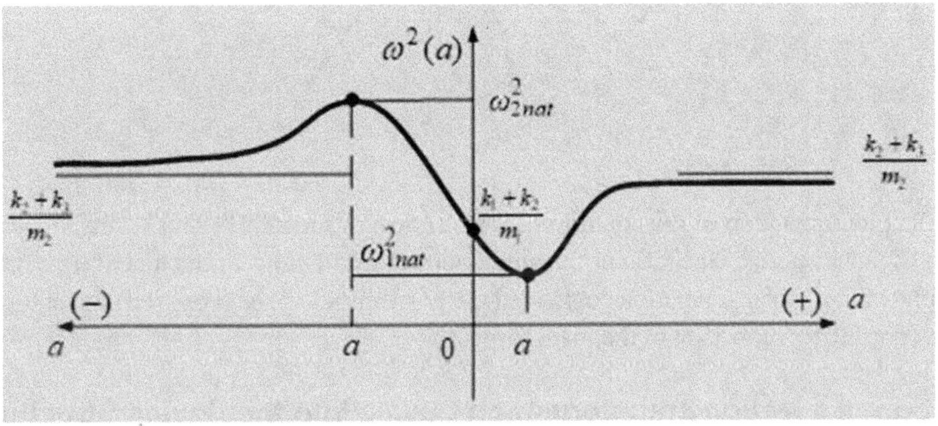

Determining the values of a and $\omega^2(a)$, i.e., natural frequencies of oscillations ω^2_{1nat} and ω^2_{2nat}, are corrected in the ratios determined by the transfer function of interpartial constraints.

Determination of Natural Frequencies Using a Graphic-Analytical Method

From the frequency characteristic equation:

$$(m_1p^2 + k_1 + k_2)(m_2p^2 + k_2 + k_3) - k_2^2 = 0 \quad (73)$$

the condition:

$$\frac{k_2}{(m_2p^2 + k_2 + k_3)} = \frac{(m_1p^2 + k_1 + k_2)}{k_2} \quad (74)$$

follows, in particular.

Figure 17. The schematic diagram of the method for determining the amplitude distribution coefficients of the main oscillations using transfer functions of interpartial constraints

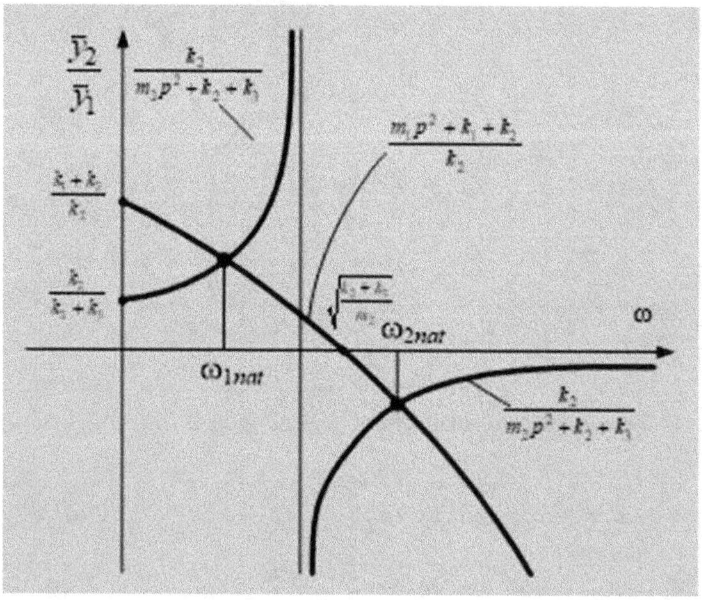

A graphical interpretation of the condition (74) is shown in Figure 17. This takes into account that the ratio $\bar{y}_2 / \bar{y}_1$ characterizes the ratio of the amplitudes of oscillations in the main forms of motion.

The results of numerical simulation confirmed the possibilities of the proposed concepts of the relations of the coordinates of motion in the main forms.

Introduction of a Motion Transformation Device Into the Device Structure

If we introduce into the computational scheme (Figure 12) a motion transformation device motion transformation device with a reduced mass L, then this can be represented in Figure 12 as the introduction of the motion transformation device parallel to the spring k_2. In this case, the structural diagram will change, and the structural mathematical model will take the form, as shown in Figure 18.

Figure 18. The structural diagram of the initial system in Figure 12 with a motion transformation device introduced parallel to the elastic element k_2

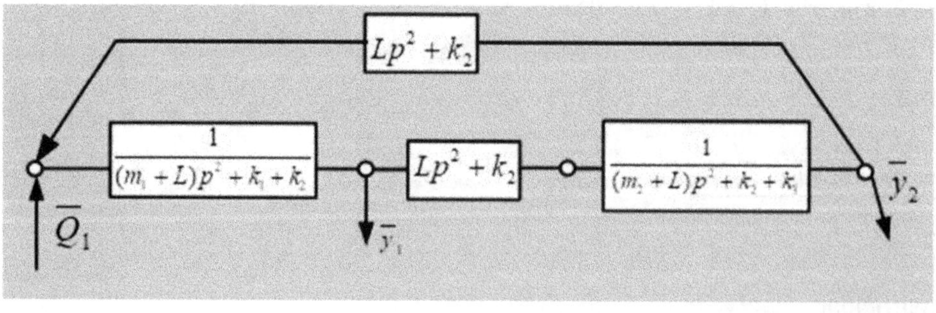

In this case, the transfer functions (64)–(66) are converted to:

$$W_1'(p)=\frac{\bar{y}_1}{\bar{Q}_1}=\frac{(m_2+L)p^2+k_2+k_3}{A_0'} \tag{75}$$

$$W_2'(p)=\frac{\bar{y}_2}{\bar{Q}_1}=\frac{Lp^2+k_2}{A_0'} \tag{76}$$

$$W_{12}''(p)=\frac{\bar{y}_2}{\bar{y}_1}=\frac{Lp^2+k_2}{(m_2+L)p^2+k_2+k_3} \tag{77}$$

Here:

$$A_0'=[(m_1+L_1)p^2+k_1+k_2)][(m_2+L)p^2+k_2+k_3)-(Lp^2+k_2)^2 \tag{78}$$

is a frequency characteristic equation.

In the reduced form, (78) takes the form:

$$p^4\left[(m_1+L)(m_2+L)-L^2\right]+p^2\left[(m_2+L)(k_2+k_3)+(m_2+L)(k_1+k_2)-2Lk_2\right]+k_1k_2+k_2k_3+k_1k_3=0. \tag{79}$$

From (79), it follows that the natural frequencies of the system will depend on the value of L. The expression for analytically determining the natural frequencies is:

$$\omega^2_{1,2\,nat}=\frac{m_1(k_2+k_3)+m_2(k_1+k_2)+L(k_1+k_3)}{2[m_1m_2+L(m_1+m_2)]}-\frac{1}{2}\sqrt{B} \tag{80}$$

where

$$B=\frac{\{m_1(k_2+k_3)+m_2(k_1+k_2)+L(k_1+k_3)\}^2-L(k_1k_2+k_2k_3+k_1k_3)(m_1m_2+L(m_1+m_2))}{[m_1m_2+L(m_1+m_2)]} \tag{81}$$

If we assume that $L=0$, then (81) is reduced to the expression (68). From (81) it can be seen that with interpartial constraints of elastic-inertial (or combined) form, the concept of partial frequencies acquires a more general meaning, since a different complex of dynamic interactions of the constituent elements is implemented in the systems.

Let us write down the expressions for the kinetic and potential energies of the system with motion transformation device in the idea that in the main oscillations the relations between the coordinates also have the form (60), that is, $y_2 = ay_1$.

In this case:

$$\mathrm{T} = \frac{1}{2}m_1\dot{y}_1^2 + \frac{1}{2}m_2\dot{y}_2^2 + \frac{1}{2}L(\dot{y}_2 - \dot{y}_1)^2 = \frac{1}{2}\dot{y}_1^2[m_1 + L + a^2(m_2 + L) - 2La] \tag{82}$$

$$\Pi = \frac{1}{2}k_1y_1^2 + \frac{1}{2}k_2(y_2 - y_1)^2 + \frac{1}{2}k_3y_2^2 = \frac{1}{2}y_1^2(k_1 + k_2 + (k_2 + k_3)a^2 - 2k_2a) \tag{83}$$

Taking into account the introduction of the motion transformation device, the kinetic and potential energy of the system acquire the full quadratic form, then:

$$\omega^2(a) = \frac{\Pi(a)}{T(a)} = \frac{k_1 + k_2 + (k_2 + k_3)a^2 - 2k_2a}{m_1 + L + a^2(m_2 + L) - 2La} \tag{84}$$

If we leave the parameters k_1, k_2, k_3, m_1, m_2 from the previous numerical experiment, then the influence of L can be illustrated, as a first approximation, with the ratio number of the mass m and L by the values: $\frac{1}{4}m$, $\frac{2}{3}m$, $5m$. The plot of $\omega^2(a)$ at $-\infty < a < \infty$ is shown in Figure 19. An increase in the reduced mass L leads to a decrease in the natural oscillation frequencies of the system, which is shown in graphs 1 and 3, which can be compared with the graph for the system in the absence of an additional constraint L the masses m_1 and m_2 "lock", which corresponds to a definite ratio between L and m. In particular, in numerical simulation, this value has the form: $L = \frac{2}{3}m$.

The transfer function of interpartial constraint (assuming that external force is applied to the object of protection m_1) has the form:

$$W_{12}(p) = \frac{\bar{y}_2}{\bar{y}_1} = \frac{Lp^2 + k_2}{(m_2 + L)p^2 + k_2 + k_3} \tag{85}$$

With a frequency determined by the expression:

$$\omega_{\text{dyn}}^2 = \frac{k_2}{L_2} \tag{86}$$

the motion along the y_2 coordinate becomes zero. At frequency:

$$\omega^2 = \frac{k_2 + k_3}{m_2 + L} \tag{87}$$

an indefinitely large increase in displacement along the y_2 coordinate is possible. With a frequency determined from the ratio:

$$\frac{k_2}{L} = \frac{k_2 + k_3}{m_2 + L} \tag{88}$$

it follows that $\frac{\overline{y}_2}{\overline{y}_1}$ turns into a $\frac{0}{0}$ type of uncertainty that can be resolved, which will be:

$$\frac{\overline{y}_2}{y_1} = \frac{L}{m_2 + L} \tag{89}$$

In this case, there must be a relation $k_2m_2+k_2L = Lk_2+Lk_3$ between L and m_2, from where $m_2 = L\frac{k_3}{k_2}$. Graphs of dependence $\frac{\overline{y}_2}{\overline{y}_1}$ on ω are shown in Figure 20.

Figure 19. The family of graphs $\omega^2(a)$ for different values of reduced masses L: 1) for L = $\frac{1}{4}m$, 2) for $L = \frac{2}{3}m$, 3) for L = 5m

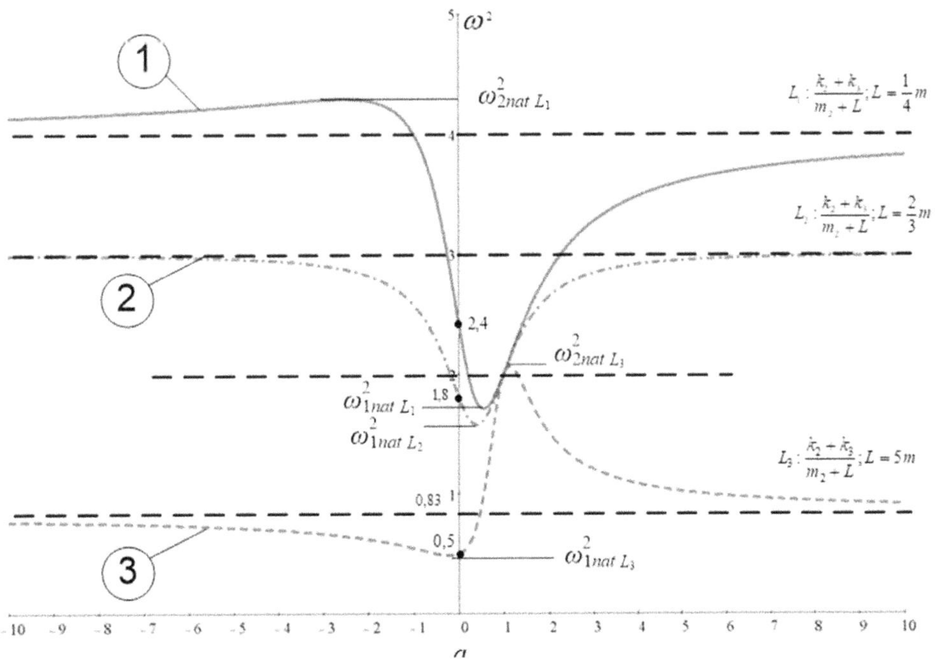

Figure 20. Plots of dependency of the transfer ratio of interpartial constraints: 1) dependency curve for dynamic damping in the pre-resonance zone, 2) graph dependencies for dynamic damping oscillations along the y_2 coordinate in the superresonance zone, 3) dependency curve in the critical case, when the system is "locked" throughout the frequency range

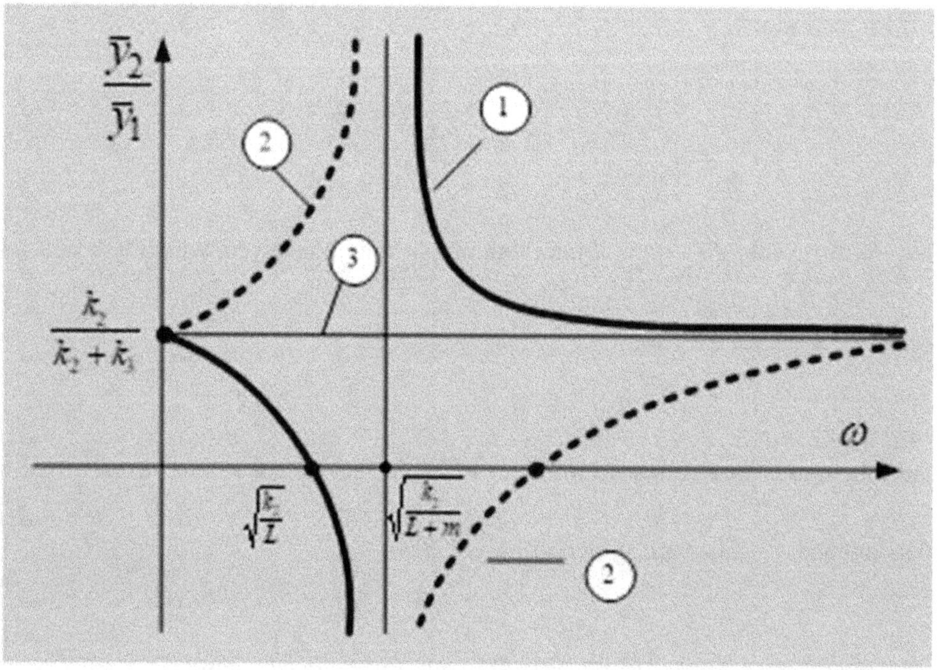

Note that in the previous sections, where the system with the motion transformation device with one degree of freedom was considered under kinematic perturbations from the support surfaces, the emergence of a critical "locking" of the system in the entire frequency range was also noted.

Consider the situation with disturbance when $\omega^2 = \frac{k_2}{L}$. In this case, from the expression for the transfer function of the form:

$$W_1(p) = \frac{\bar{y}_2}{\bar{Q}_1} = \frac{(m_2 + L)p^2 + k_2 + k_3}{[(m_1 + L)p^2 + k_1 + k_2][(m_2 + L)p^2 + k_2 + k_3] - (Lp^2 + k_2^2)} \tag{90}$$

it follows that:

$$W_1(p)\Big|_{\omega^2 = \frac{k_2}{L}} = \frac{1}{m_1 p^2 + k_1} \tag{91}$$

i.e., along the y_2 coordinate, a stop is observed ($y_2 = 0$), and along the y_1 coordinate, a motion occurs with the parameters of the system with one degree of freedom ($m_1 p^2$ and k_1).

If the frequency of external exposure is:

$$\omega^2 = \frac{k_2 + k_3}{m_2 + L} \tag{92}$$

which corresponds to the dynamic oscillation damping along the y_1 coordinate, then:

$$W_2(p) = \frac{\bar{y}_2}{\bar{Q}_1} = \frac{Lp^2 + k_2}{(Lp^2 + k_2)^2} = \frac{1}{Lp^2 + k_2} = \frac{1}{\frac{-L(k_2 + k_3)}{m_2 + L} + k_2} = \\ = \frac{m_2 + L}{-Lk_2 - Lk_3 + m_2 k_2 - Lk_2} = \frac{m_2 + L}{k_2 m_2 - Lk_3}, \tag{93}$$

which does not exclude the peculiarities of interactions at ratios $m_2 = \frac{k_3}{k_2}$. Note that such relations lead to the “locking” of the system in the entire frequency range.

The graphic-analytical method for determining the natural oscillation frequencies and the distribution coefficients of the amplitudes of motion in the first and second forms can also be applied using the ratios obtained from the frequency equation (88), which implies that:

$$\frac{Lp^2 + k_2}{(m_2 + L)p^2 + k_2 + k_3} - \frac{(m_1 + L)p^2 + k_1 + k_2}{Lp^2 + k_2} \tag{93}$$

Thus, the introduction of additional acceleration couplings, which are implemented, in the physical sense, by motion transformation devices (in particular, on the basis of non-locking screw mechanisms), significantly affect the forms of dynamic interactions of elements of mechanical oscillatory systems.

CONCLUSION

Thus, structural mathematical modeling, if we keep in mind its direction, is associated with the use of the apparatus of the theory of automatic control and the theory of mechanical circuits, in comparison with other methods of mathematical modeling, has certain advantages. They consist in the fact that the mathematical model, due to the peculiarities of its construction, allows us to introduce into consideration not only the features of the connections between typical elements, but also regulatory influences reflecting the influence of the structure. Generally, it would be possible to note a number of fundamental positions that could be developed autonomously, predetermining the possibility of obtaining new results, which is associated, in particular, with the ideas of expanding the set of standard elements, introducing additional links and transforming structural models, and more. The development of structural modeling in the theory of oscillations of mechanical linear systems is focused on the use of the apparatus of Laplace transformations, methods of frequency analysis and dynamic synthesis.

REFERENCES

Babakov, I. M. (1968). *Oscillation theory*. Nauka Publ.

Belokobylsky, S. V., Eliseev, S. V., & Kashuba, V. B. (2015). Impedance approaches as one of the forms for assessing the dynamic properties of mechanical oscillatory systems in structural mathematical modeling. Systems. *Metals Technology*, *4*(28), 7–15.

Belokobylsky, S. V., Eliseev, S. V., Kashuba, V. B., & Nguyen, D. H. (2015). Ratios of motion coordinates of elements of mechanical oscillatory systems as a form of manifestation of lever linkages. Systems. Methods. *Technology*, *3*(27), 7–14.

Belokobylsky, S. V., Eliseev, S. V., & Sitov, I. S. (2012). Interaction of links through the mass-and-inertia element in the theory of mechanical chains. Systems. Methods. *Technology*, *2*(14), 7–15.

Belokobylsky, S. V., Eliseev, S. V., & Sitov, I. S. (2013). *Dynamics of mechanical systems. Lever and inertial-elastic constraints*. Polytekhnika Publ.

Druzhinsky, I. A. (1977). *Mechanical chains*. Mashinostroenie Publ.

Eliseev, S. V., Artyunin, A. I., & Logunov, A. S. (2013a). *Dynamic interactions of machine elements: computational schemes and mathematical models of vibration states*.

Eliseev, S. V., Artyunin, A. I., & Yermoshenko, Y. V. (2013b). *Methodological approaches in system analysis and mathematical modeling of mechanical oscillatory systems*. Irkutsk: IrGUPS Publ.

Eliseev, S. V., Belokobylsky, S. V., Upir, R. Yu., & Gozbenko, V. E. (2009). Lever linkages in the problems of the dynamics of mechanical oscillatory systems: theoretical aspects. Academic Press.

Eliseev, S. V., Kashuba, V. B., Sitov, I. S., & Eliseev, S. V. (2013). The reduced stiffness of the feedback chain. Determination of dynamic reactions in a mechanical oscillatory system. Systems. Methods. *Technology*, *2*(18), 15–22.

Eliseev, S. V., & Khomenko, A. P. (2014). *Dynamic vibration cancellation: feedback concept and structural methods of mathematical modeling*. Nauka Publ.

Eliseev, S. V., Kinash, N. Zh., & Kaimov, E. V. (2015). Lever linkages of mechanical oscillatory systems. *Bulletin of the All-Russian Research and Design Institute of Electric Locomotive Engineering*, *1*(69), 112–126.

Eliseev, S. V., Kovyrshin, S. V., & Bolshakov, R. S. (2012a). Features of the construction of compact elastic elements in mechanical oscillatory systems. Interactions with elements of the system and form of the compound. Modern technologies. System analysis. *Modeling*, *4*(36), 61–70.

Eliseev, S. V., Lukyanov, A. V., Reznik, Yu. N., & Khomenko, A. P. (2006). *Dynamics of mechanical systems with additional ties*. ISU Publ.

Eliseev, S. V., Moskovskikh, A. O., Bolshakov, R. S., & Savchenko, A. A. (2012). Possibilities of integration of methods of the theory of chains and the theory of automatic control in problems of the dynamics of machines. *Science and Education, 6*. Retrieved from http://technomag.edu.ru/doc/378699.html

Eliseev, S. V., Reznik, Yu. I., Khomenko, A. P., & Zasyadko, A. A. (2008). *Dynamic synthesis in generalized tasks of vibration protection and vibration isolation of technical objects.* ISU Publ.

Eliseev, S. V., Reznik, Yu. N., & Khomenko, A. P. (2011). *Mechatronic approaches in the dynamics of mechanical oscillatory systems.* Nauka Publ.

Eliseev, S. V., Trofimov, A. N., Bolshakov, R. S., & Savchenko, A. A. (2012b). The concept of feedback in the dynamics of mechanical systems and dynamic oscillation damping. *Science and Education, 5.* Retrieved from http://technomag.edu.ru/doc/378353.html

Eliseev, S. V., Trofimov, A. N., & Kaimov, E. V. (2014). On forms of partial connectivity in collapses of mechanical systems. *International Journal of Applied and Fundamental Research*, *5*(1), 15–25.

Eliseev, S. V., Volkov, L. N., & Kukharenko, V. P. (1990). *Dynamics of mechanical systems with additional constraints.* Nauka Publ.

Ivanov, B. G. (2007). *Development of methods for calculating the dynamics and strength of units of transport equipment with lever-hinged connections* [PhD thesis]. Samara.

Kaimov, E. V., Kuznetsov, N. K., & Eliseev, S. V. (2015). Some applications of the theory of lever linkages. Academic Press.

Khomenko, A. P., Eliseev, S. V., Artyunin, A. I. & Bolshakov, R. S. (2016). Mechanical systems. *Scientific Israel- Technological Advantages, 18*(1), 43-62.

Khomenko, A. P., Eliseev, S. V., & Kaimov, E. V. (2014). Virtual lever mechanism: Dynamic oscillation damping as a form of manifestation of linkages. *News of Transsib*, *4*(20), 61–71.

Khomenko, A. P., Eliseev, S. V. & Yermoshenko, Yu. V. (2012). *System analysis and mathematical modeling in the mechatronics of vibration protection systems.* Irkutsk: IrGUPS Publ.

Khomenko, A. P., Khomenko, A. P., & Eliseev, S. V. (2012). Quasi-elements in mechanical oscillatory systems. Features of the system with the exclusion of variables of the dynamic state. Modern technologies. System analysis. *Modeling*, *2*(38), 8–17.

Kim, P. D. (2003). *The theory of automatic control* (Vols. 1–3). Fizmatlit Publ.

Kuznetsov, N. K. (2009). *Dynamics of controlled machines with additional constraints.* Irkutsk: ISTU Publ.

Lavrus, V. V. (2006). *Improvement of pneumatic lever-hinged systems in railway transport* [PhD thesis]. Orel.

Lontsikh, P. A., & Eliseev, S. V. (2014). *Dynamic quality of machinery and equipment as a tool to ensure production reliability and competitiveness of processes.* INRTU Publ.

Samara, A. A., & Mikhailov, A. P. (2005). *Mathematical modeling. Ideas. Methods. Examples.* Fizmatlit Publ.

ADDITIONAL READING

Eliseev, A. V., & Eliseev, S. V. (2019). Features of dynamic properties of diads in mesomechanics approaches. Modern Technologies. System Analysis. *Modeling*, *2*(62), 23–33.

Eliseev, A. V., Eliseev, S. V., & Borisov, B. G. (2016). Some approaches to forming the structure of a vibrational field (one-dimensional vibrational field). Actual Directions of Scientific Research of the XXI Century. *Theory into Practice*, *4*(1), 140–144.

Eliseev, A. V., Kopylov, Y. R., Nikolaev, A. V., & Eliseev, S. V. (2018). Vibrational technological processes: Mathematical models and features of dynamic properties of technical objects. Transportnoe, Gornoe i Stroitel'noe Mashinostroenie. *Nauka i Proizvodstvo*, *1*, 84–90.

Eliseev, A. V., & Vyong, K. C. (2013). Some possibilities of controlling one-dimensional vibration field of technological machine. Modern Technologies. System Analysis. *Modeling*, *1*(49), 33.

Eliseev, S. V., Bol'shakov, R. S., Kargapol'tsev, S. K., & Eliseev, A. V. (2020). Elastic-damping elements as factors in forming dynamic states of vibration technical objects. Systems. Methods. *Technologies*, *4*(48), 7–11.

Eliseev, S. V., Kargapol'tsev, S. K., Eliseev, A. V., & Nikolaev, A. V. (2019). Specific modes of dynamic states of vibrational machines: Joint damping of oscillations in two coordinates, frequency diagrams. Transportnoe, Gornoe i Stroitel'noe Mashinostroenie. *Nauka i Proizvodstvo*, *2*, 42–49.

Kashuba, V. B., Kinash, N. Zh., Eliseev, A. V., & Nikolaev, A. V. (2016). Structural mathematical modeling in the assessment of dynamic properties of mechanical vibrational systems. Systems. Methods. *Technologies*, *3*(31), 16–28.

Kashuba, V. B., Kinash, N. Zh., Sitov, I. S., Eliseev, A. V., & Nguyen, D. H. (2016). Distribution of vibration amplitudes between system elements as a manifestation of inter-partial connections. Systems. Methods. *Technologies*, *3*(31), 29–40.

Orlenko, A. I., Eliseev, S. V., & Eliseev, A. V. (2017). Features of mathematical modeling in reliability problems of current collection in collector-brush assemblies of electric traction motors. *Bulletin of Irkutsk State Technical University*, *22*(5), 46–64. doi:10.21285/1814-3520-2018-5-46-64

Afterword

Attention to the development of mathematical modeling methods during the invention and improvement of technical equipment of various purposes is quite a characteristic feature of up-to-date directions of scientific and technological progress. Serious advancements have taken place in the field of computer technologies focused on a wide range of problems that can be solved through applications, machine graphics and animation packages. Building of high-complexity technical objects, such as technical complexes, modern transportation systems, special purpose facilities, is created, controlled and monitored by appropriate software.

Essentially, modern technical objects are often automatic control systems in which the work of machine assemblies, components, mechanisms, equipment, and devices is linked into a single technological complex, working in accordance with regulations of a certain interaction of machine elements.

The most important manifestations of the quality of the new machinery being created are ensuring the equipment reliability and safe conditions of its operation.

Focusing on creating new technical means in a competitive environment and ensuring a high level of dynamic quality and competitiveness of transport and technological machines initiates attention to the development of methods, ways and means of analytical approaches and mathematical modeling at all stages of the life cycle of machines and technological complexes.

Of particular note is the importance and efficiency of the use of computer equipment in ensuring the operation of machines as a certain set of interacting fragments of technical objects. Each of them is often a fairly complex device that in turn consists of several elements.

In a certain sense, mathematical modeling of technical objects of various nature (we can consider electrical devices and systems, automatic machines, including robotic systems, etc. in this perspective) develop in parallel with the tasks of appropriately representing the complexity of interactions of elements of technical objects, which led to the formation of certain directions of mathematical modeling.

In particular, methods of structural mathematical modeling (that include circuit theory, graph theory and automatic control theory) became widely known in recent decades.

The evolvement of methods of structural mathematical modeling is closely related to the evolvement of a certain circle of ideas of physical content that reflect the principles of dynamic analogies observed in the dynamic states of electrical and mechanical chains. This allows us to introduce new approaches to the assessment of the dynamic states of technical objects.

The possibilities of adequate transformations and representations of the peculiar features of the interaction of elements of an electric and mechanical nature are widely used and are reflected in many scientific developments. Automatic control systems (if we mean technical objects, the dynamic state of which is maintained using automatic control systems) have certain specific features.

First of all, it should be pointed out that there exist so-called feedbacks in the automatic control systems. Due to these feedbacks, specific dynamic states are provided that include possible oscillatory motions relative to the equilibrium position of the system or its steady-state motion process.

Studies in the field of dynamics of technical objects for various purposes show that the methods of structural mathematical modeling, implemented on the basis of the mathematical tools of circuit theory, are appropriate to the analytical tools of automatic control theory.

The researches conducted, the results of which are considered in the monograph, are based on an understanding of the commonality of dynamic phenomena that accompany the dynamics of mechanical oscillatory systems. Representations of the dynamic states of technical objects (if their computational schemes are reduced to mechanical oscillatory systems) can be adequately represented by mechanical chains, graphs, and structural diagrams, dynamically equivalent to automatic control systems.

The commonality of structural representations is predetermined by the commonality of dynamic processes that are characteristic of technical objects for various purposes, but considered in their dynamic states as mechanical oscillatory motions.

The principal positions of methodological approaches outlined above in solving the problems of the dynamics of technical objects for various purposes allow us to develop and refine our ideas about a fairly general methodological basis that provides the formation of analytical modeling methods. These methods have the ability to consider a wide class of problems of dynamics or problems of estimation, correction and formation of dynamic states of technical objects for various purposes.

As practice shows, machines and technical devices, used in various production processes, at a certain level of consideration are reduced to physical models in the form of mechanical oscillatory systems. These systems are ultimately transformed into the corresponding computational schemes that predetermine the possibilities of structural mathematical modeling.

The emphasis on structural mathematical modeling can be considered, in a certain sense, as an attempt to formulate the necessity to take into account the jointness of the manifestation of the action of signals of a mechanical and informational character. In our opinion, this is a natural form and direction for further development of structural mathematical modeling methods.

Compilation of References

Andreevsky, B. R., & Blekhman, I. I. (2001). *Management of mechatronic vibration installations*. Nauka Publ.

Andreychikov, A. V., Khorichev, A. A., & Andreychikov, O. N. (2011). Expert system for forecasting strategic innovations. Quality. Innovation. *Education*, *12*, 26.

Antipov, V. A. (2006). *Suppression of vibration of Gordeeva and units of transport systems*. Marshrut Publ.

Atabekov, G.I. (2006). *Fundamentals of the theory of chains: a textbook*. St. Petersburg: Lan' Publ.

Averyanov, G. S., Zubarev, A. V. & Khamitov, R. N. (2008). Oscillating system dynamics with controlled vibration protection devices. *Bulletin of mechanical engineering, 6*, 27-30.

Babakov, I. M. (1968). *Oscillation theory*. Nauka Publ.

Bakalov, A. N. (2006). *Circuit Theory*. Nauka Publ.

Baranovsky, A. M. (2000). *Theoretical foundations of effective vibration isolation on ships* [PhD thesis]. Novosibirsk.

Belokobylsky, S. V. (2002). *Dynamics of systems with dry friction and its application to the problems of mining mechanics*. Mashinostroenie Publ.

Belokobylsky, S. V., & Eliseev, S. V. (2016). Dynamic analogies in problems of the theory of circuits and automatic control. Structural interpretations. *Systems. Methods. Technologies*, *32*, 7–16.

Belokobylsky, S. V., Eliseev, S. V., & Kashuba, V. B. (2013). *Applied problems of the structural theory of vibration protection systems*. Politekhnika Publ.

Belokobylsky, S. V., Eliseev, S. V., & Kashuba, V. B. (2013). *Applied Problems of the Structural Theory of Vibration Protection Systems*. Polytekhnika Publisher.

Belokobylsky, S. V., Eliseev, S. V., & Kashuba, V. B. (2013). *Applied problems of the structural theory of vibro-protective systems*. Polytekhnika Publ.

Belokobylsky, S. V., Eliseev, S. V., & Kashuba, V. B. (2015). Impedance approaches as one of the forms for assessing the dynamic properties of mechanical oscillatory systems in structural mathematical modeling. Systems. *Metals Technology*, *4*(28), 7–15.

Belokobylsky, S. V., Eliseev, S. V., Kashuba, V. B., & Nguyen, D. H. (2015). Ratios of motion coordinates of elements of mechanical oscillatory systems as a form of manifestation of lever linkages. Systems. Methods. *Technology*, *3*(27), 7–14.

Belokobylsky, S. V., Eliseev, S. V., Kashuba, V. B., Nguyen, D. H., & Tsygan, V. V. (2015). On relations between motion coordinates in mechanical oscillatory systems with lever devices. Systems. Methods. *Technology*, *2*(26), 7–13.

Belokobylsky, S. V., Eliseev, S. V., & Sitov, I. S. (2012). Interaction of links through the mass-and-inertia element in the theory of mechanical chains. Systems. Methods. *Technology, 2*(14), 7–15.

Belokobylsky, S. V., Eliseev, S. V., & Sitov, I. S. (2013). *Dynamics of mechanical systems. Lever and inertial-elastic constraints*. Polytekhnika Publ.

Belokobylsky, S. V., Eliseev, S. V., & Sitov, I. S. (2013). *Dynamics of mechanical systems: lever and inertial-elastic constraints*. Polytekhnika Publ.

Belokobylsky, S. V., Eliseev, S. V., & Sitov, I. S. (2013). *Dynamics of mechanical systems: Lever and inertial-elastic constraints*. Polytekhnika Publ.

Belokobylsky, S. V., Eliseev, S. V., & Sitov, I. S. (2013). *Dynamics of Mechanical Systems: Lever Linkages and Inertial-Elastic Constraints*. Polytekhnika Publisher.

Belokobylsky, S. V., & Sitov, I. S. (2007). Ways and means of changing the dynamic state of the technological complex for vibration smoothing. Modern technologies. System analysis. *Modeling*, (2), 46–52.

Birger, I. A. (1988). Basics of resource design. In Mechanics and scientific and technical progress: in 4 vols. (Vol. 4, pp. 174-180). Moscow: Nauka Publ.

Blekhman, I. I. (1994). *Vibration mechanics*. Fizmatlit Publ.

Bolshakov, R. S., & Eliseev, S. V. (2013). *Methodological foundations of problems of determining the reactions of constraints in mechanical oscillatory systems*. Dep. in the VINITI RAS.

Brysin, A. N. & Sinev, A. V. (2008). New approaches in solving problems of vibration isolation. *Problems of mechanical engineering and automation, 3,* 74-83.

Brysin, A. N., Shokhin, A. E., Sinev, A. V., Solovyov, V. S., Gordeev, A. B., Erokhina, T. V. & Stepanova, L. A. (2012). Improving the method of testing vibration protection systems with inertia transducers. *Problems of mechanical engineering and automation*, (4), 80-83.

Burkov, S. N. & Efremov, A. M. (2003). Theoretical study of the dynamic stability of pneumohydraulic bearings. *Scientific problems of transport in Siberia and the Far East, 2,* 124-131.

Burkov, S. N., & Efremov, A. M. (2004). Vibration strength and reliability of autonomous dielectric power plants. *Proceedings of the International conference Energy saving Technologies ISTU*, (pp. 68-74). IEA.

De Silva, C. W. (2000). *Vibration. Fundamentals and Practice*. CRC Press.

De Silva, C. W. (2000). *Vibration: Fundamentals and Practice*. CRC Press.

Dimov, A. V., & Nasnikov, D. N. (2009). Mechatronics vibration protection systems. Features of structural transformations. Modern technologies. System analysis. *Modeling, 24,* 75–85.

Domnin, D. A. (2005). *A method for improving the vibration-damping parameters of an automobile suspension by selecting rational parameters for dynamical wheel vibration dampers.* Research Gate.

Druzhinsky, I. A. (1977). *Mechanical chains*. Mashinostroenie Publ.

Druzhinsky, I. A. (1977). *The Mechanical Chains*. Mashinostroenie Publ.

Eliseev, A. V., Kuznetsov, N. K. & Moskovskikh, A. O. (2019). *The Dynamics of Machines. System conception, structural schemes and connections of elements: monograph.* Moscow: Innovative mechanical engineering.

Eliseev, A.V., Kuznetsov, N.K. & Moskovskikh, A.O. (2019). The Dynamics of Machines: System Conception, Structural Schemes and Connections of Elements. *Innovative Mechanical Engineering.*

Eliseev, S. V. & Yermoshenko, Yu. V. (2012). *Links in the dynamics of mechanical oscillatory systems.* Irkutsk: IrGUPS Publ.

Eliseev, S. V. & Zasyadko, A. A. (2004). Vibration protection and vibration insulation as a control of object vibrations. Modern technologies. *System analysis Modeling, 1*, 24-31.

Eliseev, S. V. Belokobylsky, S.V., Upyr', R.Yu. & Gozbenko, V.E. (2009). Lever linkages in the problems of the dynamics of mechanical oscillatory systems. Theoretical aspects. Irkutsk: Dep. in the VINITI RAS.

Eliseev, S. V., Artyunin, A. I., & Kaimov, E. V. (2013a). Equivalent transformations in the structures of mechanical oscillatory systems containing material points. In *Information and mathematical technologies in science and management: Proceedings of XVIII Baikal All-Russian conf. (Irkutsk, July 01–10, 2013) in 3 vol.* (*Vol. 1*, pp. 173-182). Irkutsk: ISEM of the SB RAS.

Eliseev, S. V., Artyunin, A. I., & Yermoshenko, Y. V. (2013b). *Methodological approaches in system analysis and mathematical modeling of mechanical oscillatory systems.* Irkutsk: IrGUPS Publ.

Eliseev, S. V., Artyunin, A. I., Kaimov, E. V., & Eliseev, A. V. (2013b). Pat. 136112 Russian Federation, IPC F16F 15/04. A damping device.

Eliseev, S. V., Artyunin, A. I., Kaimov, E. V., & Eliseev, A. V. (2014a). Pat. 142137 Russian Federation, IPC F16F 15/02. Device for regulating the elastic-dissipative properties of a vibration protection system.

Eliseev, S. V., Artyunin, A. I., Logunov, A. S., Nasnikov, D. N., Bolshakov, R. S., Kaimov, E. V., Mironov, A. S. & Parshuta, E. A. (2013). *Dynamic interactions of machine elements: computational schemes and mathematical models of vibration states.* Irkutsk: IrGUPS Publ.

Eliseev, S. V., Belokobylsky, S. V. & Kashuba, V. B. (2011a). Mathematical modeling in mechanical oscillatory systems: Mechatronic approaches. *Problems of Mechanical Engineering and Automation, 3*, 70-78.

Eliseev, S. V., Belokobylsky, S. V., Upir, R. Yu., & Gozbenko, V. E. (2009). Lever linkages in the problems of the dynamics of mechanical oscillatory systems: theoretical aspects. Academic Press.

Eliseev, S. V., Kovyrshin, S. V. & Bolshakov, R. S. (2012a). Features of the construction of compact elastic elements in mechanical oscillatory systems: Interactions with elements of the system and form of the compound. *Modern Technologies, System Analysis, Modeling, 4*(36), 61-70.

Eliseev, S. V., Moskovskikh, A. O., Bolshakov, R. S., & Savchenko, A. A. (2012). Possibilities of integration of methods of the theory of chains and the theory of automatic control in problems of the dynamics of machines. *Science and Education, 6.* Retrieved from http://technomag.edu.ru/doc/378699.html

Eliseev, S. V., Parshuta, E. A., & Bolshakov, R. S. (2013b). Features of vibration protection system with the object of protection in the form of a solid: dynamic responses. In *Proceedings of the International Scientific Conference "Science in the Modern Information Society"* (Vol. 1, pp. 101-116). Moscow: Academic Publishing.

Eliseev, S. V., Parshuta, E. A., & Kaimov, E. V. (2013). Vibration protection systems. Dynamic responses and methods for their determination. Engines 2013: materials of the Intern. scientific-practical conf. Omsk.

Eliseev, S. V., Parshuta, E. A., & Kaimov, E. V. (2013d). Vibration protection systems: dynamic responses and methods for their determination. In *Proceedings of the International Scientific Conference "Engines 2013"* (pp. 279-284). Khabarovsk: Pacific National University Press.

Eliseev, S. V., Reznik, Y. N., Upir' R. Yu., et al. (2009). *Mechatronics of vibration protection systems. Elements of the theory*. Dep. in VINITI.

Eliseev, S. V., Savchenko, A. A., Trofimov, A. N., Parshuta, E. A., & Artyunin, A. I. (2013). A Device for Damping. Patent RU 133232, IPC F16F7/10 F16F15/04. No. 2013105925/11, 12.02.2013; publ. 10.10.2013, Bull. No. 28.

Eliseev, S. V., Trofimov, A. N., & Bolshakov, R. S. (2015). Structural formations as a part of mechanical oscillatory systems. In *Kulagin Readings: Technique and Technology of Production Processes: XV Intern. scientific and practical conf*, (pp. 273-277). Chita: Transbaikalian State University Publ.

Eliseev, S. V., Trofimov, A. N., Bolshakov, R. S., & Savchenko, A. A. (2012b). The concept of feedback in the dynamics of mechanical systems and dynamic oscillation damping. *Science and Education, 5*. Retrieved from http://technomag.edu.ru/doc/378353.html

Eliseev, S. V., Upir', R. Yu., Nasnikov, D. N., & Logunov, A. S. (2007). The system of new model links in the structural representations of mechanical oscillatory systems. In *Innovative technologies of transport and industry: Proc. of the 45th International scientific and practical conf.* (*Vol. 3*, pp. 173-182). Khabarovsk: FEDUPS Publishing House.

Eliseev, S. V., Upir', R. Yu., Nasnikov, D. N., & Logunov, A. S. (2007). The system of new types in the structural representations of mechanical oscillatory systems. In Innovative technologies for transport and industry: *Proceed. of the 45th int. scientific conf.*, (*Vol. 3*, pp. 173-182). Irkutsk.

Eliseev, S. V., Zasyadko, A. A., Karpukhin, E. L., Kukharenko, V. P., & Metzger, P. K. (1988). Software and algorithmic complex optimization design of complex vibration protection systems. In *Proceedings of the Sixth All-Union Conference on Control in Mechanical Systems* (pp. 61-63). Lviv: Institute of Applied Mechanics of the Academy of Sciences of the Ukrainian SSR.

Eliseev, S.V. & Eliseev, A.V. (2020). Theory of Oscillations: Structural Mathematical Modeling in Problems of Dynamics of Technical Objects. *Studies in Systems, Decision and Control*, 252. Springer International Publishing.

Eliseev, S.V. & Ermoshenko, Yu.V. (2012). *Joints of units in the dynamics of mechanical oscillation systems*. Irkutsk: IrGUPS Publ.

Eliseev, S.V., Parshuta, E.A. & Bolshakov, R.S. (2012). On the possibilities of mechatronic approaches to the problems of vibration protection of technical objects. Vibration of machines: measurement, reduction, protection. *scientific and technical and manuf. magazine, 4*(31), 46-50.

Eliseev, S.V., Parshuta, E.A. & Bolshakov, R.S. (2012). On the possibilities of mechatronic approaches to the tasks of vibration protection of technical objects. *Vibration machines: measurement, reduction, protection, 4*(31), 46-49.

Eliseev, A. V., Kuznetsov, N. K., & Moskovskikh, A. O. (2019). *The Dynamics of Machines: System Conception, Structural Schemes and Connections of Elements*. Innovative Mechanical Engineering.

Eliseev, S. V. (1978). *Structural theory of vibration protection systems*. Nauka Publ.

Eliseev, S. V. (2018). *Applied system analysis and structural mathematical modeling (dynamics of transport and technological machines: connectivity of movements, vibration interactions, lever connections)*. Irkutsk State Transport University.

Eliseev, S. V. (2018). *Applied System Analysis and Structural Mathematical Modeling (Dynamics of Transport and Technological Machines: Connectivity of Movements, Vibration Interactions, Lever Connections)*. Irkutsk State Transport University.

Eliseev, S. V., & Artyunin, A. I. (2016). *Applied Theory of Oscillations in Problems of Dynamics of Linear Mechanical Systems*. Science.

Eliseev, S. V., & Artyunin, A. I. (2016). *Applied theory of oscillations in problems of the dynamics of linear mechanical systems*. Nauka Publ.

Eliseev, S. V., Artyunin, A. I., Ermoshenko, Yu. V., Moskovskikh, A. O., & Kaimov, E. V. (2013). *The method of transformation of mechanical circuits based on the introduction of intermediate devices in the compounds*. Dep. in the VINITI RAS.

Eliseev, S. V., Artyunin, A. I., & Logunov, A. S. (2013a). *Dynamic interactions of machine elements: computational schemes and mathematical models of vibration states*.

Eliseev, S. V., Bolshakov, R. S., & Kinash, N. Zh. (2015a). Leverage: Virtual leverage mechanisms, features of oscillatory processes. *News of South-West State University, 2*(59), 8–15.

Eliseev, S. V., & Eliseev, A. V. (2020). *Theory of Oscillations: Structural Mathematical Modeling in Problems of Dynamics of Technical Objects. Studies in Systems, Decision and Control* (Vol. 252). Springer International Publishing. doi:10.1007/978-3-030-31295-4

Eliseev, S. V., & Ermoshenko, Yu. V. (2010). Dynamic properties of vibration protection systems. Standing States. *Modern technologies. System analysis. Modeling., 28*, 24–30.

Eliseev, S. V., Gordeeva, A. A., & Fomina, I. V. (2011). Dynamic properties of vibrational systems at limiting values of the parameters. *Herald of the VSSUTU, 35*, 12.

Eliseev, S. V., Kargapoltsev, S. K., & Kashuba, V. B. (2016). Structural Transformations Mapping of Dynamic Interactions in Oscillatory Systems. *Bulletin of the SSUPS, 33*, 27–41.

Eliseev, S. V., Kashuba, V. B., Sitov, I. S., & Eliseev, S. V. (2013). The reduced stiffness of the feedback chain. Determination of dynamic reactions in a mechanical oscillatory system. Systems. Methods. *Technology, 2*(18), 15–22.

Eliseev, S. V., & Khomenko, A. P. (2009). Oscillation stability in two-dimensional systems with non-traditional constraints. Modern technologies. System analysis. *Modeling*, (2), 8–12.

Eliseev, S. V., & Khomenko, A. P. (2014). *Dynamic damping of oscillations: the concept of feedback and structural methods of mathematical modeling*. Nauka Publ.

Eliseev, S. V., & Khomenko, A. P. (2014). *Dynamic oscillation damping: feedback concept and structural methods of mathematical modeling*. Nauka Publ.

Eliseev, S. V., & Khomenko, A. P. (2014). *Dynamic oscillation suppression: Feedback concept and structural methods of mathematical modeling*. Nauka Publ.

Eliseev, S. V., & Khomenko, A. P. (2014). *Dynamic vibration cancellation: feedback concept and structural methods of mathematical modeling*. Nauka Publ.

Eliseev, S. V., Khomenko, A. P., & Barsukov, S. V. (2011). *Modern problems of machine dynamics. Protection from vibrations and shocks*. Dep. in the VINITI RAS.

Eliseev, S. V., Khomenko, A. P., & Logunov, A. S. (2010). Use in the schemes of pneumomechanical protection of mechanisms and motion transformation devices. Modern technologies. System analysis. *Modeling, 1*, 8–13.

Eliseev, S. V., Khomenko, A. P., & Upir', R. Yu. (2009). Mechatronics of vibration protection systems with lever linkages. Modern technologies. System analysis. *Modeling*, (3), 8–16.

Eliseev, S. V., Kinash, N. Zh., & Kaimov, E. V. (2015). Lever linkages of mechanical oscillatory systems. *Bulletin of the All-Russian Research and Design Institute of Electric Locomotive Engineering, 1*(69), 112–126.

Eliseev, S. V., Kinash, N. Zh., & Kaimov, E. V. (2015b). Lever linkages of mechanical oscillatory systems. *Bulletin of VELNII, 1*(69), 112–126.

Eliseev, S. V., Kovyrshin, S. V., & Bolshakov, R. S. (2012a). Features of the construction of compact elastic elements in mechanical oscillatory systems. Interactions with elements of the system and form of the compound. Modern technologies. System analysis. *Modeling, 4*(36), 61–70.

Eliseev, S. V., Kovyrshin, S. V., & Parshuta, E. A. (2013). Some questions of the theory of vibration isolation: Justification of structural approaches. *News of IHEA, 3*, 121–127.

Eliseev, S. V., Kovyrshin, S. V., & Parshuta, E. A. (2013a). Some questions of the theory of vibration isolation: Justification of structural approaches. *News of ISEA*, (3), 121–127.

Eliseev, S. V., Kuznetsov, N. K., & Kaimov, E. V. (2015c). On the question of the theory of lever constraints in the dynamics of mechanical oscillatory systems. *Bulletin of the Irkutsk State Technical University, 12*(107), 30–40.

Eliseev, S. V., & Logunov, A. S. (2009). Features of transient processes in pneumatic vibration protection systems. Modern technologies. System analysis. *Modeling, 1*, 8–15.

Eliseev, S. V., & Logunov, A. S. (2010). Mechatronics of pneumatic vibration protection systems. Systems. Methods. *Technologies, 7*, 9–21.

Eliseev, S. V., Lukyanov, A. V., Reznik, Yu. N., & Khomenko, A. P. (2006). *Dynamics of mechanical systems with additional ties*. Irkutsk State University Publ.

Eliseev, S. V., Lukyanov, A. V., Reznik, Yu. N., & Khomenko, A. P. (2006). *Dynamics of Mechanical Systems with Additional Ties*. Irkutsk State University.

Eliseev, S. V., Moscovskikh, A. O., Bolshakov, R. S., & Savchenko, A. A. (2012b). Possibilities of integration of methods of the theory of chains and the theory of automatic control in problems of the dynamics of machines. *Science and Education, 6*, 277–296.

Eliseev, S. V., Moskovskikh, A. O., Bolshakov, R. S., & Savchenko, A. A. (2012). The possibilities of integration of the methods of the theory of chains and the theory of automatic control in problems of the dynamics of machines. *Science and Education, 5*, 25–26.

Eliseev, S. V., Moskovskikh, A. O., & Kaimov, E. V. (2013). *Dynamics of mechanical oscillatory systems: structural analogies, mechanical circuits*. Dep. in the VINITI RAS.

Eliseev, S. V., Parshuta, E. A., & Bolshakov, R. S. (2013). Features of vibration protection system with the object of protection in the form of a solid. Dynamic responses. *Science in the modern information society: Materials of the Intern. scientific and practical conf. Moscow. Academic Publ.*

Eliseev, S. V., Parshuta, E. A., & Bolshakov, R. S. (2013). Transfer functions of mechanical oscillatory systems. Possibilities for estimating unit stiffness. Intern. *Journal of Applied and Fundamental Research, 1*, 11–18.

Eliseev, S. V., Parshuta, E. A., & Bolshakov, R. S. (2013c). Transfer functions of mechanical oscillatory systems: Possibilities for assessing reduced stiffness. *International Journal of Applied and Fundamental Research, 1*, 11–18.

Eliseev, S. V., Parshuta, E. A., Kaimov, E. V., & Kinash, N. Zh. (2014b). Mechanisms in the structure of vibration protection systems: Mathematical models, evaluation of dynamic properties. *Bulletin of VSGUT, 6*(51), 37–44.

Eliseev, S. V., Reznik, Y. N., & Khomenko, A. P. (2011). *Mechatronic approaches in the dynamics of mechanical oscillatory systems*. Nauka Publ.

Eliseev, S. V., Reznik, Yu. N., & Khomenko, A. P. (2011). *Mechatronic approaches in the dynamics of mechanical oscillation systems*. Nauka Publ.

Eliseev, S. V., Reznik, Yu. N., Khomenko, A. P., & Zasyadko, A. A. (2008). *Dynamic synthesis in generalized problems of vibration protection and vibration isolation of technical objects*. IGU Publ.

Eliseev, S. V., Reznik, Yu. N., Khomenko, A. P., & Zasyadko, A. A. (2008). *Dynamic synthesis in generalized tasks of vibration protection and vibration isolation of technical objects*. ISU Publ.

Eliseev, S. V., Trofimov, A. N., Bolshakov, R. S., & Savchenko, A. A. (2012). The concept of feedback in the dynamics of mechanical systems and dynamic oscillation damping. *Science and Education*, *5*, 25.

Eliseev, S. V., Trofimov, A. N., & Kaimov, E. V. (2014). On forms of partial connectivity in collapses of mechanical systems. *International Journal of Applied and Fundamental Research*, *5*(1), 15–25.

Eliseev, S. V., & Upir', R. Yu. (2008). Mechatronic approaches in the tasks of vibration protection of machines and equipment. Modern technologies. System analysis. *Modeling*, (4), 8–16.

Eliseev, S. V., Volkov, L. N., & Kukharenko, V. P. (1990). *Dynamics of mechanical systems with additional constraints*. Nauka Publ.

Ermoshenko, Yu. V., & Fomina, I. V. (2009). Dynamic absorbing in vibration protection systems using L-shaped lever linkages. Modern technologies. System analysis. *Modeling*, (2), 85–88.

Fomichev, P. A. (2010). *Vibration isolation of ship power plants with electropneumo-hydraulic supports* [PhD thesis]. Novosibirsk.

Fomichev, P. A., & Fomichev, E. V. (2004). *Automation of ship engines vibration protection*. Nauka Publ.

Frolov, K. V. (Ed.). (1983). Vibrations in technology: a reference book: in 6 vols.: Vol. 6. *Protection against vibrations and shocks*. Mashinostroenie Publ.

Frolov, K. V., Kraynev, A. F., & Kreinin, G. V. (2004). *Designing machines: reference book* (Vol. 1-2). Mashinostroenie Publ.

Galiev, I. I., Nekhaev, V. A., & Nikolaev, V. A. (2010). Methods and means of vibration protection of railway carriages. Moscow: GOU Publ.

Galperin, I. I. (1964). *Automation as one-sided mechanics*. Energiya Publ.

Ganiev, R. F. (2010). Problems of mechanics of machines and technologies. *Problems of mechanical engineering and machine reliability*, *3*, 3-20. Blagonravov Institute of Engineering.

Genkin, M. D., & Ryaboy, V. M. (1988). *Elastic-inertial vibration isolation systems: ultimate opportunities, optimal structures*. Nauka Publ.

Genkin, M. D., & Ryaboy, V. M. (1988). *Inert-vibration isolation systems: Ultimate opportunities, optimal structures*. Nauka Publ.

Goverdovsky, V. N. (2006). *Development of the theory and methods of designing machines with infra-low-frequency vibration protection systems*. Research Gate.

Goverdovsky, V. N., & Trofimov, A. N. (2009). Problems and perspectives of creation of vibration protection for high-speed railway transport. *Modern technologies. System analysis. Modeling*, *24*, 86–92.

Harris, C. M., & Crede, C. E. (2002). *Shock and Vibration Handbook*. McGraw — Hill Book Co.

Ivanov, B. G. (2007). *Development of methods for calculating the dynamics and strength of transport equipment aggregates with articulated kinematic relations* [PhD thesis]. Samara.

Ivanov, B. G. (2007). *Development of methods for calculating the dynamics and strength of units of transport equipment with lever-hinged connections* [PhD thesis]. Samara.

Ivashchenko, I. I. (1993). *Automatic adjustment. Theory and elements of systems*. Mashinostroenie Publ.

Kaimov, E. V., Kuznetsov, N. K., & Eliseev, S. V. (2015). Some applications of the theory of lever linkages. Academic Press.

Kamaev, V. A. (2005). *Conceptual design. Development and improvement of methods*. Mashinostroenie Publ.

Karnovsky, I. A., & Lebed, E. (2016). *Theory of vibration protection*. Springer. doi:10.1007/978-3-319-28020-2

Khokhlov, A. A. (2002). *Dynamics of complex mechanical systems*. MIIT Publ.

Khomenko, A. P. & Eliseev, S. V. (2010). Vibration protection systems with joints: Technology of construction of mathematical models. *Modern Technologies, System Analysis, Modeling, 4*(28), 8-18.

Khomenko, A. P. & Eliseev, S. V. (2013). Quasi-elements in mechanical oscillatory systems: Features of systems with the exclusion of variables of the dynamic state. *Modern Technologies, System Analysis, Modeling, 2*(38), 8-17.

Khomenko, A. P., Eliseev, S. V. & Ermoshenko, Yu. V. (2012). *System analysis and mathematical modeling in the mechatronics of vibration protection oscillatory systems*. Irkutsk: IrGUPS Publ.

Khomenko, A. P., Eliseev, S. V. & Yermoshenko, Yu. V. (2012). *System analysis and mathematical modeling in the mechatronics of vibration protection systems*. Irkutsk: IrGUPS Publ.

Khomenko, A. P., Eliseev, S. V., Artyunin, A. I. & Bolshakov, R. S. (2016). Mechanical systems. *Scientific Israel- Technological Advantages, 18*(1), 43-62.

Khomenko, A. P., Eliseev, S. V., Artyunin, A. I., Eliseev, A. V., Bolshakov, R. S., & Kaimov, E. V. (2013a). Elastic elements in mechanical systems: structural interpretations. Irkutsk: Irkutsk State Technical University Press.

Khomenko, A. P., Eliseev, S. V., Artyunin, A. I., Kaimov, E. V., & Eliseev, A. V. (2015). A damping device. Pat. 150331 Russian Federation, IPC F16F 15/04.

Khomenko, A. P., Eliseev, S. V., Artyunin, A. I., Parshuta, E. A., & Kaimov, E. V. (2013b). Mechanisms in elastic oscillatory systems: features of taking into account dynamic properties, problems of vibration protection of machines, devices and equipment. Irkutsk: Irkutsk State Technical University Press. .

Khomenko, A. P., & Eliseev, S. V. (2012). Non-traditional approaches to the construction of mathematical models of mechanical oscillatory systems with lever linkages. *Izvestia Transsiba, 4*(12), 75–86.

Khomenko, A. P., Eliseev, S. V., Artyunin, A. I., Parshuta, E. A., & Kaimov, E. V. (2013). *Mechanisms in elastic oscillatory systems: peculiarities of dynamic properties, tasks of vibration protection of machines, devices and equipment*. Dep. in the VINITI RAS.

Khomenko, A. P., Eliseev, S. V., & Kaimov, E. V. (2014). Virtual lever mechanism: Dynamic oscillation damping as a form of manifestation of linkages. *News of Transsib, 4*(20), 61–71.

Khomenko, A. P., Eliseev, S. V., & Kaimov, E. V. (2014). Virtual lever mechanism: Dynamic oscillation of oscillations as a form of manifestation of lever linkages. *News of Transsib, 4*(20), 61–71.

Khomenko, A. P., Eliseev, S. V., & Yermoshenko, Yu. V. (2012). *System analysis and mathematical modeling in the mechatronics of vibration protection systems*. Irkutsk State Technical University Press.

Khomenko, A. P., Khomenko, A. P., & Eliseev, S. V. (2012). Quasi-elements in mechanical oscillatory systems. Features of the system with the exclusion of variables of the dynamic state. Modern technologies. System analysis. *Modeling*, *2*(38), 8–17.

Khorychev, A. A. (2012). *Conceptual design of functional subsystems of cars based on heuristic classifications*. NIH.

Kim, D. P. (2003). Theory of automatic control.: Vol. 1. *Linear systems*. Fizmatlit Publ.

Kim, P. D. (2003). *The theory of automatic control* (Vols. 1–3). Fizmatlit Publ.

Kiryukhin, A. V., Tikhonov, V. A., Chistyakov, A. G. & Yablonsky, V. V. (2012). Active vibration protection – purpose, principles, state. Vibration protection and noise insulation of pipelines and experimental research. *Problems of mechanical engineering and automation, 4*, 102-110.

Kogan, A. Ya. (1997). *The dynamics of the track and its interaction with the rolling stock*. Transport Publ.

Kolovsky, M. Z. (1976). *Automatic control of vibration protection systems*. Nauka Publ.

Kopylov, Yu. R. (2011). *Dynamics of the processes of vibro-impact hardening*. Nauchnaya Kniga Publ.

Korolkov, M. V. (2010). *Development and research of analytical models of the dynamics of mechanisms with gaps in the interfaces of parts*. Research Gate.

Kuznetsov, N. K. (2009). *Dynamics of controlled machines with additional constraints*. Irkutsk: ISTU Publ.

Kuznetsov, N. K. (2009). *Dynamics of controlled machines with additional constraints*. ISTU Publ.

Lakanne, C. (2002). *Mechanical vibrations & shock. Specification development*. HPS Publ.

Lalanne, C. (2002). *Mechanical vibrations shock. Specification Development* (Vol. V). Hermes Penton Ltd.

Lavrus, V. V. (2006). *Improvement of pneumatic lever-hinged systems in railway transport* [PhD thesis]. Orel.

Lee, M. (2008). *Damping of resonant oscillations of gyroscopic systems with an active dynamic damper*. Research Gate.

Lenk, A. (1978). *Electromechanical systems. Systems with lumped parameters*. Mir Publ.

Lilov, L. K. (1993). *Modeling of related bodies*. Nauka Publ.

Lontsikh, P. A., & Eliseev, S. V. (2014). *Dynamic quality of machinery and equipment as a tool to ensure production reliability and competitiveness of processes*. INRTU Publ.

Lontsikh, P. A., & Eliseev, S. V. (2014). *Dynamic quality of machinery and equipment as a tool to ensure the reliability of production and competitiveness of processes*. NI ISTU Publ.

Lurie, A. I. (1961). *Analytical Mechanics*. GIFML Publ.

Lurie, A. I. (1961). *Operational calculus and application in technical applications*. GITTL Publ.

Makhutov, N. A., Petrov, V. P., Kuksova, V. I. & Moskvitin, G. V. (2008). Modern trends in the development of scientific research on the problems of mechanical engineering and mechanical engineering. *Problems of mechanical engineering and automation, 3,* 3-19.

Myamlin, S. V. (2002). *Simulation of rail carriage dynamics*. New Ideology Publ.

Nasnikov, D. N. (2008). Active elements as typical units in controlled vibration protection systems. Modern technologies. *System analysis Modeling, 20*, 41-49.

Nasnikov, D. N., & Logunov, A. S. (2006). Typical units in structural interpretations of mechanical oscillatory systems. Modern technologies. System analysis. *Modeling*, *12*, 78–93.

Nasnikov, D. N., Parshuta, E. A., Trofimov, A. N., & Sorin, V. V. (2010). Features of dynamic interactions in systems with an extended set of typical elements. *Bulletin of the Irkutsk Regional Branch of the Academy of Sciences of the Higher School*, *2*(17), 170–186.

Olson, G. (1947). *Dynamic analogies.* Politizdat Publ.

Panovko, G.Ya. (2006). *Dynamics of vibrational technological processes.* Institute for Computer Research.

Parshuta, E. A. (2013a). Determination of dynamic responses in mechanical oscillatory systems: structural approaches. In Information and Mathematical Technologies in Science and Management: *Proceedings of XVIII of the Baikal All-Russian Conference* (Vol. 1, pp. 216-221). Irkutsk: MESI SB RAS.

Parshuta, E. A. (2013b). Mechanisms in mechanical oscillatory systems as a form of introducing additional bonds. In Intellectual and Material Resources of Siberia: Materials of the Region Scientific-Practical Conference (pp. 42-48). Irkutsk.

Parshuta, E. A., Kaimov, E. V., & Mironov, A. S. (2012). Some Approaches to Dynamic Oscillation Damping. In *Technique and Technology: New Development Prospects: Proceedings of VII Int. Scientific-Practical Conference* (pp. 36-40). Moscow: Sputnik + Publishing.

Parshuta, E. A., & Gordeeva, A. A. (2010). Mathematical Modeling in Problems of Dynamic Oscillation Damping. Modern Technologies. System Analysis. *Modeling*, *1*(25), 149–153.

Parshuta, E. A., & Kaimov, E. V. (2013). Possible forms of changing the dynamic state of vibration protection systems. *Metallurgical Processes and Equipment*, *1*, 55–59.

Pykhalov, A.A. & Milov, A.E. (2007). *The contact problem of the static and dynamic analysis of the assembly rotors of turbomachines*. Irkutsk: IrSTU Publ.

Rakov, D. L. & Sinev, A. V. (2011). Structural analysis and synthesis of new technical systems based on the morphological approach. *Problems of machine engineering and automation, 2*, 73-80.

Rakov, D.L. & Sinev, A.V. (2011). Structural analysis and synthesis of new technical systems based on the morphological approach. *Problems of machine-building and automation, 2,* 73-80.

Reznik, Y. N. (2001). *Basics of mineral training in the development of mineral deposits.* VLADMO Publ.

Ryabov, I. M., Chernyshev, K. V., Vorobyov, V. V., & Urukov, I. N. (2009). A mathematical model of a vehicle suspension with an inertia-frictional shock-absorber. *Modern technologies. System analysis. Modeling, 21*, 29–31.

Salpin, S. N., Sinev, A. V., Lebedev, V. N., et al. (2002). *The device of angular stabilization of a suspended object on a transport vehicle.* RF No. 2181683 / 2002.

Samara, A. A., & Mikhailov, A. P. (2005). *Mathematical modeling. Ideas. Methods. Examples.* Fizmatlit Publ.

Samarsky, A. A., & Mikhailov, A. P. (2002). *Mathematical Modeling: Ideas. Methods. Examples.* Fizmatlit Publ.

Saveliev, Yu. F. (2003). The method of effective vibration protection of rolling stock and carriage based on additional mechanical devices with alternating elasticity. Omsk: OmGUPS Publ. .

Selvinsky, V. V. (2009). *Dynamics of the contact interaction of solids.*

Shedlos', A. I. (2005). *The choice of rational parameters of low-frequency articulated stands of linear displacements.* [Dissertation, Tula University].

Sobolev, V. I. (2003). *Discrete-continual mathematical models in algoritmic and programmatic resolution of problems of vibration suppression of construction and equipment*. Research Gate.

Sobolev, V. I. (2002). *Discrete-continual dynamic systems and vibration isolation of industrial screens*. ISU Publ.

Sovetov, B. Ya., & Yakovlev, S. A. (2001). *Modeling of systems*. Vysshaya Shkola Publ.

Sun, X., & Jing, X. (2015). Multi-direction vibration isolation with quasi-zero stiffness by employing geometrical non-linearity. *Mechanical Systems and Signal Processing*, *62-63*, 149–163. doi:10.1016/j.ymssp.2015.01.026

Tarasik, V. P. (2004). *Mathematical modeling of technical systems*. Dizayn PRO Publ.

Upir', R. Yu. (2009). *Dynamics of mechanical oscillatory systems taking into account the spatial forms of the connection of elementary units* [PhD thesis]. IrGUPS Publ.

Vargunin, V. N., Gusarov, V. N., Ivanov, B. G., Levchenko, A. S. (2006). Design and calculation of articulated devices and units (O.P. Mulyukin, Ed.). Samara: SamGAPS Publ.

Vargunin, V. I., Gusarov, V. N., Ivanov, B. G., & Levchenko, A. S. (2006). *Construction and calculation of artlculated means and assemblies*. Samara State Technical University Publ.

Wolfson, I.I. (2008). *Oscillations in Machines: A Textbook*. St. Petersburg: SPbGUTID Publishing.

Wolfson, I.I. (2008). *Oscillations in the machines*. St. Petersburg: SPbGUTID Publ.

Wolfson, I. I. (2013). *Dynamics of cyclic machines*. Polytekhnika Publ.

Yurevich, E. I. (2007). Theory of automatic control: educational ed. for universities (3rd ed.). St.Petersburg: BKhV – Petersburg Publ.

Zasyadko, A. A., & Nasnikov, D. N. (2008). Features of a hydraulic drive in active vibration protection systems. Modern technologies. System analysis. *Modeling*, *6*, 18 29.

Zinoviev, V. A., & Bessonov, A. P. (1964). *The fundamentals of the dynamics of machine aggregates*. Mashinostroenie Publ.

About the Author

Andrey Eliseev was born in 1975. He graduated from physics and mathematics school. In 1992 he entered Novosibirsk State University as a mechanic-mathematical faculty. He graduated from the University as a master of mathematics. He has a degree in economic mathematics. From 2011 to 2014 post-graduate student of the full-time Department, specialty "Dynamics, strength of machines, devices and equipment" in Scientific and educational center of modern technologies, system analysis and modeling (Irkutsk State Transport University). In 2014 he submitted to the consideration of thesis: "the Dynamics of a vibratory interaction of elements of technological systems subject to unilateral constraints relations" specialty 01.02.06 "Dynamics, durability of machines, devices and equipment". 2015 – defense of the candidate's dissertation at the Southwest State University, Kursk. Currently he works as an associate professor at the Department of Mathematics of Irkutsk State Transport University and an associate professor at the Department of Engineering Design and Standardization at Irkutsk National Research Technical University. The full list of publications includes more than 300 scientific papers. Over the past 10 years, the area of our research interests is the development of methods of structural mathematical modeling. The principal feature of the method is the use of the analytical apparatus of the theory of automatic control to solve the problems of the dynamics of various technical objects of transport and technical purpose, operating under intense dynamic loading, assuming that as the design schemes of technical objects will be considered mechanical oscillatory systems.

Index

M

O

P

Q

R

S

T

U

V

W

Recommended Reference Books

IGI Global's reference books are available in three unique pricing formats: Print Only, E-Book Only, or Print + E-Book.

Order direct through IGI Global's Online Bookstore at **www.igi-global.com** or through your preferred provider.

ISBN: 9781668440124
EISBN: 9781668440148
© 2022; 204 pp.
List Price: US$ **250**

ISBN: 9781799890782
EISBN: 9781799890805
© 2022; 306 pp.
List Price: US$ **260**

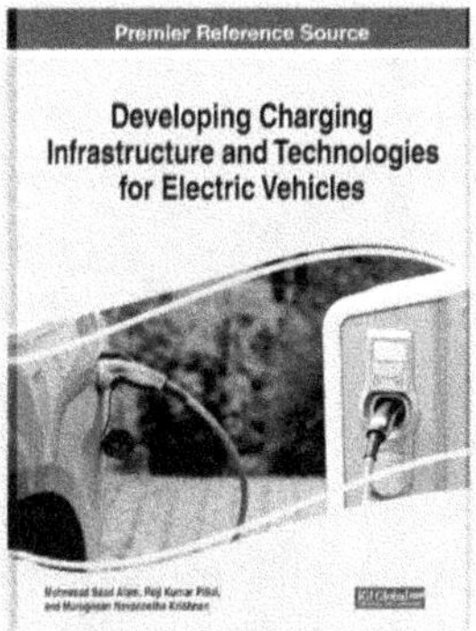

ISBN: 9781799868583
EISBN: 9781799868606
© 2022; 343 pp.
List Price: US$ **240**

ISBN: 9781799893158
EISBN: 9781799893172
© 2022; 261 pp.
List Price: US$ **250**

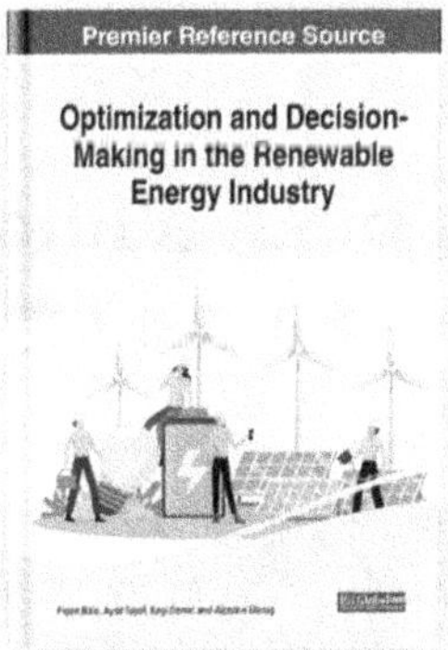

ISBN: 9781668424728
EISBN: 9781668424742
© 2022; 304 pp.
List Price: US$ **270**

ISBN: 9781799883746
EISBN: 9781799883760
© 2022; 366 pp.
List Price: US$ **215**

Do you want to stay current on the latest research trends, product announcements, news, and special offers?
Join IGI Global's mailing list to receive customized recommendations, exclusive discounts, and more.
Sign up at: **www.igi-global.com/newsletters.**

Publisher of Timely, Peer-Reviewed Inclusive Research Since 1988

IGI Global
PUBLISHER of TIMELY KNOWLEDGE

www.igi-global.com Sign up at www.igi-global.com/newsletters facebook.com/igiglobal twitter.com/igiglobal linkedin.com/igiglobal

www.ingramcontent.com/pod-product-compliance
Lightning Source LLC
LaVergne TN
LVHW062153300926
849161LV00010B/2656